STANDARDS FOR AGGREGATES

ELLIS HORWOOD SERIES IN GEOLOGY

Editors: D. T. DONOVAN, Professor of Geology, University College London, and J. W. MURRAY, Professor of Geology, University of Exeter

This series aims to build up a library of books on geology which will include student texts and also more advanced works of interest to professional geologists and to industry. The series will include translation of important books recently published in Europe, and also books specially commissioned.

RADIOACTIVITY IN GEOLOGY: Principles and Applications
E. M. DURRANCE, Department of Geology, University of Exeter
FAULT AND FOLD TECTONICS
W. JAROSZEWSKI, Faculty of Geology, University of Warsaw
A GUIDE TO CLASSIFICATION IN GEOLOGY
J. W. MURRAY, Department of Geology, University of Exeter
THE CENOZOIC ERA: Tertiary and Quaternary
C. POMEROL, University of Paris VI
Translated by D. W. HUMPHRIES, Department of Geology, University of Sheffield, and E. E. HUMPHRIES
Edited by D. CURRY and D. T. DONOVAN, Department of Geology, University College London

BRITISH MICROPALAEONTOLOGICAL SOCIETY SERIES

This series, published by Ellis Horwood Limited for the British Micropalaeontological Society, aims to gather together knowledge of a particular faunal group for specialist and non-specialist geologists alike. The original series of Stratigraphic Atlas or Index volumes ultimately will cover all groups and will describe and illustrate the common elements of the microfauna through time (whether index or long-ranging species) thus enabling the reader to identify characteristic species in addition to those of restricted stratigraphic range. The series has now been enlarged to include the reports of conferences, organized by the Society, and collected essays on specialist themes.

The synthesis of knowledge presented in the series will reveal its strengths and prove its usefulness to the practising micropalaeontologist, and to those teaching and learning the subject. By identifying some of the gaps in this knowledge, the series will, it is believed, promote and stimulate further active research and investigation.

PALAEOBIOLOGY OF CONODONTS
Editor: R. J. ALDRIDGE, Department of Geology, University of Nottingham
CONODONTS: Investigative Techniques and Applications
Editor: R. L. AUSTIN, Department of Geology, University of Southampton
FOSSIL AND RECENT OSTRACODS
Editors: R. H. BATE, British Museum Natural History, London, and Stratigraphic Services International, Guildford, E. ROBINSON, Department of Geology, University College London, and L. SHEPPARD, British Museum Natural History, London, and Stratigraphic Services International, Guildford
NANNOFOSSILS AND THEIR APPLICATIONS
Editors: J. A. CRUX, British Petroleum Research Centre, Sunbury-on-Thames, and S. E. VAN HECK, Shell Internationale Petroleum, Maatschappij
A STRATIGRAPHICAL INDEX OF THE PALAEOZOIC ACRITARCHS AND OTHER MARINE MICROFLORA
Editors: K. J. DORNING, Pallab Research, Sheffield, and S. G. MOLYNEUX, British Geological Survey, Nottingham
MICROPALEONTOLOGY OF CARBONATE ENVIRONMENTS
Editor: M. B. HART, Department of Geological Studies, Plymouth Polytechnic
A STRATIGRAPHICAL INDEX OF CONODONTS
Editors: A. C. HIGGINS, Geological Survey of Canada, Calgary, and R. L. AUSTIN, Department of Geology, University of Southampton
STRATIGRAPHICAL ATLAS OF FOSSIL FORAMINIFERA, 2nd Edition
Editors: D. G. JENKINS, The Open University, and J. W. MURRAY, Department of Geology, University of Exeter
A STRATIGRAPHICAL INDEX OF CALCAREOUS NANNOFOSSILS
Editor: A. R. LORD, Department of Geology, University College London
MICROFOSSILS FROM RECENT AND FOSSIL SHELF SEAS
Editors: J. W. NEALE, and M. D. BRASIER, Department of Geology, University of Hull
THE STRATIGRAPHIC DISTRIBUTION OF DINOFLAGELLATE CYSTS
Editor: A. J. POWELL, British Petroleum Research Centre, Sunbury-on-Thames
OSTRACODA
Editors: R. C. WHATLEY and C. MAYBURY, Department of Geology, University College of Wales

ELLIS HORWOOD SERIES IN APPLIED GEOLOGY

Published by Ellis Horwood Limited for the Institution of Geologists, Burlington House, Piccadilly London W1V 9AG

The books listed below are motivated by the up-to-date applications of geology to a wide range of industrial and environmental factors; they are practical, for use by the professional and practising geologist or engineer in the field, for study, and for reference.

A GUIDE TO PUMPING TESTS
F. C. BRASSINGTON, North West Water Authority
QUATERNARY GEOLOGY: Processes and Products
JOHN A. CATT, Rothamsted Experimental Station, Harpenden
NON-CONVENTIONAL METHODS IN GEOELECTRICAL PROSPECTING
MARK GOLDMAN, Petroleum Infrastructure Coporation Ltd, Holon, Israel
TUNNELLING GEOLOGY AND GEOTECHNICS
Editors: M. C. KNIGHTS and T. W. MELLORS, Consulting Engineers, W. S. Atkins & Partners
PRACTICAL PEDOLOGY: Manual of Soil Formation, Description and Mapping
S. G. McRAE, Department of Environmental Studies and Countryside Planning, Wye College (University of London)
STANDARDS FOR AGGREGATES
D. C. PIKE, Consultant in Aggregates, Reading
LASER HOLOGRAPHY IN GEOPHYSICS
S. TAKEMOTO, Disaster Prevention Research Unit, Kyoto University

STANDARDS FOR AGGREGATES

Editor
D. C. PIKE M.Phil.
Pike and Pike, Reading, Berks.

ELLIS HORWOOD
NEW YORK LONDON TORONTO SYDNEY TOKYO SINGAPORE

First published in 1990 by
ELLIS HORWOOD LIMITED
Market Cross House, Cooper Street,
Chichester, West Sussex, PO19 1EB, England

A division of
Simon & Schuster International Group
A Paramount Communications Company

© Ellis Horwood Limited, 1990

All rights reserved. No part of this publication may be reproduced, stored in a retrieval system, or transmitted, in any form, or by any means, electronic, mechanical, photocopying, recording or otherwise, without the prior permission, in writing, of the publisher

Typeset in Times by Ellis Horwood Limited
Printed and bound in Great Britain
by Bookcraft (Bath) Limited, Midsomer Norton, Avon

British Library Cataloguing in Publication Data

Standards for aggregates.
1. Construction materials: Aggregates
I. Pike, D. C.
624.1891
ISBN 0–13–842477–2

Library of Congress Cataloging-in-Publication Data

Standards for aggregates/editor, D. C. Pike.
p. cm. — (Ellis Horwood series in applied geology)
ISBN 0–13–842477–2
1. Aggregates (Building materials) — Standards — Great Britain.
I. Pike, D. C. II. Series.
TA441.S83   1990
620.1′91–dc20   90–4922
                                                                         CIP

Table of Contents

1 Introduction
  (D. C. Pike)..11

2 Sampling of aggregates and precision tests
  (P. M. Harris and R. Sym)
  2.1 Sampling...19
     2.1.1 Introduction..19
     2.1.2 Random sampling of ideal materials............................19
     2.1.3 Practical sampling..26
     2.1.4 Sample sizes for size analysis................................34
     2.1.5 Sample reduction..35
     2.1.6 Analysis..36
     2.1.7 Sampling *in situ*..38
     2.1.8 Sampling equipment..40
  2.2 Precision..43
     2.2.1 Variations arising during testing.............................43
     2.2.2 Standard definitions of precision coefficients................44
     2.2.3 Hierarchy of sampling terms...................................46
     2.2.4 Test result...47
     2.2.5 Definitions of repeatability and reproducibility for
            aggregate tests...48
     2.2.6 Design of a precision experiment..............................50
     2.2.7 Analysis of data from a precision experiment..................53
     2.2.8 Checking the variability caused by sampling...................55
     2.2.9 Repeatability control chart...................................56
     2.2.10 Cusum stability chart...57
     2.2.11 Inter-laboratory comparisons, using reference materials.......59
     2.2.12 Applications of precision in assessing the capability
            of text methods...61
  References..63

3 Aggregates for concrete

3.1 Marine aggregates (B. V. Brown) . . . 64
- 3.1.1 Introduction . . . 64
- 3.1.2 Environmental effects and dredging licencing . . . 65
- 3.1.3 Dredgers and discharge methods . . . 65
- 3.1.4 Use and modern history in the UK . . . 67
- 3.1.5 Geological origin and current UK dredging locations . . . 70
- 3.1.6 Processing and stockpiling . . . 71
- 3.1.7 Differences in characteristics between marine and land-based aggregates . . . 77
- 3.1.8 Effects of chloride on the corrosion of reinforcement . . . 80
- 3.1.9 Methods of testing for chloride content . . . 83
- 3.1.10 Alkali–silica reaction . . . 86
- 3.1.11 Shell . . . 87
- 3.1.12 Wear and skid resistance . . . 89
- 3.1.13 Organic impurities . . . 90
- 3.1.14 Efflorescence . . . 90

References . . . 90

3.2 Alkali–silica reaction (D. W. Hobbs) . . . 91
- 3.2.1 Introduction . . . 91
- 3.2.2 The reaction . . . 93
- 3.2.3 Sources of alkali . . . 93
- 3.2.4 Reactive silica . . . 98
- 3.2.5 Testing aggregates and cement–aggregate combinations for their reactivity . . . 102
- 3.2.6 Specifying and minimizing the risk of deleterious expansion in new construction . . . 107

References . . . 109

3.3 The drying shrinkage of aggregates (G. R. Lavers) . . . 112
- 3.3.1 Introduction . . . 112
- 3.3.2 The mechanism of aggregate shrinkage in concrete . . . 113
- 3.3.3 Experience in Scotland . . . 113
- 3.3.4 Shrinkable rock types . . . 113
- 3.3.5 The effects of shrinkage . . . 114
- 3.3.6 The measurement of drying shrinkage . . . 114
- 3.3.7 The BRE test method . . . 116
- 3.3.8 The use of *BRE Digest* 35 . . . 116
- 3.3.9 Introducing a test to British Standard BS 812 . . . 117
- 3.3.10 Precision of the methods . . . 119
- 3.3.11 Correlation between the *Digest* 35 and ten-day methods . . . 121
- 3.3.12 Factors affecting measured values of shrinkage . . . 121
- 3.3.13 Specification for shrinkage of aggregate in concrete . . . 124
- 3.3.14 Future research . . . 124

References . . . 125

4 Sands for building mortars
(N. Beningfield and T. P. Lees)
- 4.1 Types of aggregates in use, and production methods 126
- 4.2 Specification for mortar sands 127
- 4.3 The properties of mortar. 129
- 4.4 Sand properties and their effects on mortar. 134
- 4.5 Testing mortar sands 138
- References 140

5 Aggregates for bituminous materials
(G. E. Broadhead and J. F. Hills)
- 5.1 Properties and specifications of bituminous materials 142
  - 5.1.1 Introduction 142
  - 5.1.2 Bituminous materials: applications, processing, and performance 143
  - 5.1.3 Mechanical properties of bituminous mixtures 146
  - 5.1.4 The requirements for aggregates in asphalt and coated macadam specifications 149
- 5.2 Origin and classification of aggregates 169
  - 5.2.1 Introduction 169
  - 5.2.2 Geological classification of rocks 171
  - 5.2.3 Group classification of aggregates 175
  - 5.2.4 Classification and description of aggregate materials using the CADAM system 178
- 5.3 Tests and standards for aggregates in bituminous materials. 183
  - 5.3.1 Introduction 183
  - 5.3.2 Sampling of aggregates 185
  - 5.3.3 Particle size and shape 187
    - 5.3.3.1 Particle size distribution or grading 187
    - 5.3.3.2 Particle shape 192
    - 5.3.3.3 Flakiness index 193
    - 5.3.3.4 Average least dimension (ALD) 194
    - 5.3.3.5 Elongation index 195
    - 5.3.3.6 Angularity number 196
    - 5.3.3.7 Surface texture of coarse aggregate 197
    - 5.3.3.8 Shape and surface texture of fine aggregate 198
  - 5.3.4 Physical properties. 199
    - 5.3.4.1 Relative density and water absorption 199
    - 5.3.4.2 Bulk density of aggregate 201
    - 5.3.4.3 Bulk density of filler in toluene 202
    - 5.3.4.4 Voids of dry compacted filler 203
  - 5.3.5 Mechanical properties 203
    - 5.3.5.1 Aggregate crushing value 203
    - 5.3.5.2 Ten per cent fines value 206
    - 5.3.5.3 Aggregate impact value 207
    - 5.3.5.4 Los Angeles abrasion value 209
    - 5.3.5.5 Aggregate abrasion value 210

              5.3.5.6 Polished–stone value 213
      5.3.6 Soundness 222
              5.3.6.1 Soundness by the use of sodium or magnesium
                       sulphate (sulphate soundness test) 225
              5.3.6.2 Soundness by freezing and thawing 228
              5.3.6.3 Soaked ten per cent fines and modified aggregate
                       impact value tests 230
      5.3.7 Soundness of basic igneous rock aggregates 232
              5.3.7.1 Secondary mineral content by point counting 233
              5.3.7.2 Soundness by ethylene glycol 234
              5.3.7.3 Soundness by methylene blue absorption 236
      5.3.8 Adhesion of bitumen to aggregates 239
              5.3.8.1 Static water immersion tests 239
              5.3.8.2 Immersion wheel-tracking test 240
              5.3.8.3 Immersion mechanical tests 241
  References 241

6 Unbound aggregates
  6.1 Frost susceptibility   (P. T. Sherwood) 249
      6.1.1 Types of frost damage 249
              6.1.1.1 Frost shattering 249
              6.1.1.2 Frost-heave 250
      6.1.2 Measurement of frost-heave 250
              6.1.2.1 The frost-heave test 250
              6.1.2.2 Interpretation of results 251
      6.1.3 Availability of water 252
      6.1.4 Frost penetration 252
  References 253
  6.2 Stability of sub-bases   (D. C. Pike) 254
      6.2.1 Patterns of use, and specifications 254
      6.2.2 Performance requirements 255
      6.2.3 Trafficking trials 257
      6.2.4 The shortcomings of the CBR test 259
      6.2.5 The shear-box test 259
      6.2.6 A standard shear-box test 264
      6.2.7 Correlations 264
      6.2.8 Compactability tests 266
  References 269

7 International and European standards
      (D. C. Pike) 271
  7.1 Sampling 274
  7.2 Particle size distribution 274
  7.3 Shape and surface texture 275
  7.4 Particle density, water absorption value, and bulk density 275
  7.5 Resistance to crushing, abrasion, and polishing 276

Table of contents

7.6 Frost resistance and soundness.........................277
7.7 Harmful materials finer than two micrometres, especially clay
    minerals...277
References...277

Appendix 1 Status of new Parts of BS 812 at January 1990..............279

List of contributors

Neil Beningfield L.R.S.C., M.I.C.T. spent 13 years as Technical Manager of RMC Mortars Ltd. before taking his M.Phil. at South Bank Polytechnic. He is now General Commercial Manager at RMC Mortars and Chairman of the Technical Committee of the Mortar Producers Association.

Graham E. Broadhead B.Sc.(Eng.), C.Eng., M.I.C.E., F.I.H.T., F.I.A.T. received his BSc.(Eng.) and Dip.Eng. from Kings College, where he won both the Walter Smith Prize for Civil Engineering and the Tennant Medal for Engineering Geology. Before becoming a consultant in 1988, he was Technical Director for Wimpey Asphalt, where he had worked since 1959. He has sat on various BSI committees, and lectured widely.

Bevil V. Brown M.Inst.P., M.I.C.T. received his B.Sc. in Physics from London University in 1960. He has spent the last 13 years as Technical Executive for RMC Technical Services Ltd. A Chartered Physicist, he has served on several inter-industry aggregate and concrete technical and BSI committees.

Peter Harris received his B.A. in Geology from Oxford University in 1958. He is a member of the Institution of Mining and Metallurgy and has been a Principal Scientific Officer with the Mineral Intelligence Statistics and Economics Unit of the British Geological Survey since 1972.

John F. Hills M.Inst.P., F.I.A.T., F.I.H.T. received his B.Sc. in Physics from Queen Mary College, University of London in 1950 and joined Shell. He worked there in several capacities and locations until 1974, when he became Head of the Asphalt Division at Wimpey Laboratories Ltd. After a year as Technical Development Manager with Wimpey Asphalt Ltd., he became a consultant in 1986. A Chartered Physicist, he has served on several BSI and industry committees.

Donald W. Hobbs F.Inst.P. received his B.Sc. in Physics from Southampton University in 1958 and his Ph.D. in Engineering from the University of Surrey in 1973. After a spell as Senior Principal Scientist with the Cement and Concrete Association, he took up the same position with the British Cement Association in 1987. He was Secretary of the Materials and Testing Group of the Institute of Physics for 3 years and has been awarded the Henry Adams Diploma and Murrey Buxton Bronze Medal from the Institution of Structural Engineers.

Geoffrey R. Lavers F.I.Q.A., F.I.A.T., F.I.H.T. spent 21 years with Harry Stanger Ltd., materials consultants. He is now a Chief Materials Engineer with West Sussex County Council. He serves on BSI and CEN committees.

Timothy P. Lees B.Sc., F.R.S.C., M.I.C.T. obtained his degree in chemistry from the University of London in 1956. He has been Chief Chemist at Taylor Woodrow Construction since 1987, having previously spent 15 years as Chief Chemist of the Cement and Concrete Association. From 1984 to 1986 he served as Chairman of the Road and Building Materials Group of the Society of Chemical Industry and is Chairman of the BSI technical committee on mortars.

Philip T. Sherwood B.Sc., F.R.S.C. was Principal Scientific Officer with the Royal Commission on Environmental Pollution from 1971 to 1974, and held the same post at the Transport and Road Research Laboratory from 1974 to 1984. He is now a consultant and serves on BSI committees.

Roger Sym received his B.Sc. in Mathematics in 1966 and his Ph.D. in Statistics in 1969, both from Imperial College. He then worked as Principal Statistician with the British Cement Association (formerly the Cement and Concrete Association) until 1987, in which year he was awarded the SCI Road and Building Materials Group's Jubilee Award. He is a Fellow of the Royal Statistical Society and now works as a consultant.

1
Introduction

D. C. Pike

In 1988, the annual demand for aggregates for use in construction in Great Britain was over 290 million tonnes (BACMI 1989). Although this is the largest tonnage of bulk mineral worked in the country, having an annual value when delivered to its points of use exceeding £1000 million, there is a rather limited general literature on the use of aggregates. This is possibly because the subject lies at the fringes of several disciplines. The study of aggregates can absorb the full attention of a geologist or engineer, but some people have made, and continue to make, valuable contributions to knowledge about aggregates without having had any formal training in engineering or geology. Much of the relevant research is published in papers not usually read by many producers and users of aggregates, and some of the useful findings of practical work in the industry may not be published at all.

This book is intended to fill a gap in the literature by giving information relevant to the production of standards for aggregates used in construction; it consists of chapters on six selected aspects of technology that have been particularly prominent in the past five years in the UK, written by leading contributors to knowledge in their fields. It is not intended to provide a uniform and comprehensive treatment of the whole subject of aggregates but, taken with the references cited, the book will be useful to students and teachers who have a background in civil engineering, construction management, or geology, as well as providing authoritative guidance to commercial and technical specialists in the quarrying industry, and to other technologists in the field of construction materials.

The word 'aggregates' embraces a variety of materials; these are mostly crushed rocks or sands and gravels, but industrial byproducts and recycled waste materials used as substitutes for quarried materials are included too. The prime function of aggregates is to provide bulk, whether as filling materials, for example in road embankments, or in composites such as concrete, mortars, and bituminous paving mixtures. At the lower end of quality there is necessarily some overlap between aggregates and engineering soils; for the purposes of this book, aggregates are materials that are sold in commerce, whereas soils are encountered *in situ* on construction sites, and are not commodities of trade.

An unusually high price may be paid for a special aggregate that has particularly good quality in a given application; for example, calcined bauxite is sometimes specified in road surfacings required to provide very high levels of skidding resistance. This material may cost over £200 per tonne, but the general run of aggregates cost say £3 to £8 a tonne at the point of production. Therefore they are essentially low-cost minerals. It is important to have this in mind when considering the appropriate levels of technology to apply to aggregates.

In Great Britain in 1985, roughly 75 million tonnes of aggregates were used in concrete; about 54 million tonnes were used as unbound road materials or for fill; about 19 million tonnes were used in bituminous mixtures (excluding asphalt sand); about 18 million tonnes were used for building sands (including asphalt sands); and the balance went to a whole range of smaller uses (*Business Monitor* 1985).

The change in annual production of natural aggregates in Great Britain from 1955 is shown graphically in Fig. 1.1. The general pattern of these statistics was one of steady growth at an approximate rate of about 6 per cent compound until 1973, when a total output of some 256 million tonnes was reached. Subsequently, as a result of a rapid rise in fuel prices and a rapid reduction in public funds available for construction, annual demand reduced sharply, falling eventually to 182 million tonnes in 1981. Since then an increase in spending on construction has led to another period of growth in aggregates production at an average annual rate of over 6.5 per cent. To what extent this will continue into the future is a matter for speculation, and, naturally enough, some speculators will be luckier than others.

It is always difficult to predict demand accurately. A view taken in 1986 from within the quarrying industry, formed from an analysis of basic economic trends, was that total annual demand for aggregates would not then rise by more than 10 million tonnes by 1995, nor by more than 15 million tonnes by 2005 (Singleton 1986). Sometimes attempts are made to fit mathematical models to past demand curves and to extrapolate them into the future. By this means it has been predicted that demand will reach 400 million tonnes a year by 2000 (Gribble 1990). These forecasts produce widely different results even in the short term.

Despite major falls in 1987 in international stock markets, demand for aggregates in the UK has continued to grow. Furthermore, it is clear that spending on construction, in real terms, is still at a fairly modest level. The energy crisis of the 1970s, insofar as it increased transport costs and held back growth in demand, has been largely overcome; and there is a growing perception that seriously increased traffic congestion now affects most major modes of transport in many parts of the UK. Thus, in addition to large individual projects, like the Channel Tunnel, an increase in general construction associated with transport can be expected. A major increase in spending on roads has been announced, and other transport systems, as well as other parts of the UK's infrastructure (e.g. water treatment works), are known to be targeted for substantial capital works. All this suggests that aggregate consumption will be fairly buoyant in the immediate future, provided that the economy remains sound. The market share held by different types of aggregates is also subject to change.

Since 1950 the annual output of crushed rocks has grown from about half that of

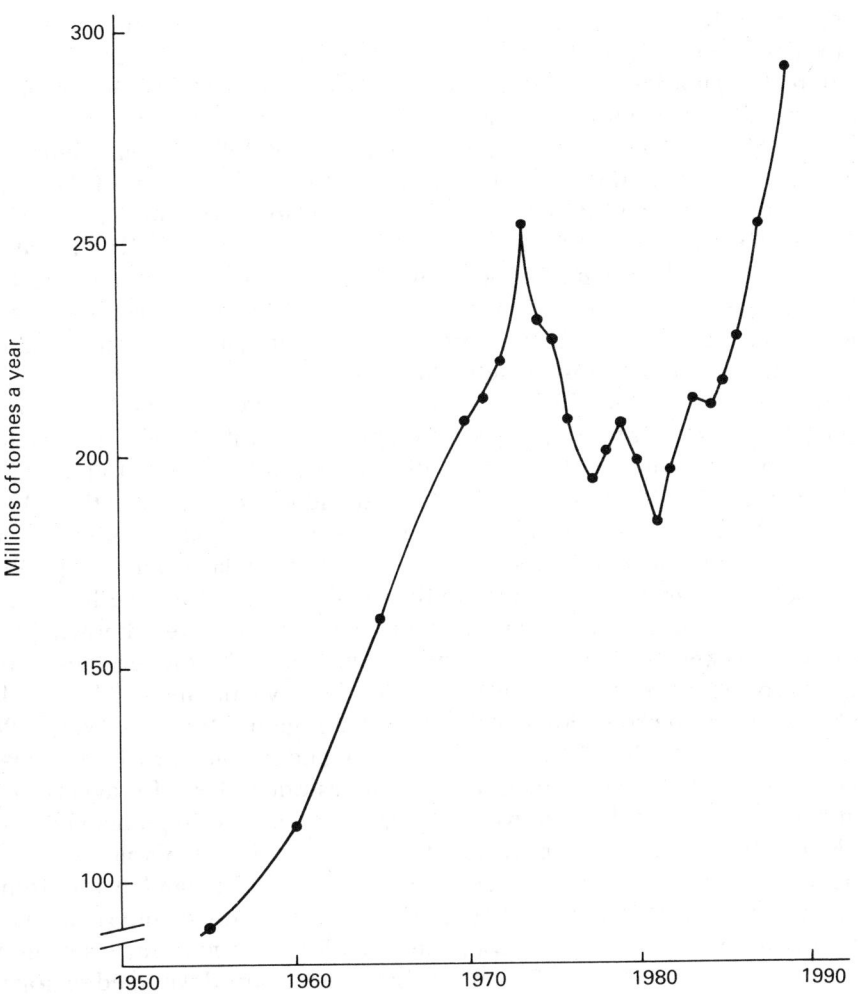

Fig. 1.1 — Growth in annual production of natural aggregates in Great Britain. *Source*: BACMI.

sand and gravel, to rather more than an equal share of total production. In 1950, 7 million tonnes of manufactured aggregates, mainly metallurgical slags, were sold. By 1970, this trade had increased to 23 million tonnes, but it has since fallen back to about 5 million tonnes a year, mainly because of the drop in the production of slag caused by changes in the size and methods of the steel industry. The current output of manufactured aggregates includes purpose-made lightweight aggregates for special applications. In 1985, the *per capita* annual demand for aggregates was 3.8 tonnes in the UK, 4.7 tonnes in France, and 5.4 tonnes in the USA. In 1988, these figures had grown to 5.1 tonnes per head in the UK, 6.7 tonnes in France, and 7.0 tonnes in the USA. Differences in demand between countries obviously reflect variables such as the pace of infrastructure construction and renewal.

Within the UK, and excluding the occasional large projects that can greatly distort local markets, regional distribution of demand could be expected to follow distribution of population quite closely. The population densities of the southeast of England and of the conurbations in the Midlands and northern England are substantially higher than the national average, and it follows that demand for aggregates is high, per unit of area, in those parts. Conversely, parts of the country having low population densities show consequential low demand for aggregates. Scotland has 34 per cent of the area of Great Britain, but only about 10 per cent of its population. Its overall demand for aggregates, per unit of area, is therefore lower than the average for the UK. These differences in intensity of demand strongly influence the commercial activities of producers of aggregates, and the regulatory policies of central and local government planners.

Because aggregates generally have a low value per unit weight, and because they are required in large quantities, it follows that producers will choose, if possible, to site their production units within a short distance of their intended markets (the average haulage distance from a land-based sand and gravel quarry in the UK is of the order of 20 miles). But, in highly-developed areas, it becomes increasingly difficult to obtain planning permission to work minerals. In particular, the problems of finding sufficient aggregates in the southeast of England to serve local needs, without causing unacceptable environmental damage, are well known. Those problems were acknowledged by a committee appointed by Government fifteen years ago to consider the supply of aggregates (Advisory Commitee 1975), and are still causing concern to professional planners in that region (Deakin & Waite 1988). One example of the results of the pressures of demand in southeast England is the serious examination of a deep-mined source of limestone in Kent for aggregates.

As it becomes more difficult to work aggregates in areas of high demand, some materials are brought in over much greater distances. For example, substantial quantities of crushed rocks are imported to the southeast of England by rail from the Mendips and from the Midlands, and by ship from various British and Irish quarries and from across the English Channel (Tidmarsh 1988). Large tonnages of marine sands and gravel are also landed. This means that users in any developed region may have to become accustomed to materials that they have not previously used, and each new material may raise different technical questions. This trend and its effects will intensify and spread with time.

The wide variety, in terms both of age and type, of rocks that outcrop in Great Britain results in a profusion of aggregate types. For example, in the southeast of England, only small quantities of crushed rocks are produced owing to the general lack of outcrops of older, indurated rocks; the predominant local aggregates are flint sands and gravels derived originally from nodules of flint in the Chalk. Substantial quantities of quartzite sands and gravels are worked in the English Midlands. Carboniferous Limestone is worked extensively, e.g. in the Mendip Hills, to the north of the Bristol coalfield, in south and north Wales, in Derbyshire, on the borders of the Lake District, in North Yorkshire and Lancashire, and to a minor extent in Scotland. Other kinds of limestone and a wide range of other sedimentary, igneous, and metamorphic rocks are worked in other parts of the country. Trade catalogues (e.g. BACMI 1986, SAGA 1983) give some useful information about the locations of quarries and, in some cases, the petrology of the products. Those who seek more detailed information will consult the specialized literature.

The theme of this book was chosen because of the increasing importance of standards to users and producers of aggregates. It would be fair to say that, only fifteen years ago, standards for aggregates in the UK had limited significance. Some of the output fell nominally within the scope of various British Standards, but a lot did not, and, in practice, those standards were often ignored. A series of events then caused producers and users to seek improvements to British Standards. In the early 1970s, severe corrosion of metal reinforcements in concrete was discovered and was attributed to the presence of excessive levels of chlorides used as additives. The concrete industry had recently suffered the consequences of the misuse of high alumina cement, and was not prepared to tolerate another blow to the confidence of specifiers and users in concrete as a structural material. It was decided to exclude the use of calcium chloride in reinforced concrete, although it had been used for many years to achieve especially early strengths. There had been no evidence in the UK of any problems of corrosion of reinforcement associated with the small amounts of salts in marine aggregates, but some specifiers also then decided that levels of chlorides in marine aggregates must be maintained at very low levels.

Having failed at first to obtain agreement in the drafting of limits on chloride contents for British Standards, the producers of marine sand and gravel decided to take a leading role. They had to prove that the extremely tight limits being proposed by some specifiers were unnecessary and commercially damaging. Research and standards development by the Sand and Gravel Association (Jordan & Pike 1976) resulted in acceptance of limits on chlorides that were tight enough to prevent corrosion but which avoided unnecessary extra processing costs. These have stood until the present, although another aspect of chlorides — their role in the alkali–silica reaction — is now coming to the fore.

The next significant event was the revision of the British Standard for concrete aggregates, BS 882 (BSI 1973). Before the work started it was not thought that the existing standard needed much alteration, although some sand producers had asked for modifications to be made to the grading specifications because their materials did not fit any of the four permitted zones of particle size distributions. These had been settled some 25 years earlier, following tests on about 500 samples of sand (Newman & Teychenne 1954). The grading zones were constructed arbitrarily so that there were no overlaps at the 600 μm size: Zone 1 (the coarsest) allowed 15 to 34 per cent finer than 600 μm; Zone 2 allowed 35 to 59 per cent; Zone 3 allowed 60 to 79 per cent; and Zone 4 allowed 80 to 100 per cent. Limits at other sizes were allowed to overlap, and were set at values that included most of the samples tested in the original study. An additional tolerance allowed some marginal materials to comply. Some research on crushed rock sands had suggested that changes ought to be made to the grading zones, and the producers of natural sands sought changes too.

The sand producers found it hard at first to convince the specifiers of their needs, so the Sand and Gravel Association undertook a major study. Over 16 000 gradings were examined this time, and, when they were analysed, an undeniable need for change emerged. About 20 per cent of individual samples did not fall into any zone, and only about one in twenty sources had gradings that consistently fell into a single zone. It was concluded that, although the zoning system was helpful to some users in designing concrete mixes with a particular sample of sand, it was not appropriate for a compliance standard. Following a lengthy debate, a more flexible system of sand gradings was agreed (Pike & Harrison 1984). This permitted the majority of sources of concrete sands in current use to be described by BS 882, which now includes three overlapping gradings. These gradings (coarse, medium, and fine, abbreviated to C, M, and F) have ranges that are approximately twice the width of the former zones.

At about the same time as these changes in BS 882 were being developed, attempts were in progress to update the standard for building sands (BSI 1976). Yet again there was an initial difference of opinion between specifiers and producers, and then a period of research by the producers. This centred on the need to clarify the basic sieve test method (Pike & Limbrick 1981). Having eventually agreed that all test samples of building sands be washed before being sieved, the BSI comittee then agreed to changes in the limits applied to the results of tests so that many more of the sources of sand in commerce can now meet the requirements of BS 1200. These kinds of development have helped to focus attention on the need to specify by reference to the standards. This trend has not yet fully run its course: it was recently argued by a County Council at a planning inquiry that it is the performance of the sand that the purchaser is concerned with, and not so much whether it complies with any particular British Standard grading. Even so, a senior official responsible for building control from a City Council in that county has publicly spoken about costly problems caused by the use of probably non-standard sands. Current investigations by the BSI committee suggest that closer regulation of harmful fines may be more important than grading.

Any serious assessment of standards for aggregates must include a review of test methods. In 1981 a major revision of the British Standard for test methods for aggregates (BSI 1975) was begun. BS 812 is a very widely used standard and it is important in commerce. The annual sales of the 1975 edition averaged 600 a year for each of its four parts, and many other standards make reference to it. When the revision was started, it was soon concluded that substantial new work was required to include recent developments in standard tests, to draw in additional, non-standard tests, and to obtain and publish reliable precision data. In view of the breadth of the subject, a decision was taken early on to break the revised BS 812 into over twenty parts. The results of the first eight years of work are summarized in Appendix 1.

The 1975 edition of BS 812 included some precision estimates, but these were inadequate because they had not been based on the correct procedures of BS 5497 (BSI 1979). With the passage of time, more attention has been paid to testing for compliance: more tests are being applied more frequently by more people. This has led to growing numbers of disputes over differences between laboratories in results of tests on nominally identical materials. In turn, this has brought about the cynical, but perhaps understandable, practice among some producers of sending samples to several laboratories in the expectation of obtaining at least one quotable, favourable result. It has been common to hear complaints that the test methods are inadequate; or that the laboratories do not carry out the tests properly; or that specifiers were not accepting the obvious limitations of the tests. To solve these problems it is vital to obtain reliable precision data and, wherever possible, to improve precision. The work that has been completed in this field is one of the two main topics covered in Chapter 2, which also deals with the special problems of sampling aggregates.

The three authors of Chapter 3 deal with aspects of concrete aggregates that have caused considerable controversy over the past ten years: marine aggregates, alkali–silica reactivity (ASR), and drying shrinkage. Chapters 4, 5, and 6 deal with aggregates for building mortars, bituminous mixtures, and unbound road sub-bases.

The end-uses covered by this book cover about three quarters of the total output of aggregates in the UK. The remainder includes railway track ballast, thought to consume about 10 million tonnes a year, and a multitude of minor uses. The final chapter considers overseas and international standards for aggregates. The UK now has to take a serious interest in Europe because, within a few years, existing British Standards must be replaced with European (i.e. CEN) standards. If the UK does not play a full part in the process of drafting those new standards, there could be severe problems for its users and producer. The UK has been awarded the chairmanship and secretariat of the committee (CEN/TC154) that will write European standards for aggregates. This will create new interest in many features of the subjects covered in this book.

REFERENCES

Advisory Committee on Aggregates (1989) *Aggregates: the way ahead.* HMSO.
British Aggregate Construction Materials Industries (1989) *Statistical Year Book.*
British Standards Institution (1975) *Testing aggregates.* BS 812. Now being superseded in many Parts.
British Standards Institution (1973) *Specification for aggregates from natural sources for concrete (including granolithic).* BS 882, 1201. Now superseded by the 1983 edition.
British Standards Institution (1976) *Building sands from natural sources.* BS 1198, 1199, and 1200 (with subsequent amendments).
British Standards Institution (1979) *Precision of test methods.* BS 5497.
Deakin, H. & Waite, C. (1988) Planning for aggregates in the south east. The Planning Officer's viewpoint. *Quarry Management.* January.
Department of Trade and Industry (1985) *Business Monitor PA1007.* HMSO.
Gribble, C. (1990) Predicting future UK aggregate production. *Quarry Management.* January, p 27.
Jordan, J. P. R. & Pike, D. C. (1976) Limits on chloride contents of marine aggregates. *Sand and Gravel Association Technical Paper TP11.*
Newman, A. J. & Teychenne, D. C. (1954) A classification of natural sands and its use in concrete mix design. *Symp. on mix design and quality control of concrete.* Cement and Concrete Association.
Pike, D. C. & Harrison, R. (1984) Revision of BS 882: background and changes. *Chemistry and Industry.* February.
Pike, D. C. & Limbrick, A. J. (1981) A study of sieve tests for building sands. *Chemistry and Industry.* September.
Singleton, D. F. (1986) Construction material provision. The market situation — demand and supply characteristics. *Quarry Management.* April.

Sand and Gravel Association (1983) *Products and services of producer and associate members*.

Tidmarsh, D. (1988) Rail- and sea-borne aggregates for the south-east. *Quarry Management*. February.

2

Sampling of aggregates and precision tests

P. M. Harris and **R. Sym**

2.1 SAMPLING

2.1.1 Introduction

In the aggregate industry, sampling procedures have traditionally been less rigorous than is usual in mining. This is primarily because aggregates, unlike metallic ores, have tended to be fairly uniform in composition and not critically dependent on the content of a small amount of a valuable constituent. Hence quality control during aggregate production often depends on little more than recognizing, and avoiding, very obvious variations in the mineral composition of the rock and trusting to the latitude of the specification to accommodate smaller variations.

As specifications for aggregates become more rigorous, however, the need to test becomes more important, and questions increasingly arise about how representative samples are, and how much reliance can be placed on the outcome of a particular test.

In addition to variations caused by sampling, substantial variations also occur during sample testing. It is important to know how large these variations are so that undue risks are not imposed on producers and purchasers of aggregates when tests are called on in specifications, and so that test methods showing excessive variations can be identified and improved. Knowledge of the normal level of variation that occurs within a test can be used by laboratories to check that their results are reliable. Hence 'precision', the term used to describe these variations, is the subject of the second part of this chapter.

2.1.2 Random sampling of ideal materials

The concept of the 'representative sample' is frequently encountered in testing aggregates, but all too often this turns out to be nothing more than an average-looking sample which is deemed, without proper justification, to be representative of a quarry or a larger batch of aggregate, and suitable for testing to determine properties that are held to be typical of the original source. It is often not appreciated that, without a knowledge of how closely the properties of the sample relate to those of the original, quantitative data derived from the sample will be of doubtful relevance to the whole

Samples selected by the hand and eye of a competent geologist might well be typical of a given source, but they are always suspect to a degree, unless there is a quantitative way of expressing the relationship between a particular property in the sample and the same property in whatever it is supposed to represent.

In contrast, samples taken on a random basis, in which each unit in the original lot has an equal chance of selection, can be subjected to a mathematical analysis which allows them to be regarded as representative of the whole within known limits.

An appreciation of some of the theoretical aspects of random sampling can be gained by considering 'ideal' material consisting of equally sized and shaped particles. If we take random selections of five particles from such a material in which 19 per cent of the particles are black and 81 per cent white, then the chance of selecting all the possible combinations is determined by the formula

$$\frac{n!}{(n-r)!\,r!}(0.19)^r(0.81)^{n-r} \tag{2.1}$$

where n is the total number of particles in the selected sample (and is small relative to the original material) and r is the number of black particles, as shown in Table 2.1.

Table 2.1 — Probability of selecting different combinations in 5 particle samples from a mixture of 19 per cent black and 81 per cent white particles

| No. of particles | | Probability % |
Black	White	(to nearest whole number)
5	0	0
4	1	1
3	2	4
2	3	19
1	4	41
0	5	35

Clearly it would be unwise to assume that any randomly selected sample of five particles was typical of the original. It will also be clear that the chances of selecting a sample containing 60 per cent or more of black particles, which might be thought of

Sec. 2.1] Sampling 21

as being wildly unrepresentative, is quite small. In fact, there is an overwhelming probability (95 per cent) that the sample selected will contain 0–40 per cent black particles (0, 1, or 2 particles) which is some sort of approximation to the average black content of 19 per cent.

If larger samples are taken, the spread of values decreases and the chances of selecting highly unrepresentative combinations becomes very much smaller. This can be illustrated by considering selections of samples containing 10, 20, and 40 particles and calculating the probability of various combinations as previously. The results, excluding combinations with a very small probability of occurrence, are shown in Table 2.2.

Table 2.2 — Probability of selecting different combinations of 10, 20, and 40 particle samples from the same mixture considered in Table 2.1

Sample of 10			Sample of 20			Sample of 40		
Black	White	Probability %	Black	White	Probability %	Black	White	Probability %
0	10	12	0	20	1	0	40	0
1	9	29	1	19	7	1	39	0
2	8	30	2	18	15	2	38	1
3	7	19	3	17	22	3	37	3
4	6	8	4	16	22	4	36	6
5	5	2	5	15	16	5	35	10
6	4	0	6	14	10	6	34	14
			7	13	4	7	33	16
			8	12	2	8	32	15
			9	11	1	9	31	13
			10	10	0	10	30	9
						11	29	6
						12	28	3
						13	27	2
						14	26	1
						15	25	0

From the foregoing, the changes that occur in the probability of making various selections as the sample size increases can be summarized as in Table 2.3.

Table 2.3 — Probability of obtaining certain combinations in samples of 5, 10, 20, and 40 particles from a mixture of 19 per cent black and 81 per cent white particles

Sample size (no. of particles)	5	10	20	40
Chances of zero black	35	12	1	0
Chances of 10–30% black	41	77	85	93
Chances of >40% black	5	3	1	0

Hence it will be clear that the chances of making highly unrepresentative selections, such as 0 black, or more than 40 per cent black, become vanishingly small as the number of particles in the sample increases. Also, the chances of obtaining a sample of composition within a certain specified range containing the true value (i.e. 10–30 per cent black) increases significantly as the sample size increases. The situation is illustrated in Fig. 2.1 in the form of histograms.

As larger samples are selected, the chances of obtaining a perfect sample increase steadily until the ultimate limit is reached when the whole of the original lot comprises the sample. Hence, the only way to be completely sure of obtaining a truly representative sample by a random method is to take the whole of the original lot. For practical purposes this is clearly impossible, although it serves to illustrate the point that there is always a possible error involved in taking a sample to represent an original lot. The smaller the sample, in terms of the number of particles, the greater will be the potential error. Practical sampling entails taking a sample large enough to be representative within acceptable and known limits, but small enough for convenience in handling.

The probabilities associated with the larger samples (e.g. Fig 2.1(d)) become increasingly difficult to calculate because of the relatively large number of particles. Consequently, where large numbers of particles are concerned (as indeed they are in most real sampling situations) the problem is more conveniently dealt with by using the Normal or Gaussian distribution. In this case, where p is the percentage of the constituent of interest in the original population, n is the number of particles in the sample, a is the percentage of the constituent of interest in the sample, and s is the standard error, it can be shown to an acceptable degree of approximation that a will fall within the range $p \pm ks$ with a probability that can be calculated, and the standard error $s = \sqrt{\dfrac{p(100-p)}{n}}$. The normal distribution is characterized by certain important properties:

(1) 99.7 per cent of the values deviate from the mean by not more than three times the standard error ($k = 3$).
(2) 95 per cent of the values deviate from the mean by not more than twice the standard error ($k = 2$).
(3) 67 per cent of the values deviate from the mean by not more than one standard error ($k = 1$).

For those not familiar with the term, the 'standard error' is a measure of the scatter likely to occur in the estimated value a.

Hence the general binomial formula for the probability of selecting samples of a given composition on which Fig. 2.1 is based,

$$\frac{n!}{(n-r)!\,r!} \cdot \frac{(p)^r (100-p)^{n-r}}{100^n} \tag{2.2}$$

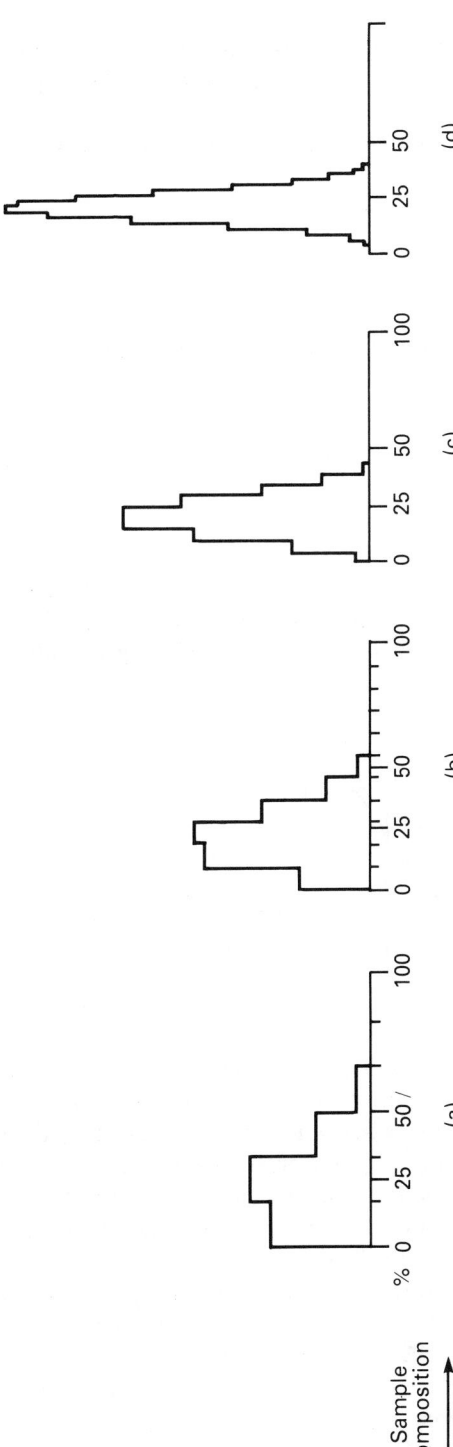

Fig. 2.1 — The probability of making different selections from a lot containing 19% black and 81% white particles when samples containing 5(a), 10(b), 20(c), and 40(d) are taken.

where n = total number of particles in the sample,
r = number of particles of the constituent of interest in the sample, and
p = concentration of the constituent of interest in the original population, as a percentage,

can be replaced by the general equation of the normal distribution:

$$y = \frac{1}{s\sqrt{(2\pi)}} e^{-(x-p)^2/(2s^2)} \tag{2.3}$$

By taking the mean as the origin and making measurements in units of standard error, this equation reduces to:

$$y = \frac{1}{s\sqrt{(2\pi)}} e^{-x^2/2} \tag{2.4}$$

Equation (2.4) defines the normal curve shown below, in which the x axis is now graduated in units of standard error and the areas under the curve show the proportion of the distribution associated with the values given on the x axis.

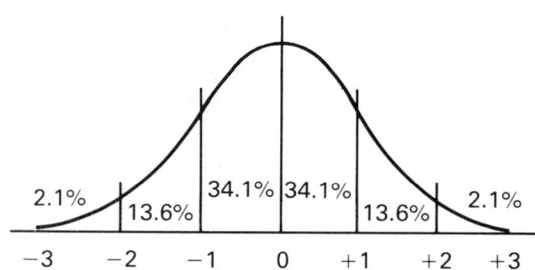

By using the normal distribution as an approximation to the binominal distribution considered above, it is possible to consider the effects of sample size on the probability of making certain selections when the number of particles becomes very large. For example, by using the same theoretical aggregate containing 19 per cent black particles, we can now consider what the chances are of selecting a sample that falls within ± 10% of the true composition (i.e. 19 ± 1.9%) as we increase the size of the sample from 50 to 3200 particles (Table 2.4).

From Table 2.4 it will be clear that a sample containing 1600 particles has a 95 per cent chance of having a composition lying within ± 10 per cent of the true value, and a sample of 3200 particles is over 99 per cent certain of having a composition within 10 per cent of the true value. In practice, 95 per cent is usually regarded as close enough for sampling purposes, and consequently a sample of some 1600 particles would be regarded as adequately representative of the composition of the original within ± 10 per cent. In other words, an analysis of the sample would be said to be subject to a possible error of ± 10 per cent within 95 per cent confidence limits.

Table 2.4 — Relationship between sample size and chances of obtaining a representative sample for a mixture of 19 per cent black and 81 per cent white particles

Size of sample (no. of particles)	% probability of selecting a sample having a composition within ± 10% of the true value
50	27
100	38
200	50
400	67
800	83
1600	95
3200	99

These errors and confidence limits, however, depend on the composition of the mixture, which in this case contained 19 per cent black particles. Had it contained 50 per cent black particles, then a smaller sample, in terms of number of particles, could have been selected for the same error and confidence limits. On the other hand, if the mixture had contained only a small proportion of black particles then a much larger sample would be required for the same error and confidence limits.

The size of sample needed to ensure representativeness of ± 10 per cent of the true value within 95 per cent confidence limits for various compositions has been calculated, and some typical values are given in Table 2.5.

Table 2.5 — Relationship between sample size and composition for the proportion of the constituent of interest to be within ± 10 per cent of the true value with a probability of 95 per cent

	Average composition (% black particles)							
	1	2	5	10	20	40	60	80
No. of particles in sample	38 000	18 800	7300	3457	1537	576	256	96

From Table 2.5 it will be clear that the size of sample necessary for a given accuracy, in the measurement of a particular constituent, depends on the concentration of that constituent. It will also be obvious that very large samples are necessary if the constituent of interest is present at a very low level of concentration.

At this stage it is important that the means of expressing error or accuracy is fully understood. In the above tables the error referred to has been a relative error, that is to say the value 80 ± 10 per cent means a range of 72–88 per cent and the value of 5 ± 10 per cent means a range of 4.5–5.5 per cent. In some instances, however, an error can be expressed in absolute terms, in which case the value 80 ± 10 per cent means a range of 70–90 per cent and 20 ± 10 per cent means a range of 10–30 per cent.

Hence 'relative error' $= 200\dfrac{s}{p}$ and 'absolute error' $= 2s$.

The significance of the two types of error is illustrated in Fig. 2.2. The absolute error is given by the solid lines, and the relative error is given by the broken lines. For example, the values given in Table 2.5 can be verified to some extent by extrapolation between the broken lines corresponding to 8 per cent and 12 per cent. The graph conveniently summarizes the relationship between accuracy, size of sample, and composition. It can be used to design sampling procedures for aggregates that have particles of approximately the same size and that can be defined in terms of a simple two-component mixture, such as, for example, a mixture of limestone and basalt particles. However, for very small values of p and large values of n, it is preferable to use a graph where the relative error is plotted against the proportion of mineral and the number of particles (Fig. 2.3).

2.1.3 Practical sampling

When very large numbers of particles are to be considered, counting becomes impractical and it is preferable to measure sample size on a weight basis. In such cases, the number of particles can be converted to a weight, depending on the particle size, by assuming a nominal particle density and a spherical particle shape (Fig. 2.4). In this case the sample weight (w) is given by the formula:

$$w = \frac{n\pi\rho d^3}{6} \quad (2.5)$$

where $n =$ no. of particles, $\rho =$ specific gravity, $d =$ particle diameter. For most aggregates a particle density of 2.65, on which Fig. 2.4 is based, will be sufficient, but if the material has a significantly different density, allowance may be made by making a *pro rata* adjustment to the sample weight. For example, in the case of a dolorite or basalt aggregate, the sample weight can be increased by 10 per cent, but smaller adjustments would not be worth making in view of the overall accuracy of the method. A working graph applicable to samples of single-size aggregate with a 10 per cent error is included (Fig. 2.5).

If the aggregate has a wide size distribution, it is possible to design separate sampling routines for each size fraction or to accept different errors for each particle size range within the sample. However, it will be clear from Fig. 2.4 that small

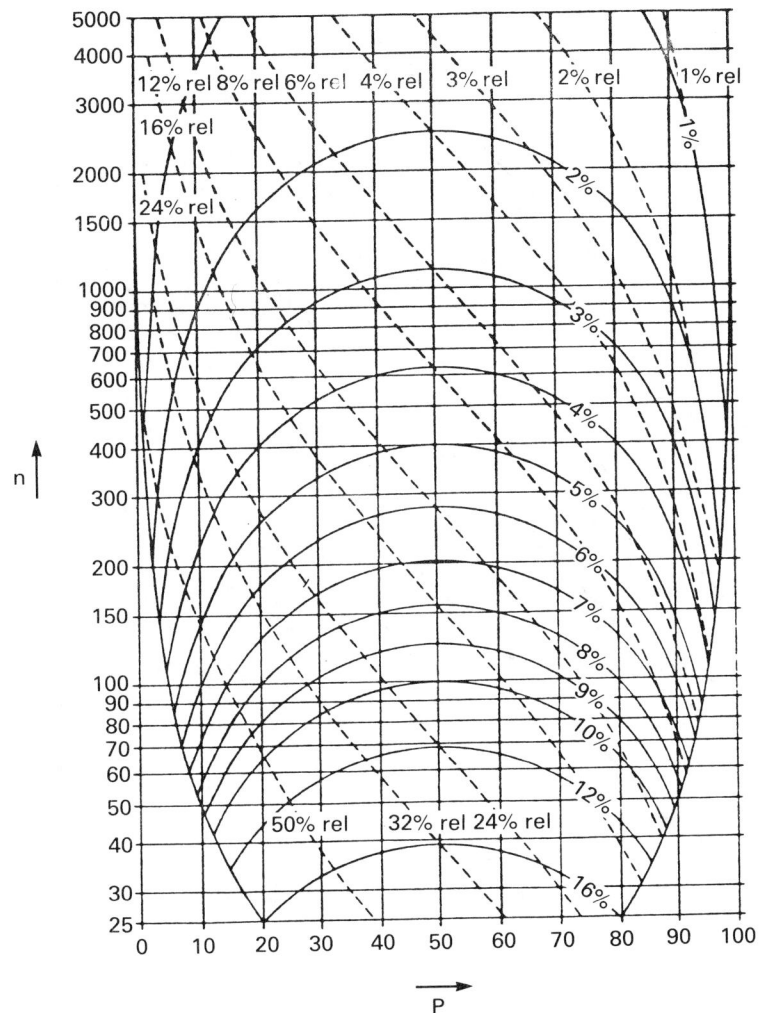

Fig. 2.2 — Relationship between size of sample (n), composition (p), relative error, and absolute error. The graph can be used to control the accuracy of point counting or the selection of samples of uniformly sized, fully-liberated particles.

increases in particle size require very large increases in sample weight to maintain the same accuracy, hence the overall error associated with a sample will depend critically on the proportion of coarse material. (Formula (2.5) shows a third power relationship between the mass of a particle and its diameter.)

It should be noted that the proportion of mineral in the sample (Figs 2.2 and 2.3) refers to any mineral or rock type occurring as discrete particles that can be clearly recognized. For example, it can refer to particles of granite in a gravel but not to the quartz grains in the granite unless the material has been crushed sufficiently for the constituent minerals such as quartz to be substantially free.

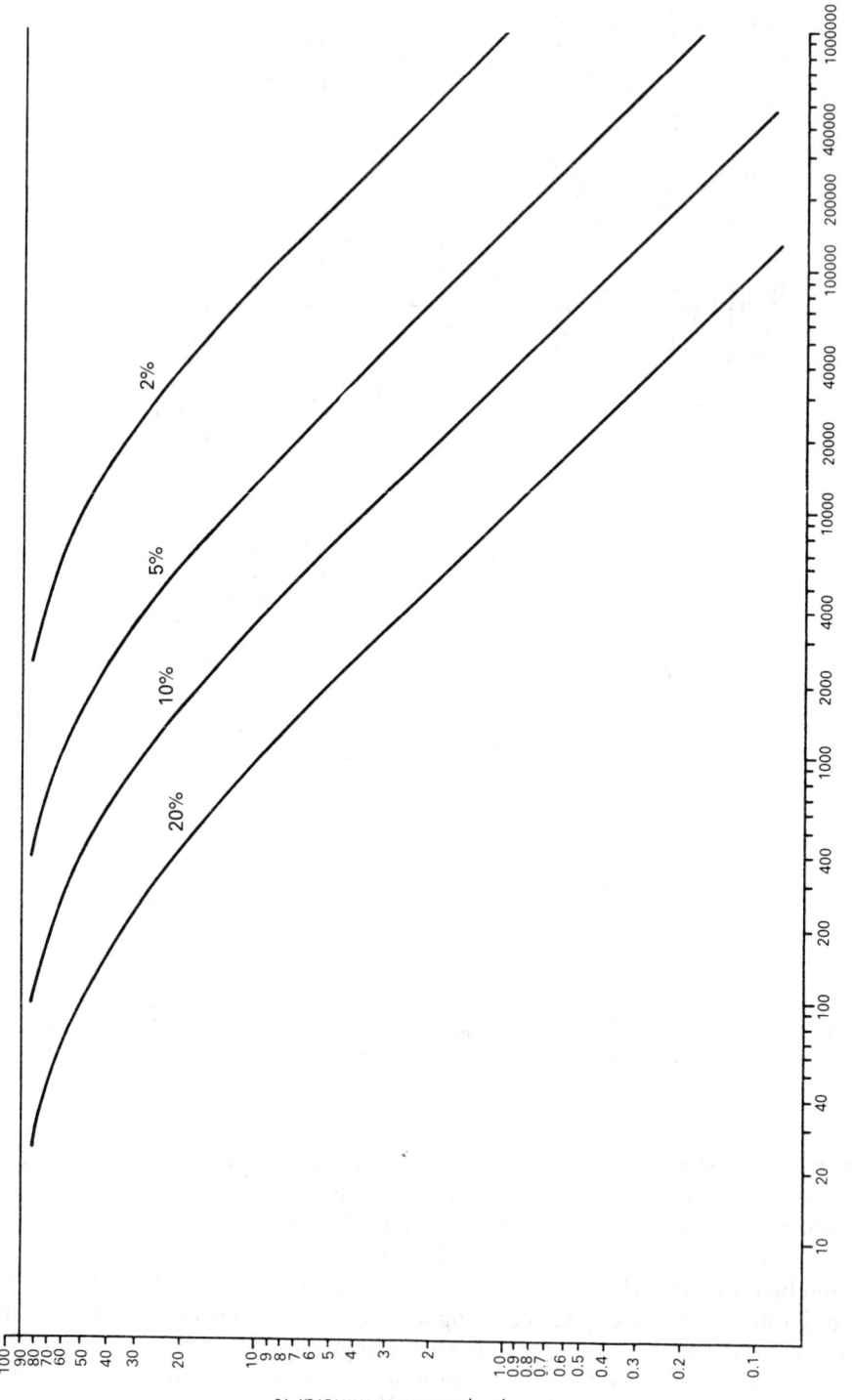

Fig. 2.3 — Number of particles vs mineral proportions for various relative sampling errors (based on single-sized, fully-liberated particles).

Sec. 2.1] **Sampling** 29

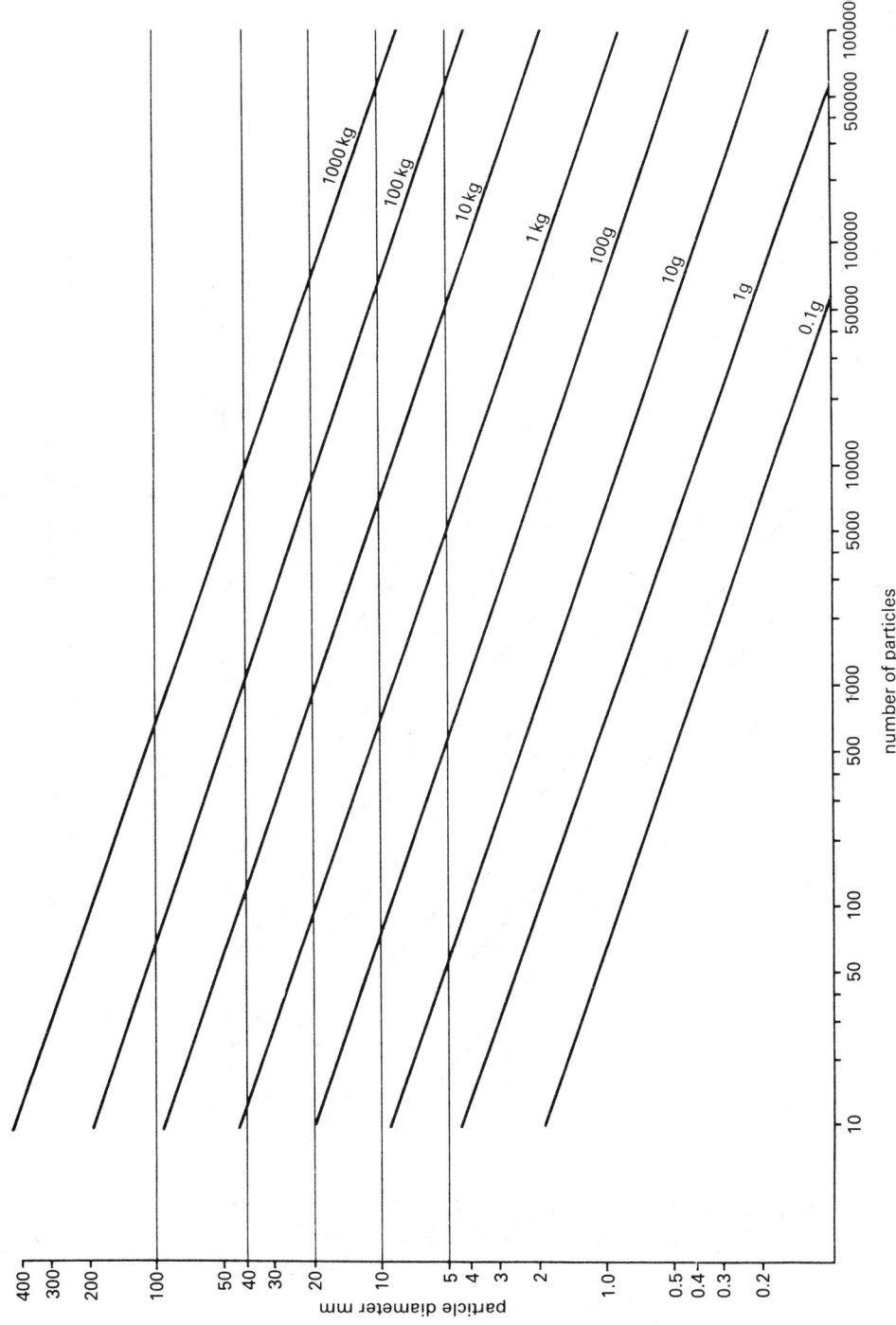

Fig. 2.4 — Relationship between size of particles, number of particles, and weight of sample.

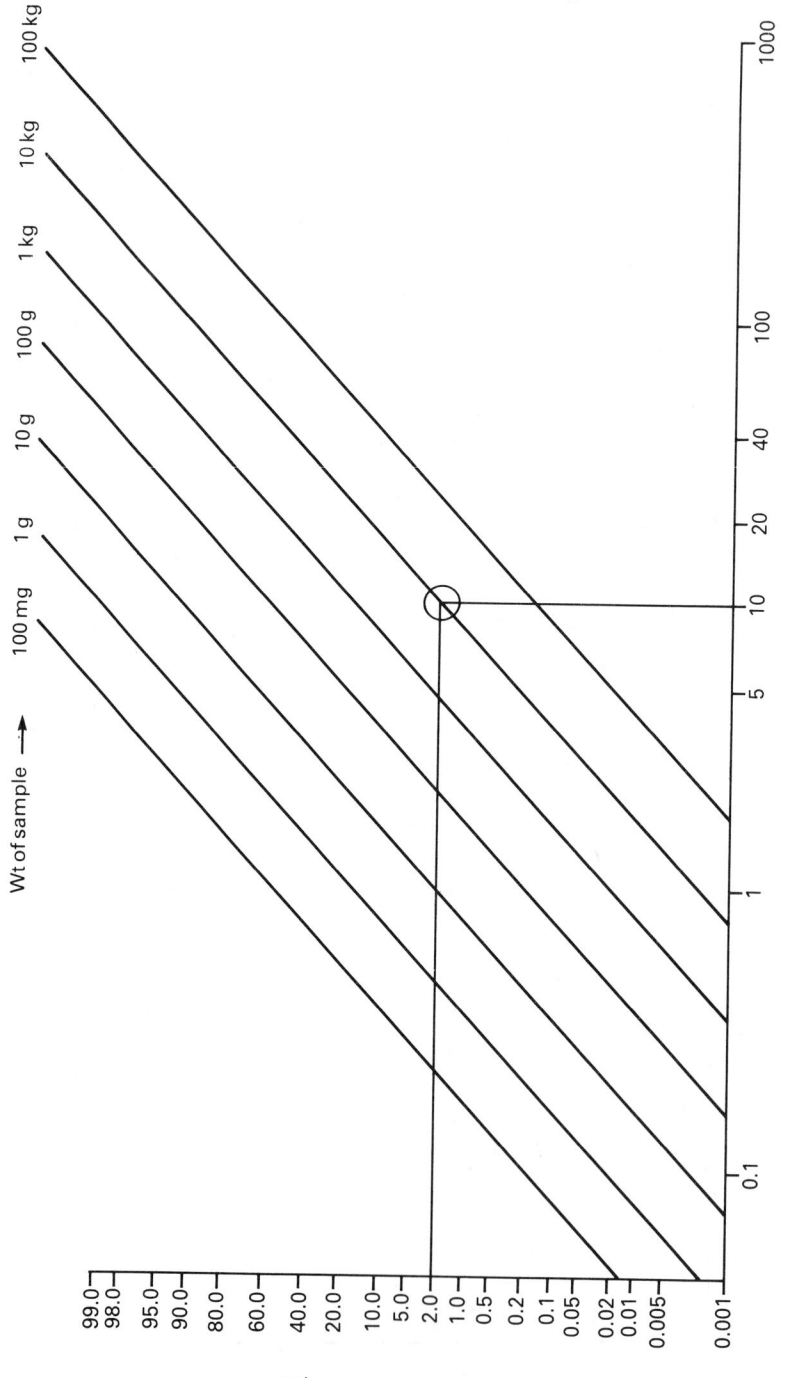

Fig. 2.5 — Weight of test portion necessary to achieve an accuracy of ± 10% relative for a given constituent (based on single-sized, fully-liberated particles).

In cases where the mineral is not free, the number of particles (or weight of sample) required for a given sampling error can be reduced according to the ratio between the size of the mineral and the size of the enclosing particle. Thus, if the diameter of the particle is on average 10 times larger than the mineral it contains, then the weight of sample can be reduced by a factor of $\sqrt{10}$. If the particle is 100 times larger than the mineral, the sample can be reduced by $\sqrt{100}$ times, and so on. However, such reductions are valid only when the mineral is distributed randomly within the host. In cases of doubt it is better to assume complete liberation.

To accommodate these and other factors present in practical sampling situations, a comprehensive sampling method has been developed by Pierre Gy (1982) which is applicable to all mineral products whether used for aggregates or other purposes, and is widely used in the mining industry. It has been briefly described by Ottley (1966).

The fundamental relationship is:

$$M = \frac{Cd^3}{u^2} \tag{2.6}$$

where d = diameter of the largest particle present, conventionally taken as the sieve size on which 5 per cent of the sample is retained and 95 per cent passes.

u = relative standard error = s/p
M = weight of sample
C = a constant for the material being sampled, which can be defined in terms of four parameters: f, g, L, m, (that is, $C = fgLm$), where
f = shape factor which can be taken as 0.5 in all practical cases likely to be met with in sampling aggregates
g = particle size distribution factor which usually has a value of 0.25, except for closely-sized materials where it is 0.5
L = liberation factor which has values between 0 and 1. Where particles of different minerals are completely separate and liberated from each other, $L = 1$. Where the material is completely homogeneous $L = 0$.
m = mineralized composition factor which can be calculated from:

$$m = \frac{1-p}{p}[(1-p)r + pt] \tag{2.7}$$

where r is the density of the mineral of interest, t is the mean density of the other minerals, p is the proportion of the mineral of interest in the material being sampled.

Of the above factors, f can be taken as a constant for aggregates. The particle size distribution factor (g) can be assumed to be 0.25 in crushed but unsized material and 0.5 in very closely-sized material produced by screening. The liberation factor is difficult to apply without a detailed knowledge of the nature of the material. Where the material is completely segregated into particles of different composition, as was the case with our earlier examples concerning separate black and white particles, then $L=1$. When the material is completely homogeneous, as might be the case with a glass, then L would approach zero, C would also approach zero, and in consequence the required sample weight would become vanishingly small. This would indeed be the case with a pure crystal or perfect glass when even one molecule could adequately represent the composition of the whole. In other cases where a particular mineral constituent is attached to other minerals, it is necessary to estimate the ratio of the maximum size (d') of the mineral grains under analysis to the maximum size (d) of the particles in the sample. If the material were to be crushed to the extent needed to give complete liberation of the particles of interest, then d' would be the size of the material in this state. Gy gives the following rule of thumb for calculating the liberation factor:

$$L = \sqrt{(d'/d)} \tag{2.8}$$

Some values obtained by using this formula are shown in Table 2.6.

Table 2.6 — Values of the liberation factor L

d/d'	100	40	10	4	1
L	0.10	0.16	0.32	0.50	1.00

The mineralized composition factor, m, is greatly simplified where all the particles have the same density, as is the case with most aggregates; it then depends on the composition and mean density of the particles.

Pierre Gy's formula is not easily expressed graphically, although slide rules and other calculating devices have been developed to facilitate its application. At present programmable calculators probably offer the best means of applying the formula on a routine basis.

Gy's formula is intended to be used to derive an approximate sample size — it does not give an exact prescription. Later sections of this chapter describe how sampling variability can be checked experimentally. If it is found that sampling errors are larger in some cases than in others, then Gy's formula can be used to investigate the reasons.

One example of the application of Gy's formula occurs in draft Part 104 of BS 812 (BSI 1988) which is concerned with petrological examination. The minimum masses for test portions currently proposed for draft Part 104 are larger than those given in some other Parts of BS 812. This is because Part 104 is concerned with the quantitative analysis of materials which may be present at low concentrations and in a fully-liberated condition. Hence large samples are necessary to ensure that such analyses are not subject to serious errors. Other Parts are concerned with properties which are less likely to depend on liberated material present in low concentrations, hence smaller samples have been allowed in the expectation that the associated errors will be manageable. Another example where large samples would be necessary arises when it is required to detect the presence of deleterious contaminants, for example clay, which may be present in small amounts as liberated consitutents.

To use Gy's formula to select the correct sample mass, it is necessary to know the composition of the material a — a chicken and egg dilemma! Draft Part 104 overcomes the problem by using preliminary estimates of the proportions, and it also simplifies the procedure by using likely values for the physical parameters f, g, L, r, and t in equations (2.6) and (2.7). Thus it contains the following instructions:

Prepare a representative test portion, using the sample reduction procedures given in Clause 6 of BS 812: Part 102. If a constituent of particular interest is known to be present and if the approximate proportion of that constituent has been estimated by preliminary examination or has been inferred from previous results, calculate the minimum mass of the test portion, using the following formula:

$$M = 0.0002\,(100-p)\,d^3/p$$

where

M = minimum mass of test portion, kg
p = estimated proportion of the constituent of interest, per cent by mass
d = nominal maximum particle size, mm

Otherwise prepare the test portion for routine examination, using the appropriate mass given in the following table.

Minimum quantities of test portions

Nominal maximum particle size	Minimum mass of test portion
40 mm	50 kg
20 mm	6 kg
10 mm	1 kg
5 mm and smaller	100 g

The masses given by the formula and the table relate to an accuracy level of plus or minus 10 per cent relative, and were calculated by substituting in Gy's formula (equation (2.6)) the following values of the physical parameters:

$f = 0.5$ (the shape factor likely to be appropriate when sampling aggregates)
$g = 0.4$ (particle size distribution factor for fairly closely sized materials)
$L = 1$ (liberation factor for fully liberated materials)
$r = t = 2650 \times 10^{-9}$ kg/mm^3 (likely density of aggregates)

together with the relative standard error corresponding to the chosen accuracy level:
$u = 0.05$.

These values give the coefficient of 0.0002 in the formula in Part 104. The masses in the table in Part 104 were obtained from the formula by using the estimated proportion of the constituent of interest of 20 per cent (that is, $p = 20$) and rounding the calculated values somewhat.

2.1.4 Sample sizes for size analysis

The method given in equations (2.6) and (2.7) for calculating sample sizes is not appropriate when the object of the analysis is to determine the particle size distribution. Gy gives a method in his book for this situation which may be simplified to the following formula when the particle densities are similar.

$$M = 10^{-9}[(1/p - 2)v + \Sigma p_i v_i] r/u^2 \qquad (2.9)$$

where
M = sample mass in kg
r = particle density in kg/m^3
u = relative standard error (usually 0.05)

as before, and

p = mass proportion of the fraction of interest
p_i = mass proportion of fraction i
v = average particle volume of fraction of interest
v_i = average particle volume of fraction i

and the summation is over all size fractions.

The average particle volume of a fraction defined by two sieve sizes d_1 mm and d_2 mm is given by:

$$v = (d_1^3 + d_2^3)/4 \text{ mm}^3 \qquad (2.10)$$

As an example, consider the calculation of the sample size needed for a size analysis of a coarse material containing particles ranging in size from 63.0 mm to 5.00 mm.

Applying equation (2.10) to the sieve sizes in Table 2.7 gives the average particle sizes in the table. To use the method you need to either make a guess at the particle size distribution of the material or to have the result of a preliminary assessment: for Table 2.7 it is assumed that the material is evenly graded, giving the mass proportions shown in the table. The value

$$\Sigma p_i v_i = 19976.6 \text{ mm}^3$$

may now be calculated. If the particle density

$$r = 2650 \text{ kg/m}^3$$

and a relative error of ±10 per cent is required in the measurement of the percentage of material in the range 63.0 mm to 37.5 mm, so

$$u = 0.05 \quad v = 75695 \text{ mm}^3 \quad p = 0.205,$$

then equation (2.9) gives a sample mass of

$$M = 10^{-9} \cdot [(1/0.205 - 2) \, 75695 + 19976.6] \cdot 2650/0.05^2$$
$$= 252 \text{ kg}.$$

The sample masses required to give the same relative error in the determination of the percentage of material in the other size ranges may be calculated similarly, and are given in Table 2.7. In this example though the situation is dominated by the sample size required by the 63.0 to 37.5 mm fraction, so the sample size would need to be 250 kg.

For fine material the method can give very small sample sizes, so that in practice the sample size will be determined by the need to use a standard sampling device and an adequate number of increments.

2.1.5 Sample reduction

It will be clear from Figs 2.4 and 2.5 that, where aggregates are coarse and the proportion of material under analysis is small, then very large sample weights will have to be taken to maintain an acceptable error. For example, reference to Fig. 2.5 and simple extrapolation will show that, if the particle size is about 100 mm and the composition at about the 2 per cent level, then a sample size of the order of 10 000 kg will be required to maintain an accuracy of ±10 per cent. Even when it is practical to take a sample of such a weight it would be difficult, if not impossible, to subject it to any sort of analysis. In such cases, it is necessary to crush the sample to a smaller size, then take a sub-sample by a random method and, if necessary, repeat this exercise several times until a conveniently sized sample is obtained for analysis. In the more extreme cases, where for example a sample is required for chemical analysis, a sample of several tonnes may have to be reduced to a few grams of fine powder.

Table 2.7 — Calculation of sample mass required for size analysis when the material has the size distribution shown in the table and a particle density of 2650 kg/m³

Fraction	Size range	Average particle volume	Mass proportion	Required sample mass
i	mm	v_i mm³	p_i	kg
1	63.0 to 37.5	75695	0.205	252
2	37.5 to 20.0	15184	0.248	54
3	20.0 to 10.0	2250	0.274	25
4	10.0 to 5.0	281	0.274	22

Provided that successive samples are taken by random methods and appropriate sample weights are selected at each stage according to a suitable formula, then the relative error in sampling by this method can be calculated as follows from:

$$\Sigma e = \sqrt{(e_1^2 + e_2^2 + \ldots e_n^2)} \qquad (2.11)$$

where e_1 is the relative error of the 1st sampling operation, e_2 is the relative error of the 2nd sampling operation, and e_n is the relative error of the nth successive sampling operation. For example, if a sample is taken from an original lot which is representative to within ±10 per cent, and this sample is then crushed to a finer size and a further sub-sample is taken which is also representative to within ±10 per cent, then the sub-sample is representative of the original lot to within $\pm\sqrt{(10^2+10^2)}$ or approximately ±14 per cent. Hence, when making successive sampling operations, it is necessary to bear in mind the overall error inherent in the whole sample reduction sequence and to design the process accordingly. It is also necessary to recalculate the liberation factor (L) for each size at which sampling is carried out until the maximum value ($L=1$) is reached.

A useful tool for designing a sample reduction operation is to draw the equation

$$M = C(100-p)\,d^3/p \qquad (2.12)$$

on log/log paper, where it appears as a straight line. The value of the constant C is that used to determine the sample mass. The sample reduction stages should be chosen so that the sample always stays on the 'safe' side of the line. For example, suppose that 240 kg of 40 mm materials is to be reduced to 2.36 mm material. With $p=5\%$, $C=0.0002$, the equation corresponds to the line shown on Fig. 2.6. From the figure, a convenient procedure is to crush to 10 mm, reduce to 4 kg, crush to 2.36 mm and reduce to about 50 g. Other schemes could be devised, depending on the equipment available for crushing.

2.1.6 Analysis

When a suitable sample has been selected by the methods outlined above, it may be analysed chemically or physically, and the results will be relevant to the original lot within the errors and confidence limits arrived at during the sampling process.

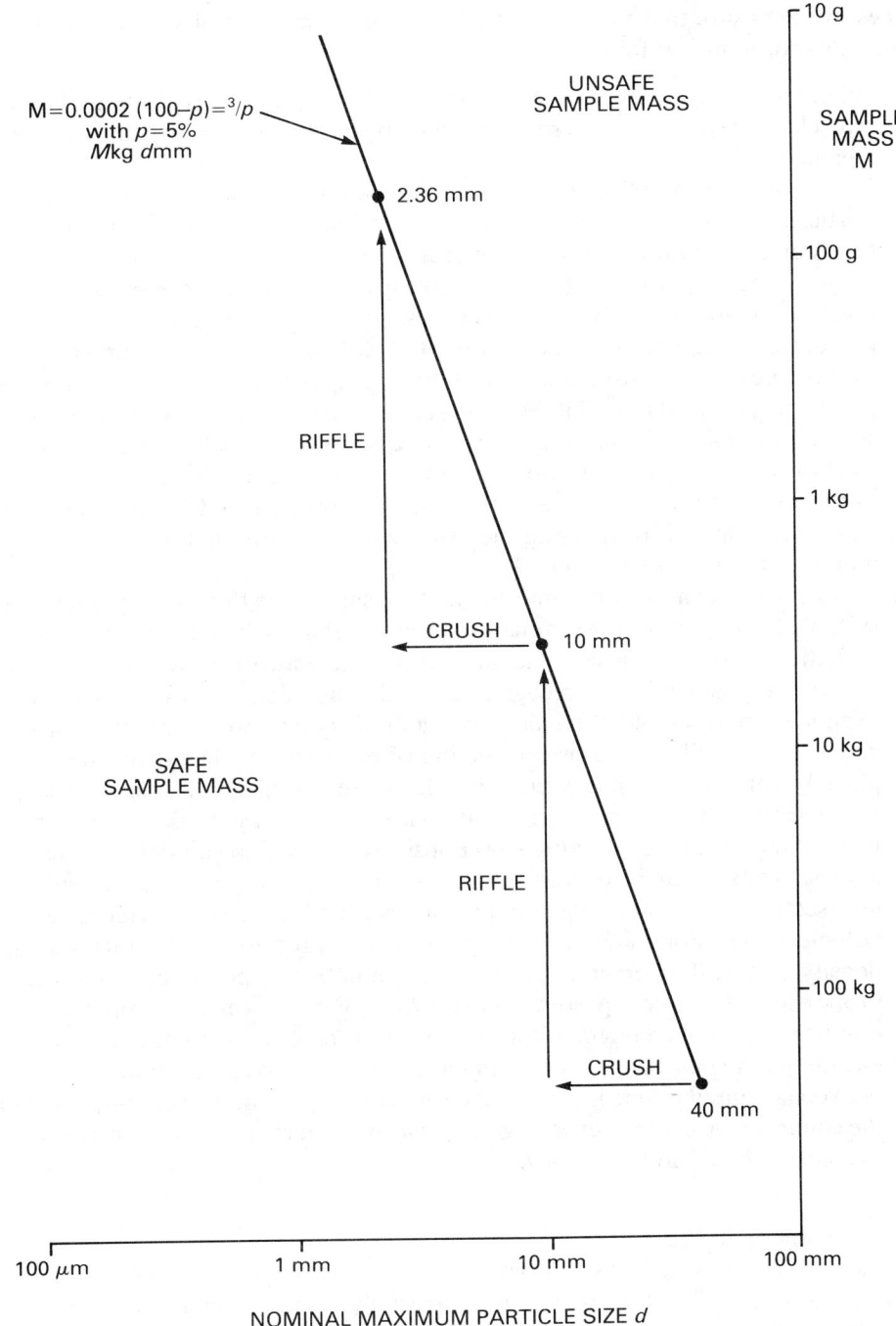

Fig. 2.6 — Graph for the control of sample reduction operations.

If the sample is to be subjected to mineralogical or petrographic analysis, it is necessary to ensure that the results are reported on a weight or volume proportion basis, by proceeding as follows:

(1) Where the constituents of a sample can be identified as discrete liberated particles, it is preferable to separate them by hand into separate lots that can be weighed.
(2) Where hand separation becomes tedious or impractical, as with very fine particles, it is necessary to take a sub-sample that is representative within known limits and to count the constituent grains, by eye or under a binocular microscope, depending on the grain size. Because the number of particles counted must be converted to a weight, it is necessary to be able to estimate the volume and density of the particle. The density follows easily from the composition, but the volume must be estimated separately for each particle, which is obviously not practical in unsized material. However, if the particles are screened into closely-sized fractions, an ideal volume can be assumed for each particle in a given fraction and a weight proportion can be readily arrived at. Hence, grain counting as a method of analysis can be applied only to closely-sized fractions, samples of which are selected by applying the usual sampling formula for a given error to material that has been screened.
(3) In other cases examinations may be made of samples in thin or polished sections suitable for microscopic examination or as polished slabs that can be examined with the naked eye. The use of a microscope is necessary only when a constituent cannot be resolved and recognized by the unaided eye. For this type of examination, a sample of particulate material is cast into a resin block which is cut so as to exhibit an unbiased section of particles. An imaginary net is then placed over the section in which the mesh size is larger than the largest grain present. When microscopic examination is necessary, a special stage that moves in fixed increments on rectangular coordinates is used to simulate the net. The analysis consists of recording the nature of each mineral present under the intersections of the net. The proportions recorded can be converted directly to volume proportions and ultimately to weight proportions by introducing a density factor. The error depends on the number of points counted and the proportion of mineral present, as shown in Fig. 2.2 where n represents the number of points counted. However, in this type of analysis it is necessary to ensure that no part of a section is traversed more than once and that the distance between counted points is larger than the largest grain in the section, otherwise the count can be made over as large an area and as many sections as are necessary to achieve the desired accuracy.

2.1.7 Sampling *in situ*

One advantage of sampling small batches of processed aggregate, rather than the rock *in situ*, is that the whole of a batch can usually be moved and passed through a sampling routine that to a greater or lesser extent mixes and randomizes the material before a single sample is recovered that is representative within known limits. Although these procedures can be expensive for large tonnages, and thus cannot always be justified financially, they are nevertheless available for use in difficult sampling situations where extreme segregation is expected, and they should always be used for smaller batches and laboratory samples.

In contrast, sampling *in situ* is by definition concerned with sampling material as it occurs without any facility for mixing before samples are taken. Such situations occur when deposits in the ground, or large stockpiles that cannot be moved, are to be sampled. In these circumstances a wide spread of values is likely to occur, hence the selection of a single representative sample may be much more difficult. The main problem is that all parts of the deposit are unlikely to be equally accessible and there might well be a tendency to select samples from a surface outcrop, the face of a quarry, or the surface of a large stockpile, resulting in a biased sample because the less accessible parts have been neglected. This situation can be overcome only by careful design of sampling procedures, in most cases it means resorting to drilling or digging sample pits that pass through the whole deposit and are spaced at regular intervals.

Before going further with procedures there are some basic concepts that should be clarified. The concept of a single representative sample may not be relevant to an *in situ* deposit in all cases. For example, if a deposit changes uniformly in composition in the same general direction that quarrying is progressing, then a single average sample may represent the output of the quarry only over a very short time interval. In these circumstances it is necessary to consider what the sample is meant to represent and whether or not a plan showing the distribution of values might not be more useful than a single sample. In addition, if the material is to be extensively processed after quarrying, it will be necessary to take account of the changes brought about by the processing before any *in situ* analysis can be related to product composition.

In situ sampling can be based on two quite different principles. The first is that of random sample selection, in which each part of the deposit has an equal chance of being sampled. The second is based on the use of an expert sampler, usually a geologist who, on the basis of his knowledge of the geology of the deposits, selects samples which, in his best judgement, will be representative of the whole. The second method will probably be quicker and cheaper as the geologist will try to keep the analytical work to a minimum and in many cases might be able to avoid taking samples from less accessible parts of the deposit. For example, the geologist might well confine an investigation to a quarry face because his knowledge of the local geology leads him to expect that the face will be sufficiently representative of the whole deposit. Whereas, with a random sampling method, any part of the deposit might be sampled, a situation which would almost certainly necessitate some drilling into the deposit behind the quarry face.

The main disadvantage of using the expert sampler method is that the accuracy of the samples cannot be quantified. Their status depends essentially on the reputation and reliability of the geologist. In many cases this is entirely acceptable, but where there is the possibility of a dispute, a less subjective and demonstrably impartial method may be preferred.

With any form of *in situ* sampling, however, the first requirement is to define the entity that is to be sampled. This may seem straightforward in the case of a static stockpile that is clearly visible and obviously of finite extent. However, with a deposit in the ground it is essential that the material exposed to sampling comprises the entirety of the reserve that is to be worked. Hence, the first step before any samples are taken is to define and quantify the reserve in terms of surface area, depth, and estimated *in situ* tonnage. At this stage it is also important to be aware of any interbedded or other inferior materials that can be avoided by selective working, and to ensure that any such materials are excluded from the quantified reserve.

In a new aggregate working it is usual to sink some boreholes or trial pits to evaluate the reserve. If this operation is properly designed it can form a most effective method of evaluation as it allows a range of samples to be collected over a large reserve under statistically controlled conditions well in advance of quarrying operations. Subsequently the data are applicable to the product throughout the life of the reserve, and its availability might well obviate the need for further testing.

There is no optimum sampling interval that can be universally applied in advance to an aggregate reserve in the ground. Much will depend on the inherent variability of the deposit, the nature of which is apparent only after sampling has been carried out. However, it is essential that boreholes or trial pits are distributed on a regular or random pattern in such a way that each part of the previously defined reserve has an equal chance of being selected for sampling.

For many assessment purposes a pattern of sampling points spaced at approximately 200 m intervals, with samples taken at every 2 m of depth, will be sufficient. Such a system would give approximately one sample for every 200 000 tonnes extracted. However, with sand and gravel in deposits less than 3 m thick it would be preferable to space the holes at 100 m intervals and take samples at depth intervals of 1 m. Much will depend on the observed variability between the samples, the latitude allowed in the specifications, and the funds available for detailed sampling.

The UK domestic mining industry has abundant experience of sampling UK reserves on very intensive patterns where close quality control is required. For example, the former ironstone workings at Scunthorpe, opencast coal workings, and silica sand operations, have all utilized close drilling patterns and extensive sampling to ensure a thorough evaluation of reserves before they are worked, and close control of composition during the extraction operation. The aggregate industry traditionally has not needed to go to such lengths, although there is no technical reason for not doing so where there is a financial justification.

Having assembled a number of random samples from a resource by the methods outlined above, the next step is to make sure that each sample represents the same volume of deposit, or otherwise make appropriate adjustments. Subsequently a mean and standard deviation for the whole deposit can be calculated. However, if there is a significant difference between material from different sampling locations — that is to say, if all the material from one borehole is significantly different from all the material from another — then it might be better to calculate an average from each borehole and subsequently calculate an average for all the boreholes weighted according to the depth of strata that each hole represents.

2.1.8 Sampling equipment

The fundamental problem in the mechanics of sampling lies in the recovery of a sample by truly random methods.

If the problem is conceived in terms of a population consisting of identically sized and shaped particles, such as billiard balls, in which some are red and others are white, then random samples are easily selected. This is because the red balls are identical in every respect, save colour, to the white balls and hence any selection method, provided that it takes no account of colour, should produce a random sample. Blindfold selections from well-mixed balls spread out on a table top, arranged in a heap or contained in a bag, could be expected to provide equally random samples with errors, depending on the size of sample, that could be easily calculated.

Unfortunately, particles of natural aggregates are not uniform; they may differ in shape, angularity, surface texture, and to some extent in size and density. These properties affect the velocity at which the particles move in air or water, and the rate at which they roll down inclined surfaces and over the surfaces of other particles. The practical effect of these differences is that the various components often tend to unmix, segregate or stratify when they are tipped on to a conical heap, or when they are loaded into a drum or other container. In the case of the drum, particularly if it is subjected to much vibration, the coarser, more dense, and less angular particles will tend to report toward the base, while the finer material remains at the top. In stockpiles, the coarse material often tends to accumulate around the base. Similar sorts of segregation can also occur on conveyors and in associated handling plant, sometimes with a superimposed cyclical variation in the product brought on by the operating characteristics of the equipment.

All such tendencies toward segregation should be considered and allowed for in any sampling operation which, as far as is practicable, should aim to ensure that each particle has an equal chance of selection. For example, if samples are to be taken from a conveyor, it is desirable to stop the belt and take all the material within a narrow strip extending right across the belt. Samples from the belt may then be taken in the same way at regular or random intervals to ensure that all the material has an equal chance of selection. If the material is present in a stockpile it is desirable, although not perhaps always practicable, to pass the whole of the stockpile along a conveyor and take samples as described above.

Small stockpiles and other lots of aggregate that can be assembled into piles can be sampled by the time-honoured technique known as cone and quartering. This requires the aggregate to be formed into a conical heap by piling all the sample on to the apex so that it is distributed with a radial symmetry; then bisecting the pile through its apex, and finally bisecting one of the halves through the apex to form a sample that is one quarter of the original. Further reduction may be carried out by assembling the quarter sample into a smaller symmetrical conical pile and repeating the operation. When the conical piles are of known size, a cruciform structure made up of steel plates can sometimes be used advantageously to quarter the sample. Aggregates in vehicles, drums, or other containers should generally be expected to have segregated to some extent, hence the whole of the container or vehicle should be emptied into a conical pile or onto a conveyor, and sampling should be carried out by the appropriate method already described.

For the accurate division of materials when relatively small weights are entailed, special equipment for dividing samples has been developed that requires only simple routine procedures on the part of the operative. The most commonly used device of this type is a riffle such as that described in BS 812: Part 102. A riffle essentially comprises a hopper of rectangular cross-section that feeds material directly into a series of adjacent uniform slot-shaped apertures arranged so that alternate apertures discharge in opposite directions. Material passing through the device is then divided into equal parts and collected in separate trays on opposite sides of the riffle (Fig. 2.7). Subsequently the contents of one tray can be reduced further by repassing

Fig. 2.7 — Laboratory scale riffle.

as many times as is necessary to achieve the desired sample size. In riffles the width of the slot-shaped apertures, and hence the size of the whole apparatus, should be proportional to the diameter of the largest particle to be handled, and in general the aperture width should exceed the maximum particle diameter by a factor of 1.5. As large samples of coarse material are needed to maintain a given accuracy, the advantage of large riffles with a high throughput in handling coarse material will be readily appreciated. Similarly in the preparation of fine materials for analysis, where amounts are small, the advantages of a riffle that allows a small sample to be distributed uniformly across the apertures will be obvious.

When small samples are to be recovered from large lots on a routine basis, it is often convenient to make use of a rotary sampler. There are several variations of the basic design, but the machine consists essentially of a series of at least 16 identical hoppers, arranged around the periphery of a horizontal rotating disc. Each hopper delivers a sample into a separate glass or metal container to which it is attached, often by means of a screw thread (Fig. 2.8). A stream of particles is delivered through a fixed vibrating feeder and is allowed to fall vertically into the hoppers, which rotate slowly, allowing each hopper and container to collect an equal part of the material. The speed of rotation will depend on the diameter of the disc, and the size of the original sample. It should be fast enough to ensure that each hopper passes under the particle stream many times, and slow enough to avoid strain on the equipment or air turbulence that might prevent the particles from falling freely into the hoppers. A small sample can be recovered from the rotary sampler by taking the contents of one container. Larger samples can be obtained by combining the contents of several containers, as required. To avoid the tedium of cleaning out a large number of containers, some machines include a close-fitting base to the rotating disc, which collects the particles when the containers are not in place beneath the hoppers. With the use of such an attachment, the sampling operation can proceed with only one or two containers in place to collect small samples, while the remaining material is recovered in one batch.

Using the above principle, rotary samplers have been developed for continuous operation by arranging for the sample collected in one hopper to be continuously discharged from the machine while the remaining material passes through the other hoppers, and is then recombined and discharged separately. In some instances the machine may be reduced to a single hopper that intersects a falling stream of particles once in each revolution.

As with riffles, the size of a rotary sampling machine depends primarily on the maximum size of particle that it is intended to handle, and the hoppers should be bigger than the maximum size of particle they are to handle by a factor of at least 1.5.

2.2 PRECISION
2.2.1 Variations arising during testing
In addition to the variation that arises during sampling and sample reduction, variation can also occur during the testing of a sample. Table 2.8 shows the results of repeat determinations of drying shrinkage (see section 3.3 in the next chapter). Each row of results in Table 2.8 shows the drying shrinkage observed on three prisms made from the same batch of concrete, so the three prisms in a row were as alike as can be achieved in practice. Nevertheless, variation between the results obtained on individual specimens is evident. Table 2.9 shows the results of repeat measurements of flakiness index. Here each row contains the results of tests on two test portions of the same sample, and, again, variation between the test portions can be seen. Table 2.10 shows some measurements of sulphate content where duplicate samples were examined. In this example there appears to be variation between the samples, even though the duplicate samples were taken from the same stockpile of aggregate. These examples illustrate different ways in which variation in the test results can occur.

There is also the probability that results can differ between individual laboratories on a consistent basis. One cause of these variations could be heterogeneity of the aggregates, giving rise to differences between samples and differences between test portions of the same sample. Other causes of variation could be a failure of an

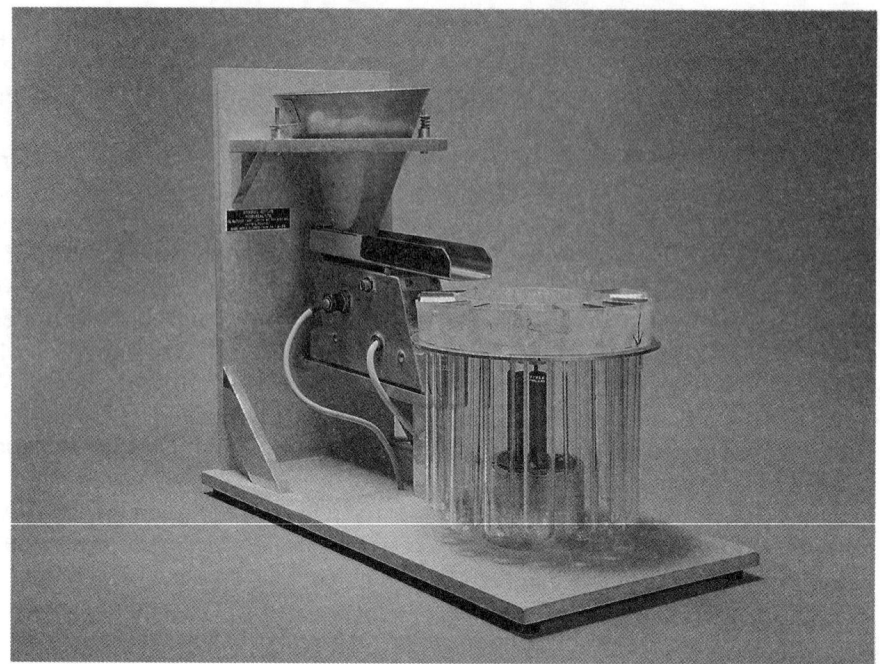

Fig. 2.8 — Laboratory scale rotary sampler with 16 hoppers.

Table 2.8 — Drying shrinkage data — microstrain

Test portion	Specimen 1	Specimen 2	Specimen 3	Between-specimen range
1	600	620	610	20
2	560	580	580	20
3	700	650	660	50
4	610	570	600	40
Average between-specimen range				33

operator to follow a test method rigorously (either through carelessness or a lack of understanding of the method), or differences in operator technique, or in the equipment or method used in different laboratories.

2.2.2 Standard definitions of precision coefficients

The general term used to describe the variation between repeated tests is 'precision', and two measures of precision have come into common use. 'Repeatability' is used to

Table 2.9 — Flakiness index data

Laboratory	Laboratory sample	Test portion 1	Test portion 2	Between-test-portion range
1	1	18.0	20.3	2.3
	2	16.7	16.9	0.2
2	1	14.0	13.6	0.4
	2	15.9	15.4	0.5
3	1	23.9	20.3	3.6
	2	21.3	20.3	1.0
Average between-test-portion range				1.3

Table 2.10 — Sulphate content — % total sulphate ion

Laboratory	Laboratory sample 1			Laboratory sample 2			Between-lab-sample range
	Test portion 1	Test portion 2	Sample mean	Test portion 1	Test portion 2	Sample mean	
1	1.16	1.12	1.14	0.89	0.89	0.89	0.25
2	0.45	0.45	0.45	0.62	0.62	0.62	0.17
3	0.85	0.86	0.86	0.87	0.88	0.88	0.02
4	1.21	1.27	1.24	1.05	1.08	1.07	0.17
Average between-laboratory-sample range							0.15

describe the variation that occurs when tests are performed under conditions that are as constant as possible, that is when tests are performed during a short interval of time in one laboratory by one operator using the same reagents and equipment. 'Reproducibility' is used when tests are repeated in different laboratories with different operators and with differing equipment and supplies of reagents. These terms thus refer to two extreme conditions, repeatability measuring the minimum variation that occurs between test results, and reproducibility the maximum.

These terms are defined in ISO 5725, and in its equivalent BS 5497 (BSI 1987), and are discussed in that Standard in some detail. However, the definitions therein suppose that tests can be repeated on identical samples of material. With aggregates, there will always be differences between samples, either in their grading or in the proportions of the constituents. The discussion of sample sizes, sampling, and of sample reduction methods earlier in this chapter was aimed at obtaining samples with properties within an acceptable tolerance of the notional true value for the material being sampled. Nearly all aggregate tests are destructive, so variation of the material under test will almost inevitably contribute to the variation of the test results, and must be considered when applying the concepts of precision to aggregates.

2.2.3 Hierarchy of sampling terms
It is important to define a hierarchy of sampling terms on which unambiguous definitions of precision terms can be based. The definitions given here are taken from BS 812: Part 102 (BSI 1984). They are illustrated by Fig. 2.9.

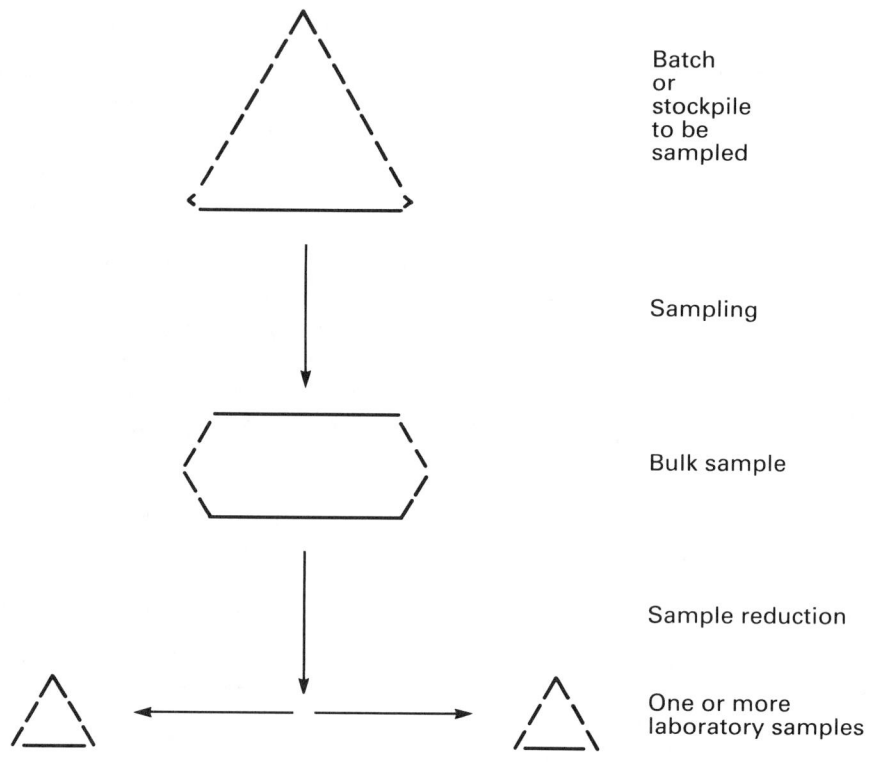

Fig. 2.9 — Sampling terms.

Bulk sample. A bulk sample is obtained from the material being sampled as the aggregation of the sampling increments. The size of a bulk sample is determined by the number and size of the sampling increments: the size of the sampling increment being a specified function of the particle size, and the number of increments depending on the heterogeneity of the material being sampled, as discussed earlier in this chapter. The bulk sample should also contain sufficient material for all the tests for which it is to be used.

Laboratory sample. The material sent to a laboratory is described as the laboratory sample, and may either be the whole of the bulk sample or may be obtained from it by sample reduction. The size of the laboratory sample is a function of the particle size and the liberation of the constitutents of interest, as discussed earlier, and again must be sufficient for all the specified tests.

Test portion. At the laboratory, the laboratory sample is reduced by one or more reduction operations to the quantity required by a particular test method. The material produced at the final stage of sample reduction, and used as a whole in the test, is referred to as the test portion.

Specimens. A particular test may require two or more specimens to be made from a test portion.

As an example, suppose that it is required to assess the drying shrinkage property of the 20 to 10 mm aggregate produced by a quarry. A bulk sample is obtained by taking 40 sampling increments of about 3 kg each. This is reduced by riffling to a 25 kg laboratory sample. At the laboratory this is further reduced by several riffling operations to the 1.5 kg test portion needed to make a batch of concrete from which the three prisms required by the drying shrinkage test method can be made.

The value of this hierarchy of sampling terms is that it avoids arbitrariness in the definitions of precision coefficients and in their applications. Differences between test portions will relate to the size of test portion required by a test method. Differences between bulk or laboratory samples will relate to the sizes of samples required by the sampling standard.

2.2.4 Test result

In the above example a 'test result' could be understood to be either the drying shrinkage of a single prism, or the average drying shrinkage of the three prisms. The precision of the average will not be the same as the precision of the single determination. There are many other similar examples where there is a need to be clear about the meaning of a 'test result'. The most useful definition of a test result is the result obtained by applying the test method to a test portion. On this basis, when more than one determination is made on a test portion, the test result will be calculated as the average of a number of determinations, as shown in Fig. 2.10. All

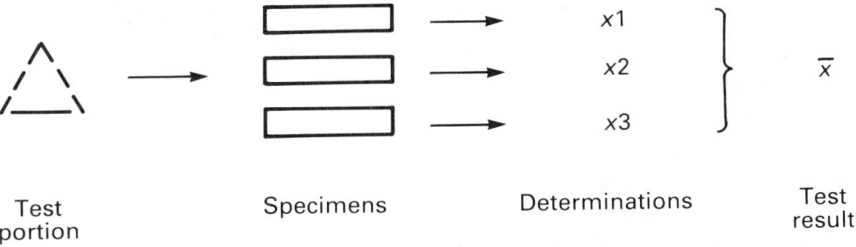

Fig. 2.10 — Test result.

variations that occur in the test method will then show up in variations between test results. In the example, a test result should be defined as the average drying shrinkage of the three prisms. If a test result were to be defined as the result obtained from a single prism, then differences between test portions or variations in the batching and mixing of the concrete would possibly fail to be seen in differences between prisms.

Further to this, sample reduction is always needed to obtain a test portion from a laboratory sample, so the possibly substantial variations that can occur in the sample reduction operation are an unavoidable part of the test method. To ensure that these variations are seen in differences between repeat tests on a laboratory sample, it is important that the test portions for repeat tests are separated at the first stage of sample reduction, as depicted in Fig. 2.11.

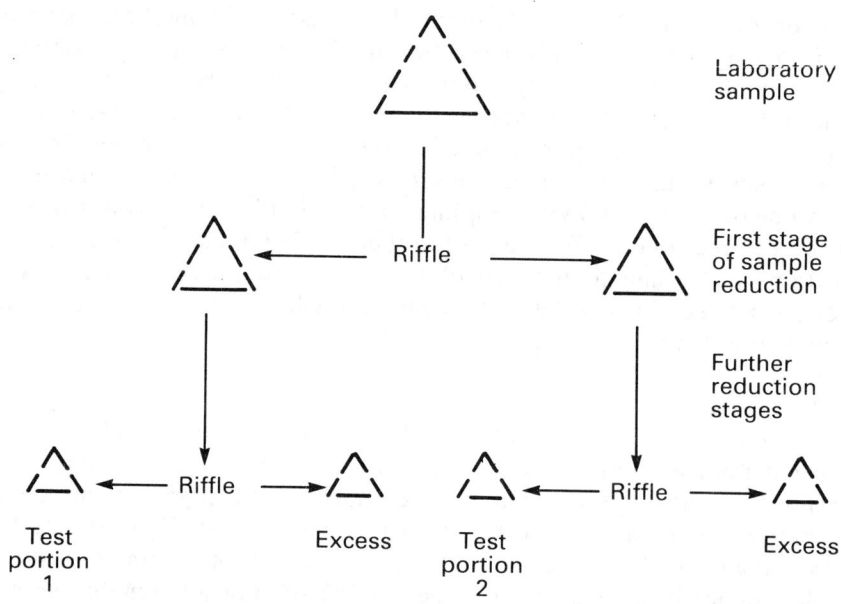

Fig. 2.11 — Sample reduction.

2.2.5 Definitions of repeatability and reproducibility for aggregate tests
Repeatability conditions arise when test results are obtained:
(a) on different test portions,
(b) of the same laboratory sample,
(c) with the same test method,
(d) in the same laboratory,
(e) by the same operator,
(f) using the same reagents and equipment,
(g) within short intervals of time.

Reproducibility conditions arise when test results are obtained:
(a) on different test portions,
(b) of the same laboratory sample or of different laboratory samples,
(c) with the same test method,
(d) in different laboratories,
(e) with different operators,
(f) using the respective laboratories' reagents and equipment,

Repeatability value (r_1) The difference between two test results obtained under repeatability conditions may be expected to exceed r_1 on one in twenty occasions.

Reproducibility value (R_1 or R_2). The difference between two test results obtained under reproducibility conditions may be expected to exceed R_1 or R_2 on one in twenty occasions.

The precision coefficients are given subscripts to indicate that they differ from the standard definitions which are given in terms of tests on identical samples. R_1 applies when the test results are obtained on the same laboratory sample, R_2 when they are obtained on different laboratory samples. (Both circumstances may arise in practice.) With these definitions, variation between laboratory samples contributes to R_2, so that inadequacies in the sampling procedure will increase R_2. Variation between test portions contributes to r_1, R_1 and R_2: in particular, if the value for r_1 for a test method is exceptionally large the reason could be that the mass of test portion is too small.

The precision coefficients may also be defined by using a mathematical model:

$$y = m + B + G + e \tag{2.13}$$

where
    y is a test result
    m is the general average test result
    B allows for variation between laboratories
    G allows for variation between laboratory samples
    e allows for variation between test portions.

This model is similar to the model given in ISO 5725 but with the addition of the term allowing for variation between samples. Each of 'B', 'G', and 'e' in the model represents populations of possible values which may be taken to have zero means, and variances:

    V_L = variance between laboratories
    V_S = variance between laboratory samples
    V_r = variance between test portions.

The precision coefficients may now be defined by:
$$r_1 = 2.8\sqrt{V_r} \qquad (2.14)$$

$$R_1 = 2.8\sqrt{(V_r + V_L)} \qquad (2.15)$$

$$R_2 = 2.8\sqrt{(V_r + V_L + V_S)}. \qquad (2.16)$$

These equations are of interest in showing how the variances contribute to the precision coefficients. They are also of value because they can easily be modified to give coefficients that can be used when averages of test results are being compared instead of individual test results. For example, if averages of 'n' test results are being compared, replace V_r by

$$V_r^* = V_r/n \qquad (2.17)$$

in the above formulae. This procedure is valid only if test portions are separated at the first stage of sample reduction.

2.2.6 Design of a precision experiment

The experimental design given in ISO 5725 needs to be extended to allow variation between laboratory samples to be measured. Fig. 2.12 shows schematically how the samples should be prepared. In this example the soundness of a 14 to 10 mm road surfacing aggregate was tested by repeated cycles of soaking in saturated magnesium sulphate solution and then drying. The measurement of soundness used is the percentage of the test portion that passes a 10 mm sieve at the end of the soaking and drying cycles. The experiment included other test methods, so 100 kg laboratory samples were appropriate. The samples were taken from a stockpile and then allocated randomly to the participants so that each participant received two laboratory samples. Random allocation of laboratory samples to laboratories is important in that it helps prevent individual laboratory's results being adversely affected by any systematic variation between the samples, and it ensures that the model for sample variation in equation (2.13) is valid. Each laboratory then riffled out two 450 g test portions from each laboratory sample, as shown in Fig. 2.13, and obtained the test results given in Table 2.11.

When there is a possibility of variation between the laboratory samples, this design is valuable because it allows the variation between the samples to be estimated, and this in turn allows a reproducibility coefficient (R_1) to be calculated which measures between-laboratory variability free from between-sample variability. If only one laboratory sample were to be tested per laboratory, it would not be possible to separate between-laboratory and between-sample variability in the analysis, and one would not know how much of the reproducibility coefficient was due to sampling variability.

A full precision experiment would entail carrying out the above design at a number of levels chosen to represent the range of materials, material combinations, or concrete mixes to which the test method is applied in practice. The recommended number of participating laboratories is 15.

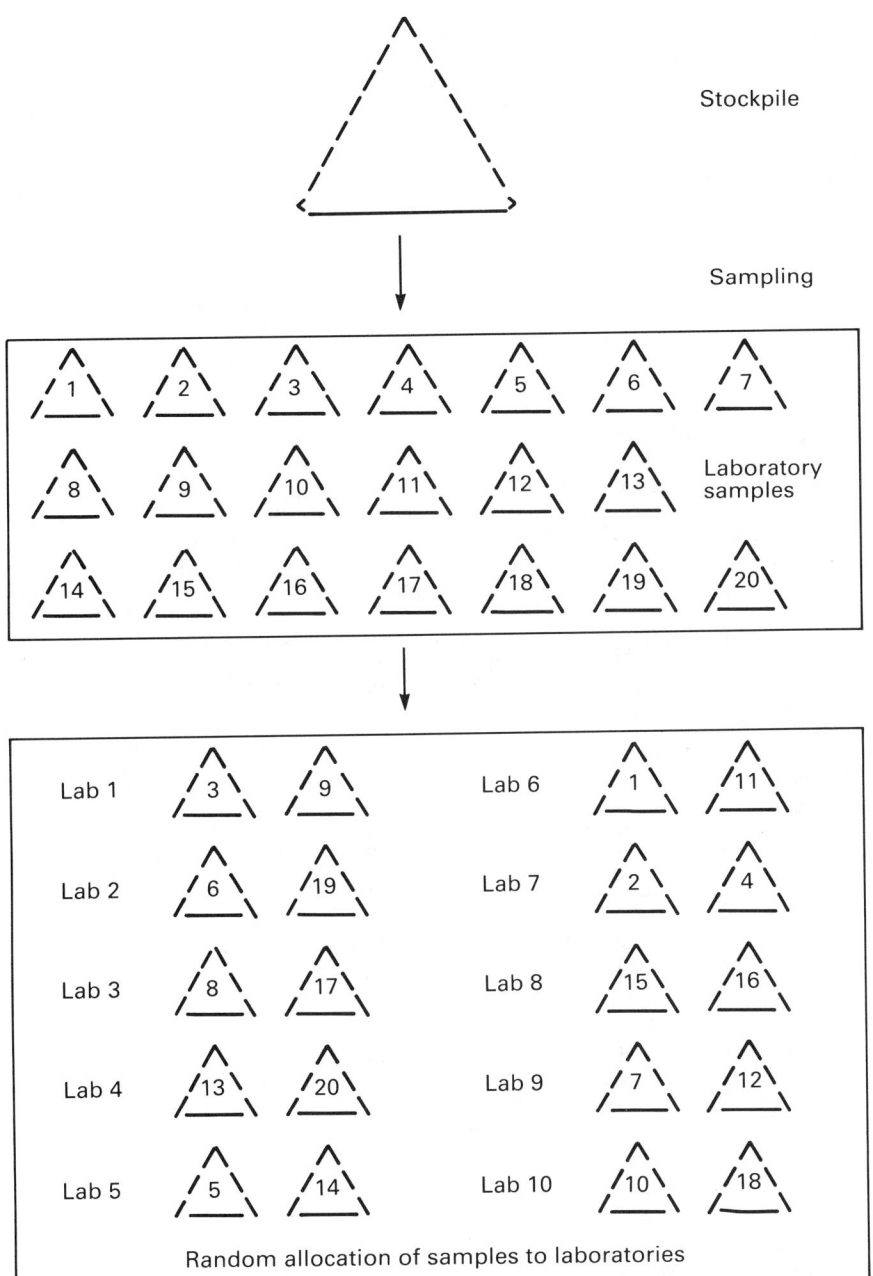

Fig. 2.12 — Samples for a precision experiment.

Sampling of aggregates and precision tests [Ch. 2

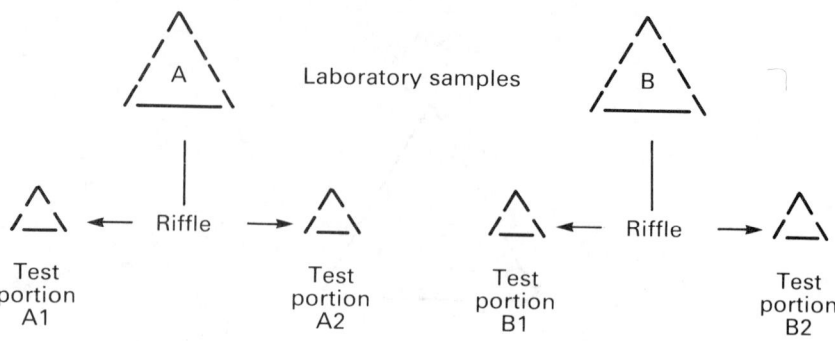

Fig. 2.13 — Test portions for a precision experiment.

Table 2.11 — Calculation of precision coefficients

Magnesium sulphate soundness of 14 to 10 mm aggregate
Data units — % passing 10 mm sieve

Laboratory	Laboratory sample	Test portion 1	Test portion 2	Sample mean	Laboratory mean	Between-lab-sample range	Between-test-portion range
1	1	31.0	28.9	29.95			2.1
	2	25.1	23.3	24.20	27.07	5.75	1.8
2	1	16.7	15.4	16.05			1.3
	2	20.3	13.5	16.90	16.47	0.85	6.8
3	1	20.4	17.2	18.80			3.2
	2	21.1	14.0	17.55	18.17	1.25	7.1
4	1	13.0	15.5	14.25			2.5
	2	11.7	13.8	12.75	13.50	1.50	2.1
5	1	18.8	22.6	20.70			3.8
	2	22.7	13.8	18.25	19.47	2.45	8.9
6	1	25.3	22.3	23.80			3.0
	2	22.3	18.9	20.60	22.20	3.20	3.4
7	1	13.9	12.6	13.25			1.3
	2	13.8	16.2	15.00	14.12	1.75	2.4
8	1	17.8	16.6	17.20			1.2
	2	22.1	16.8	19.45	18.32	2.25	5.3
9	1	24.7	24.2	24.45			0.5
	2	27.9	34.0	30.95	27.70	6.50	6.1
10	1	12.0	18.9	15.45			6.9
	2	16.4	8.7	12.55	14.00	2.90	7.7
General average					19.11%		
Variation between test portions							S1=10.6
Variation between laboratory samples						S2=5.6	
Variation between laboratories					S3=26.4		

Gives $V_r = 10.6$ $V_S = 0.3$ $V_L = 23.6$
and $r_1 = 9.1\%$ $R_1 = 16.4\%$ $R_2 = 16.4\%$

Sec. 2.2] **Precision** 53

The between-test-portion ranges, between-sample ranges, and laboratory averages for all the data obtained in the experiment are shown in Figs 2.14, 2.15, and 2.16.

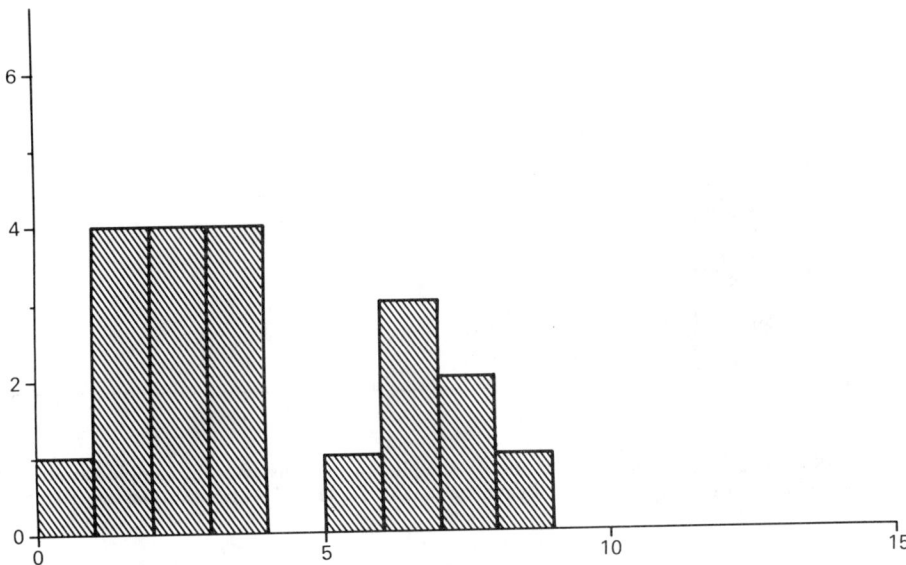

Fig. 2.14 — Histogram of between-test-portion ranges. Illustrates the repeatability of the magnesium sulphate soundness test, using the data from Table 2.11 — per cent passing 10 mm sieve.

2.2.7 Analysis of data from a precision experiment
The data given in Table 2.11 may be analysed as follows. Let:

$S1$ = the sum of the squares of the between-test-portion ranges shown in Fig. 2.14, divided by 40 (twice the number of these ranges),
$S2$ = the sum of the squares of the between-sample ranges shown in Fig. 2.15, divided by 20 (twice the number of these ranges),
$S3$ = the variance of the laboratory averages shown in Fig. 2.16.

Then:

$S1$ estimates V_r (2.18)

$S2$ estimates $V_S + (V_r/2)$ (2.19)

$S3$ estimates $V_L + (V_S/2) + (V_r/4)$, (2.20)

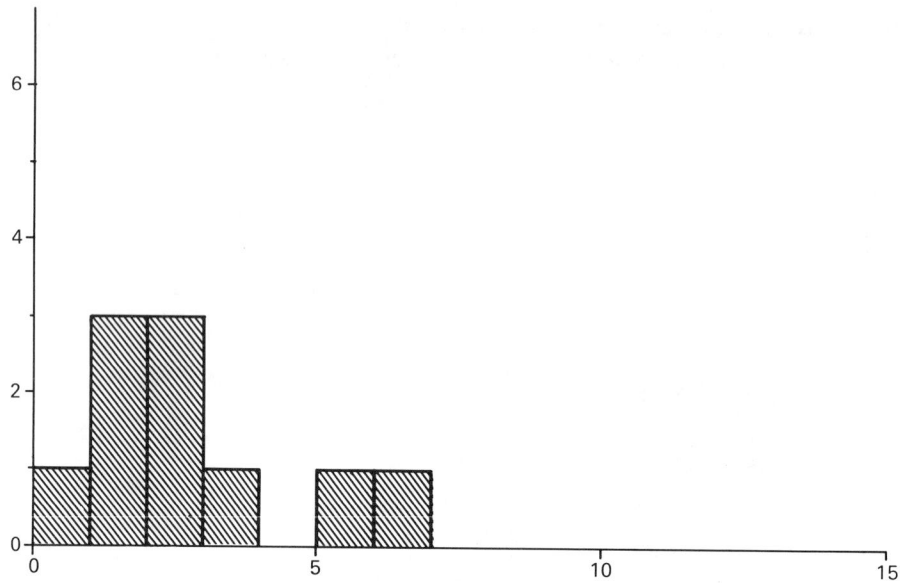

Fig. 2.15 — Histogram of between-laboratory-sample ranges. Illustrates the sample variation found in the precision experiment that gave the data in Table 2.11 — per cent passing 10 mm sieve.

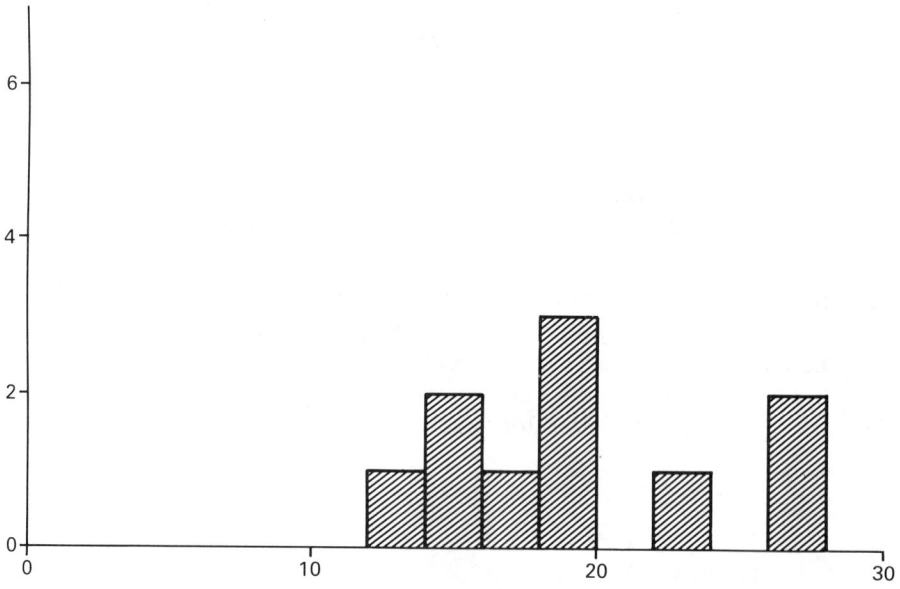

Fig. 2.16 — Histogram of laboratory means. Illustrates the reproducibility of the magnesium sulphate soundness test, using the data from Table 2.11 — per cent passing 10 mm sieve.

so estimates of the variances may be calculated by using

$$V_r = S1 \qquad (2.21)$$
$$V_S = S2 - (S1/2) \qquad (2.22)$$
$$V_L = S3 - (S2/2) \qquad (2.23)$$

and r_1, R_1, and R_2 may then be calculated by using equations (2.14) to (2.16). Table 2.11 sets out the calculations. The analysis is more complicated when the data are incomplete.

A complete analysis of the data from a precision experiment requires the data to be examined for 'outliers' and 'stragglers', using the statistical tests given in ISO 5725, and deciding which of these should be excluded from the calculation of the precision coefficients. The reasons for the occurrence of outliers and stragglers should be sought in case they indicate a fault in the test method or in the participants. When outliers and stragglers do occur, it is important to remember that they have occurred, otherwise in some applications the precision coefficients could suggest that the precision of the test method is better than it really is.

2.2.8 Checking the variability caused by sampling

The theoretical calculation of sample sizes given earlier requires estimates to be made of the various factors involved (shape, liberation, etc., see section 2.1.3). It may be difficult to obtain accurate estimates of these factors, in which case the calculated sample size will not be totally reliable. The advice given in standards, such as BS 812: Part 102 (BSI 1984), can only be general in nature and may give unsatisfactory results with some materials or in some sampling situations. Further, an individual tester may not fully appreciate the importance of following the standard sampling procedures and may consequently obtain unreliable results.

With the hierarchy of sampling terms defined in section 2.2.3 it is a straight forward matter to obtain data that will allow the variability caused by sampling to be checked. All that is required is that a number of duplicate bulk samples be taken in such a way that the two samples in a pair represent the same batch of material, and then that duplicate tests be carried out on each sample. The data can be tabulated as shown in Table 2.11, and analysed in the same manner. The only difference is that the analysis will yield an estimate of the between-batch variation instead of the between-laboratory variation. Such procedures are described in BS 812: Part 102 and ISO 1988 (ISO 1975), and in ASTM E877 (ASTM 1982), for example.

The samples could be subjected to a grading analysis, or to petrographical examination, or to any other test method. A histogram of between-sample ranges should be drawn (as in Fig. 2.15) to assist in the identification of stragglers and outliers and to illustrate the variability being assessed.

As for assessing the acceptability of the sampling variation, it would be desirable for the sampling variance to be small compared with the between-laboratory variance:

$$V_S < 0.25 V_L \qquad (2.24)$$

In the context of quality control, it would also be desirable for the sampling variance to be small compared with the between-batch variation:

$$V_S < 0.25 V_B \qquad (2.25)$$

If the sampling variability is found to be excessive, compliance of the increment size and other details of the procedure with BS 812: Part 102 should be checked and followed by examination of the factors which contribute to equation (2.6) to see if any explanation can be found. It should be possible to reduce the sampling variability by increasing the number of sample increments. To achieve a substantial reduction in the sampling variability the number of increments must be increased by a factor of at least two.

2.2.9 Repeatability control chart

A responsible laboratory will want to ensure that it is achieving a satisfactory standard of testing, and one way to do this is to enter between-test-portion ranges on a control chart. Many test methods call for duplicate tests, so that data are often available to check the repeatability in this way. If the recommendation that duplicate test portions should be separated at the first stage of sample reduction is followed, then checking repeatability also checks the variability caused by sample reduction. When the repeatability of the test method depends on the level of the results or on other factors, the data should be divided up into classes within which the repeatability ought to be the same, and a separate control chart run for each class.

An example of a repeatability control chart is shown in Fig. 2.17. The data are

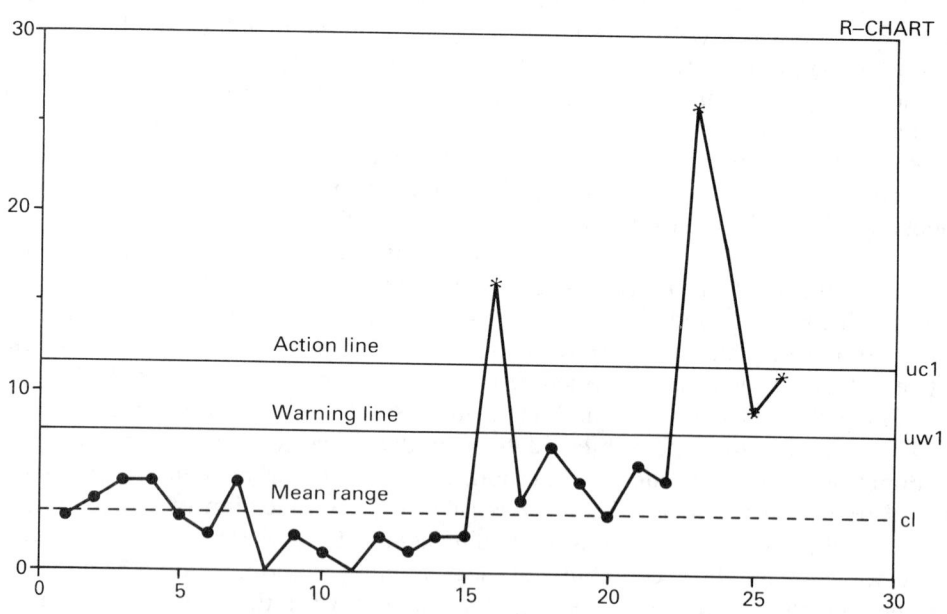

Fig. 2.17 — A repeatability chart for ten per cent fines values. Between-test-portion ranges in Table 2.12 plotted against record number.

given in Table 2.12. The chart shows between-test-portion ranges plotted against record number, also action and warning lines, and a line corresponding to the mean range. A deterioration in repeatability is signalled if a single point exceeds the action line or if two successive points exceed the warning line, and an improvement in repeatability is signalled by ten successive points below the mean range. In the example, signals are given by Points 17 and 18 and by Point 25. The action and warning lines shown in Fig. 2.17 correspond to the 'S2' scheme of BS 5703: Part 3 (BSI 1981) for ranges of two test results, and are calculated as:

$$\text{action line} = 3.61\,\bar{w}$$
$$\text{warning line} = 2.45\,\bar{w}$$

where $\bar{w}$ = the mean range.

The mean range should be calculated from at least 20 between-test-portion ranges, using the laboratory's own data, and recalculated whenever a signal of a change in the repeatability is given.

A repeatability control chart is best used by an operator as a learning tool; when the chart signals that a deterioration in repeatability has occurred the operator should be encouraged to try to identify the cause of the problem and then report it, together with a proposal for its solution, to his supervisor. Used in this way, the chart will encourage a positive attitude amongst laboratory staff. The charts will also improve confidence in the results reported by the laboratory, both among the laboratory's own staff and by its users, and provide a means by which the laboratory can demonstrate the quality of its testing to its customers.

2.2.10 Cusum stability chart

Repeatability measures the variability of tests in the most favourable conditions, in particular when tests are carried out within a short interval of time. There are causes of variation that will not show up in a repeatability check; for example, changes in the calibration of instruments, differences between deliveries of reagents, wear in equipment, and changes in the environment. It is common practice for chemists to keep stocks of 'private' reference materials, and to test them from time to time to check that the level of their results has not changed. It is equally important for aggregate testers to do the same, although there are practical difficulties — the size of test portion is usually several kilograms for an aggregate test, so large quantities of material need to be stored, and some tests take so long that regular testing of a private reference material would occupy too much of the laboratory's facilities. Even so, the stability of aggregate tests should be checked by using private reference materials when it is practical to do so. The data should be used to draw a cusum chart because this is the best method of detecting changes in testing level from small quantities of data.

Fig. 2.18 shows a cusum chart drawn up from the data in Table 2.13. The target value T has been set at the average of the first four test results in this example. Changes in the slope of the cusum indicate changes in the level of the results obtained on the private reference material. On Fig. 2.18 the upward slope of the points from Test 15 onwards suggests that the level of the test results increased at that time. With

Table 2.12 — Data for repeatability chart

Ten per cent fines values. Data units: kN

Record number	Test portion a	Test portion b	Between-test-portion range |a–b|
1	149	146	3
2	152	148	4
3	156	161	5
4	155	160	5
5	151	154	3
6	160	158	2
7	150	155	5
8	155	155	0
9	167	165	2
10	164	165	1
11	150	150	0
12	153	151	2
13	141	140	1
14	137	139	2
15	142	144	2
16	139	155	16
17	148	152	4
18	145	152	7
19	163	168	5
20	153	150	3
21	149	155	6
22	150	145	5
23	216	242	26
24	218	236	18
25	140	149	9
26	159	148	11

the cusum, a V-mask is used to decide when significant changes in level have occurred. A suitable V-mask for this application is given by the C2(b) scheme described in BS 5703: Part 3, and is shown placed on Test 17 in Fig. 2.18: there are several points below the lower limb of the V-mask indicating that the change in slope is significant, so that the data signal that a change in the level of testing has occurred at some time before this result was obtained. The later data confirm the trend.

It will prove useful to keep a log of events such as changes in operators, equipment, reagents, and of recalibrations, equipment overhauls, and maintenance. These can be used to help identify the cause of a change in the level of the results. For

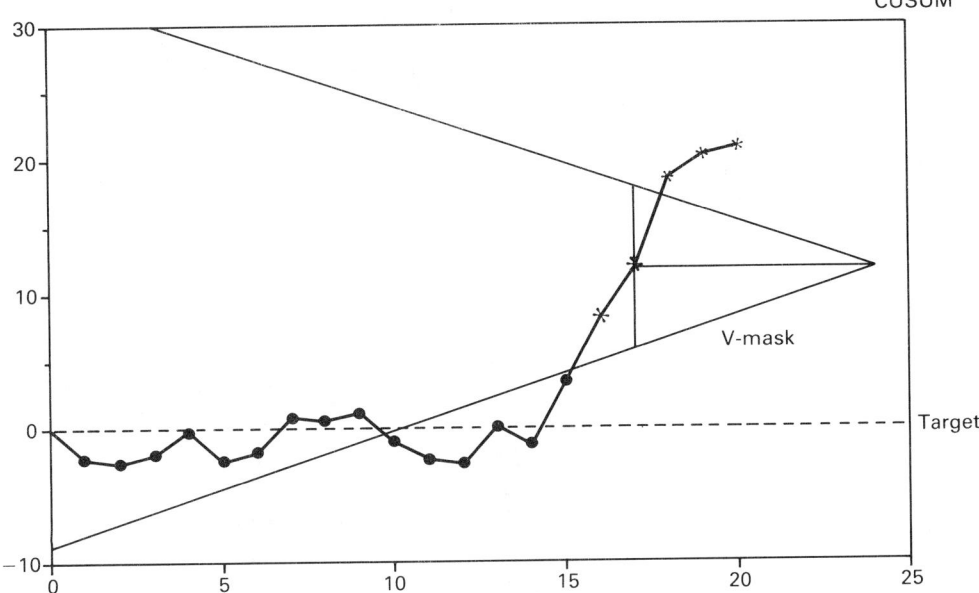

Fig. 2.18 — A cusum stability chart for acid solubility tests. Cusum in per cent from Table 2.13 plotted against test number.

further advice on the construction of cusum charts the reader should refer to BS 5703.

Cusum stability charts should be used in the same way as repeatability control charts, primarily as means by which the operators can measure and improve the quality of their testing. Their use will add further to confidence in the laboratory's results.

2.2.11 Inter-laboratory comparisons, using reference materials

The repeatability control charts and cusum stability charts both provide only internal checks: there is a need for a laboratory to check that its testing level is consistent with those of other laboratories. In precision experiments it is often found that between-laboratory variability is the largest source of variation in a test. Ideally inter-laboratory checks would be done, using reference materials having properties that have been established by following the principles set out in ISO Guide 35 (ISO 1985), but suitable reference materials are not available for the majority of aggregate tests. The alternative is to organize a testing programme similar in nature to a precision experiment, including a sufficient number of laboratories (15 or more according to ISO Guide 35) so that the properties of the 'reference material' used in the programme are established by the programme itself. Then the results of the programme can be used to check the testing level of the participating laboratories.

Criteria are needed to decide if a laboratory's results are acceptable or not, and these should be based on the established precision of the test method. At present

Table 2.13 — Cusum chart data

Acid solubility. Data units — per cent		
Test number	Test result $x\%$	Cusum $\Sigma(x-T)\%$
1	26	−2.3
2	28	−2.6
3	29	−1.9
4	30	−0.2
5	26	−2.5
6	29	−1.8
7	31	0.9
8	28	0.6
9	29	1.3
10	26	−1.0
11	27	−2.3
12	28	−2.6
13	31	0.1
14	27	−1.2
15	33	3.5
16	33	8.2
17	32	11.9
18	35	18.6
19	30	20.3
20	29	21.0

$T = (26+28+29+30)/4 = 28.3\%$

there is no standard to give guidelines on how to derive such criteria, although a future revision of ISO 5725 should provide the necessary help. Where both repeatability and stability are being checked internally, a laboratory need take part in inter-laboratory comparisons only occasionally, but when internal stability checks are not practical, there may be a case for more frequent inter-laboratory comparisons.

It is clear that a laboratory that carries out repeatability checks, stability checks, and takes part in inter-laboratory test programmes, would have to undertake a considerable amount of testing and administrative work. A similar situation arises in manufacturing if a producer attempts to ensure the quality of a product by extensive sampling and testing schemes. He may achieve his objective, but at a heavy cost. As a general rule, the economical way to achieve quality is to identify the causes of variations and to learn how to prevent them occurring. In the context of laboratory work, the product is the test result, and the prevention of variations in the product requires a disciplined approach to testing combined with the statistical techniques described above. The amount of effort expended on these techniques should depend on the degree of control achieved by the laboratory: if the control charts show that a high standard is being achieved, then the effort can be reduced.

2.2.12 Applications of precision in assessing the capability of test methods

In manufacturing, when the variability of a characteristic of a product is measured by a standard deviation 's', and the specification for the product requires the characteristic to fall between upper and lower specification limits of 'U' and 'L', it is common to describe the relationship of the variability to the specification as the capability of the process:

$(U-L)/s \ll 6 \Rightarrow$ low relative capability
$(U-L)/s \approx 6 \Rightarrow$ moderate relative capability
$(U-L)/s \gg 6 \Rightarrow$ high relative capability.

The same concept can be applied to testing when a test is used to decide into which grade a material falls. If a test method is required to have 'high relative capability', then the precision coefficients must satisfy:

$$r_1, R_1, \text{ and } R_2 \ll (U-L)/2.1 \tag{2.26}$$

where U and L are now the grade boundaries, $2.1=6/2.8$, and 2.8 is the factor in equations (2.14) to (2.16). Given the results of a precision experiment, this criterion can be used to assess the 'capability' of the test method.

For example, the available data for the BS 812: Part 120 aggregate shrinkage test (BSI 1988) indicate a reproducibility of

$$R_1 = 0.22\,\bar{x} \text{ microstrain} \tag{2.27}$$

where $\bar{x}$ is the level of shrinkage of interest. If a shrinkage class is defined by the boundaries:

$L = 450$ microstrain $U = 650$ microstrain

so

$\bar{x} = 550$ microstrain

and

$$R_2 = 121 \text{ microstrain}. \tag{2.28}$$

But

$(U-L)/2.1 = 95$ microstrain,

so the test method has low relative capability.

If the criterion is not satisfied, then one way to improve the precision of the test method is to increase the number of determinations: the improvement can be calculated by using equation (2.17). This will lead to a reduction in the repeatability coefficient, but, when the between-laboratory variance is as large (or larger than) the between-test-portion variance (as is often the case), it will not reduce the reproducibility coefficients very much. It may be possible to achieve an improvement in reproducibility by modifying the test method, but it is more likely in the case of established methods that an improvement will be achieved by the widespread use of the statistical methods described earlier. Apart from increasing the number of tests, there is no simple practical way of satisfying equation 2.26 if the between-laboratory variability is too large. Until such time as an improvement in the reproducibility can be demonstrated, the rule then has to be interpreted as the minimum interval between specified grades that is realistic.

In the above example, with the reproducibility as shown in equation 2.28, the interval between grades for the test method to have a moderate relative capability is

$$U-L \approx 254 \text{ microstrain}.$$

When a specification contains only a one-sided limit on the characteristic of a product the above considerations do not apply, and one cannot calculate an upper limit to the precision coefficients in the same way. With only a specified lower limit, for example, the producer has the option of allowing for the variability of the testing by providing material of sufficiently high quality that the risk of failing to meet the specification is small. There is still a need to ask if the precision of a test is good enough. Increasing the quality of the product is not an economical solution to the problem of poor precision. Disputes between producers and specifiers can be caused if they can come to different conclusions about the quality of a product, and there are risks that poor-quality material will be judged acceptable and good quality material rejected. The uncertainty that is caused by between-laboratory, sampling, and between-test-portion variabilities can be expressed by using 95 per cent confidence limits. Given a test result 'x' these can be calculated as:

$$\text{95 per cent confidence limits} = (x-d) \text{ to } (x+d) \tag{2.29}$$

where

$$d = R_2/\sqrt{2}. \tag{2.30}$$

If, in a particular instance, the test result is within 'd' of the specification limit 'L' then one cannot be confident that the material either complies or fails to comply with the specification. In general, if the interval

$$(L-d) \text{ to } (L+d)$$

represents a substantial difference in the cost of the material, or a substantial range of the available materials, then there is a need to improve the reproducibility of the test method.

For example, suppose that an upper limit is imposed on the percentage passing the 10 mm sieve with the magnesium sulphate soundness test of

$$U = 20 \text{ per cent} \qquad (2.31)$$

According to the data in Table 2.11

$$R_2 = 16.4 \text{ per cent} \qquad (2.32)$$

so

$$d = 12 \text{ per cent} \qquad (2.33)$$

and the above interval becomes 8 per cent to 32 per cent. Although one would expect the precision to be better than indicated by equation (2.32) when the percentage passing is small, this interval appears to be large, and there is a possibility of many materials giving results within it.

REFERENCES

American Society for Testing and Materials (1982) *Standard method for sampling and sample preparation of iron ores*. E 887.

British Standards Institution *Testing aggregates*. BS 812:
  Part 102. *Methods for sampling*. 1984.
  Part 104 (draft). *Procedure for quantitative and qualitative petrographic examination of aggregates*. 1988.
  Part 120. *Methods for the determination of drying shrinkage*. 1988.

British Standards Institution (1987) *Precision of test methods*. Part 1. Guide for the determination of repeatability and reproducibility for a standard test method by inter-laboratory tests. BS 5497.

British Standards Institution (1981) *Guide to data analysis and quality control using cusum techniques*. Part 3. Cusum methods for process/quality control by measurement. BS 5703.

Gy, P. M. (1982) *Sampling of particulate materials. Theory and practice*. Elsevier. Oxford.

International Standards Organisation (1975) *Hard coal — sampling*. ISO 1988.

International Standards Organisation (1985) *Certification of reference materials. General and statistical principles*. ISO Guide 35.

International Standards Organisation (1986) *Precision of test methods — Determination of repeatability and reproducibility for a standard test method by inter-laboratory tests*. ISO 5725.

Jones, M. P. (1987) *Applied mineralogy — a quantitative approach*. Graham & Trotman. London.

Ottley, D. J. (1986) Gy's sampling slide rule. *Mining and Minerals Engineering*. October pp 390–395.

Van der Plas, L. & Tobi, A. C. (1965) A chart for judging the reliability of point counting results. *Am. J. Sci.* **263** 87–90.

Zussman, J. (1977) *Physical methods in determinative mineralogy*. (2nd edition) Academic Press. London.

3

Aggregates for concrete

B. V. Brown (Section 3.1)
D. W. Hobbs (Section 3.2)
G. R. Lavers (Section 3.3)

This chapter deals with three selected topics from the broad field of concrete aggregates. The first subject discussed is marine aggregates; the second is alkali-silica reaction; and the third is drying shrinkage.

Each of these topics has been extensively debated in the UK over the past ten years, and British Standards have been drafted or amended to take account of research and commercial information.

Each of the three sections concludes with its own list of literature references.

3.1 MARINE AGGREGATES

3.1.1 Introduction

Marine aggregates are predominantly obtained by dredging deposits in the sea and in river estuaries, but the description also includes a limited quantity derived from beach deposits. Beach sands are often single-sized and need blending to produce a graded material. Dredging has occurred around much of the continental shelf of the United Kingdom and the lower reaches of the Rhine and the North Sea. Supplies are, however, most abundant off the coast of the UK, and these now make a significant contribution toward meeting the demand for aggregates in the British Isles. Several million tonnes have also been exported to Continental Europe.

The geological nature and origins of marine aggregates are similar to those of land-based sands and gravels. Thus, general performance and characteristics can be expected to be similar; any differences that do exist are associated with the location from which they are won. Marine aggregates are typically smooth-textured and rounded in shape, and the method of their extraction and processing results in aggregates that are relatively free from dust, clay, and silt. The presence of sea water introduces potentially harmful marine salts, but, as with contaminants in land-derived material, these can be adequately controlled by proper processing. Another principal difference arises from the presence of marine creatures in the sea and their remains such as sea-shells. These, and the other differences, are now well understood, and satisfactory controls exist to allow marine aggregates to play their rightful role as a natural resource.

3.1.2 Environmental effects and dredging licensing

In 1976 the Advisory Committee on Aggregates, chaired by Sir Ralph Verney, recommended that the use of marine aggregates should be encouraged to make good shortfalls in land-based aggregates in the south east of England. This recommendation was adopted in Government guidelines issued in 1982. (Department of the Environment, 1982).

The winning of marine aggregates can have less direct environmental impact on the population than quarrying operations on land. Transport from deep water wharves, for instance to delivery sites in London and other conurbations, may use routes that are shorter than those from the nearest land pit, leading to a reduction in lorry traffic. The quantity of land made available for aggregate extraction can be reduced; it has been estimated that this equates to about $1\frac{1}{2}$ square miles per year at 1988 rates of supply.

On the other hand, it has been suggested that fishing interests may be harmed, but this matter is well understood now, and a code of practice exists between the dredging and fishing industries. The code covers four areas of major concern: the protection of herring spawning beds; sand eel banks; oyster beds; and physical interference between dredgers and fishing vessels and gear.

Permission to dredge is covered by the 1967 Continental Shelf Act which vests ownership of the sea-bed in the Crown. The Crown Estate Commissioners are responsible for the issue of licenses for both prospecting and extracting sand and gravel. This control provides a safeguard against coastal erosion, damage to fishing interests, and protection of pipelines and cables laid across the sea-bed.

Spruell & Uren provide much useful information on the costs of dredging aggregate and licensing arrangements (Spruell & Uren, 1986). They show that, despite the heavy initial investment in ships, wharves, and licenses, the economics of supply are viable for many of the major areas where aggregate demands exist. This view is supported from data supplied by the industry (Nunny & Chillingworth, 1986).

3.1.3 Dredgers and discharge methods

The history and development of dredging and discharge methods is succinctly summarized in a recent report. (Spruell & Uren, 1986). In the early days, only shallow-water deposits were worked via a grabbing crane mounted on a pontoon or small barge. These early methods were overtaken by the development in the late 1930s of centrifugal pumps which were mounted on the decks of ships with dredge pipes lowered overboard. Fig. 3.1 shows the MV *Cambourne* with such a dredge pipe mounted along the side of the dredger.

Today there are two principal types of dredger, known as anchor and trailer dredgers. The principles of their operation are shown in Figs 3.2 and 3.3.

The anchor dredger works, as its name implies, at anchor with the dredge pipe lowered into the deposit. Sand, gravel, and water are pumped aboard. As the hold is filled, water flows over and drains back into the sea. Periodically the dredger is

66 Aggregates for concrete [Ch. 3

Fig. 3.1 — MV *Cambourne*.

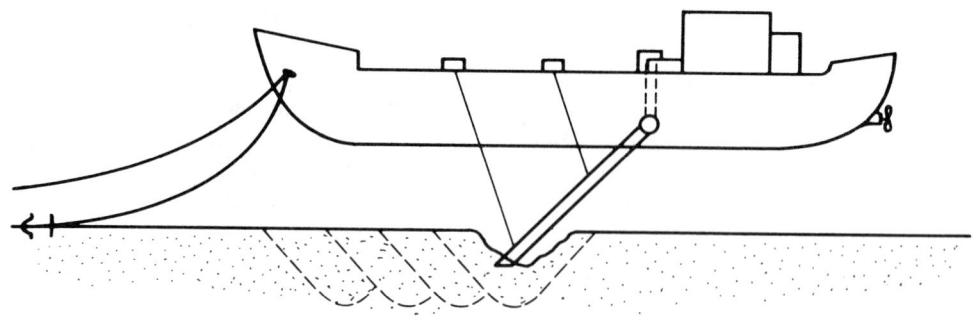

Fig. 3.2 — Principle of anchor dredging.

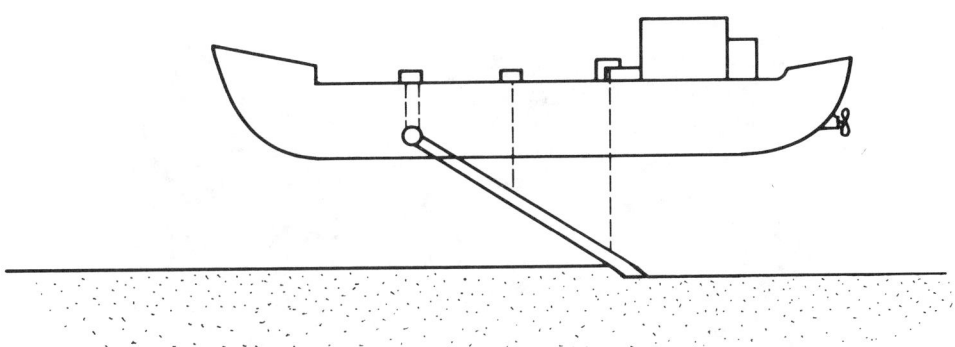

Fig. 3.3 — Principle of trailer dredging.

winched a few feet forward to replenish the supply of aggregates. This type of dredger can work a deep deposit with very little movement, but can leave a series of holes in the sea-bed.

The trailer dredger, on the other hand, moves slowly and continuously, removing a thin layer some 25 mm deep in a groove across the sea-bed. This can be used to dredge shallow deposits, and it leaves a smooth sea-bed, but obviously dredging covers a wide area.

The capacity of sea-going dredgers has continuously increased, and a range of ships, from as small as 150 tonnes up to 4500 tonnes, is now regularly in operation. Even larger units are occasionally used in protected deep harbour waters, but are not yet suited for sea-going dredgers. Many of the larger sea-going ships have some primary screening ability.

Dredger discharge methods are varied and often demonstrate the ingenuity of engineers. Initially, shore-based cranes were used to grab material from the holds. This was very slow, and the need for greater speed led firstly to the installation of drag-scraper buckets. These were winched through the cargo up a ramp, and then discharged onto a feed conveyor (see Fig. 3.4).

Another development now used is a bucket wheel mounted on a travelling carriage, winding along through the cargo and being progressively lowered into the cargo as it moves forwards and backwards through the hold. The lower part of the hold is shaped to allow complete removal of the cargo (see Fig. 3.5).

The fastest method yet developed is to pump the cargo ashore, using a specially modified dredger pump. However, this requires extra facilities at the receiving wharf to provide collecting and settling tanks or lagoons.

3.1.4 Use and modern history in the UK

Marine aggregates are predominantly used in concrete production, but they are also found in all the other end uses of aggregates including drainage, fill, mortars, sub-base material, block-making, asphalt, and coated products. However, it should be remembered that their smooth texture and rounded particle shapes make them less appropriate where particle interlock characteristics are required, such as in maca-

Fig. 3.4 — Drag-scraper buckets.

dams and sub-base materials. This deficiency may be overcome by blending with other material and-or crushing. An example of the magnitude of some of these other end uses is the substantial quantities of fill supplied occasionally. One site alone, Thamesmead in Greater London, was supplied with over 900 000 tonnes in the late 1970s and early 1980s.

It is often said the best assessment of a material is the test of time, and marine aggregates have now had well over half a century of use during which their satisfactory performance has been demonstrated. H. J. Greenham Limited, now part of the Taylor Woodrow Group, dredged the lower reaches of the Thames as early as 1918, using barges with grabs which worked the exposed sand and gravel banks at low tide. One of the first reports of the use of marine sand and gravel dredged from offshore areas goes back to 1926 when such aggregates were used for

Fig. 3.5 — MV *Cambourne*: bucket wheel discharge system mounted on a travelling carriage.

making concrete for Spillers Silos in Cardiff Docks. In 1948 Kingston Gas Works at East Cowes was built, using 100 000 tonnes of marine aggregates, and in 1949 the Ocean Terminal building at Southampton Docks used 15 000 tonnes. A major expansion in use occurred from the mid-1960s onwards.

Table 3.1 shows the increasing role of marine aggregates in the United Kingdom

Table 3.1 — Production of marine-dredged sands and gravels in the United Kingdom

	Production (million tonnes)				
Year	1955	1965	1975	1985	1987
Total	3.4	7.2	14.5	16.3	18.8

Sources: Crown Estate Commissioners and the Department of the Environment and Business Statistics Office.

with time. The contribution has been largest in the southeast of England because of the environmental pressures in this area. Table 3.2 indicates the significant quantity of such sand and gravel landed in the southeast from 1969–1987. Some 86 per cent of the landings in the southeast in 1983 was used for concrete, corresponding to around $3\frac{1}{2}$ million cubic metres of concrete. This order of percentage use in concrete has

Table 3.2 — Landings of marine dredged sands and gravels in the south east of England

	Landings (million tonnes)			
Year	1969	1983	1985	1987
Total	5.3	8.6	10.4	11.3

Sources: 1969: Crown Estate Commissioners and the Department of the Environment and Business Statistics Office. 1983-7: South East Region Aggregate Working Party.

continued since that time, resulting in an increasing concrete production *pro rata* to landings. Many of London's landmarks contain marine aggregates, notable examples include the Barbican development, the National Theatre, and the Thames Barrier (Fig. 3.6). Further, it is estimated that some 45 per cent of all the concreting aggregates in London used in 1985 were marine dredged. This reflects both their satisfactory performance and the economics of supply.

Fig. 3.6 — The Thames Barrier under construction.

3.1.5 Geological origin and current UK dredging locations

The geology of marine deposits is a specialist subject, and the following provides only a rudimentary appreciation of the origin of marine aggregates. The reader is referred to other more comprehensive references for detailed information (Collis & Fox, 1985).

Much of the continental shelf bordering the United Kingdom is covered by a mixture of deposits, resulting from several glaciations in the Quaternary era. Each of these glacial periods contributed to the sand and gravel deposits. Glaciers advancing from Scandinavia deposited large quantities of material off the coasts of the British Isles. This was supplemented by materials derived from erosion of the land masses by rivers swollen with water from the melting ice sheets as they retreated. This erosion carved out the bedrock, leaving channels which were subsequently filled with sand, gravel, and clay when the sea level rose. The geological origin of marine aggregates is thus similar to land-won material, that is, river and glacial deposition. This is demonstrated in a comparison of petrological examinations of materials taken from the Vale of St Albans and the southern part of the North Sea, see Table 3.3. The

Table 3.3 — Comparison of petrology of typical land-based and marine aggregates

Rock type	Percentage	
	Land-based (Vale of St Albans)	Sea-dredged (southern North Sea)
Flint	77	72
Flint cortex (i.e. porous flint)	12	2
Quartz/quartzite	10	12
Sandstone	10	13
Others	1	1

similarity of materials is also shown in Fig. 3.7, which illustrates a section cut through concrete in which the left-hand side of the figure contains Thames Valley aggregates, and the right-hand side contains sea-dredged aggregate from the southern North Sea.

These gravel deposits have accumulated far more extensively off the coast of the British Isles than elsewhere on the Continent Shelf and Western Europe. The major dredging areas are listed in Table 3.4, and their locations are illustrated in Fig. 3.8. Supplies of coarse and fine aggregate in commercially appropriate proportions are available off most of the major sectors of the UK coast, except in the Bristol Channel, where the deposits are predominantly sand, and the Humber Banks, which contain an abundance of coarser materials. In the Mersey area, the demand has mainly been for sand, but gravel is also available in Liverpool Bay.

The principal petrographic compositions of the aggregates from the main marine sources are set out in Table 3.5.

3.1.6 Processing and stockpiling
Many gravel aggregates require washing, and the emphasis does not change, irrespective of whether they contain chlorides, clay, or silt. Whatever the make-up of

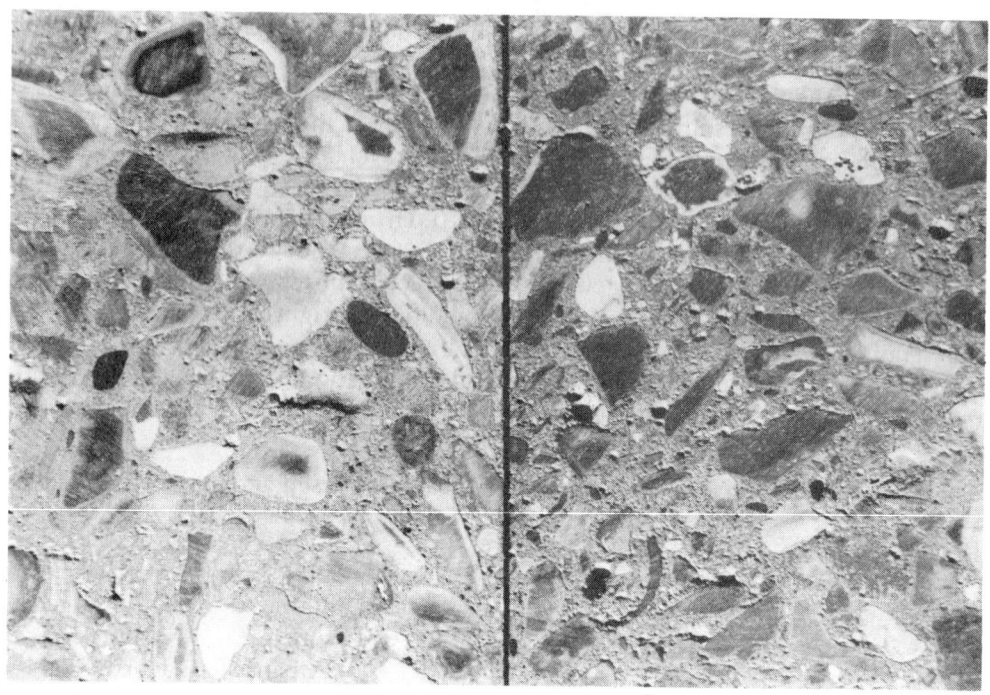

Fig. 3.7 — Sections through concrete made with Thames Valley aggregates (left) and sea-dredged aggregates (right).

Table 3.4 — Dredging grounds around the United Kingdom in 1987

Area	Dredging grounds/banks
East coast	East and West Humber
Norfolk coast	Lowestoft, Yarmouth, and Southwold
Thames Estuary and southern North Sea	Shipwash, Kentish Knock, Sunkhead Towers, Long Sands, and Gabbard
South coast	Owers, Horse Tails, Area A, Pot, and Solent
West coast	Bristol Channel, Liverpool Bay, and Mersey Estuary

the contaminating content, reduction to acceptable levels by washing is generally necessary.

Sea-water contains dissolved cations (sodium, potassium, magnesium, calcium, etc.) and anions (chloride, sulphate, etc.). Although ion concentrations vary around the World, there is typically a chloride ion content of some 1.78 per cent in UK waters. The chloride ion, by its nature, is readily soluble in water, and the content in

Sec. 3.1] **Marine aggregates** 73

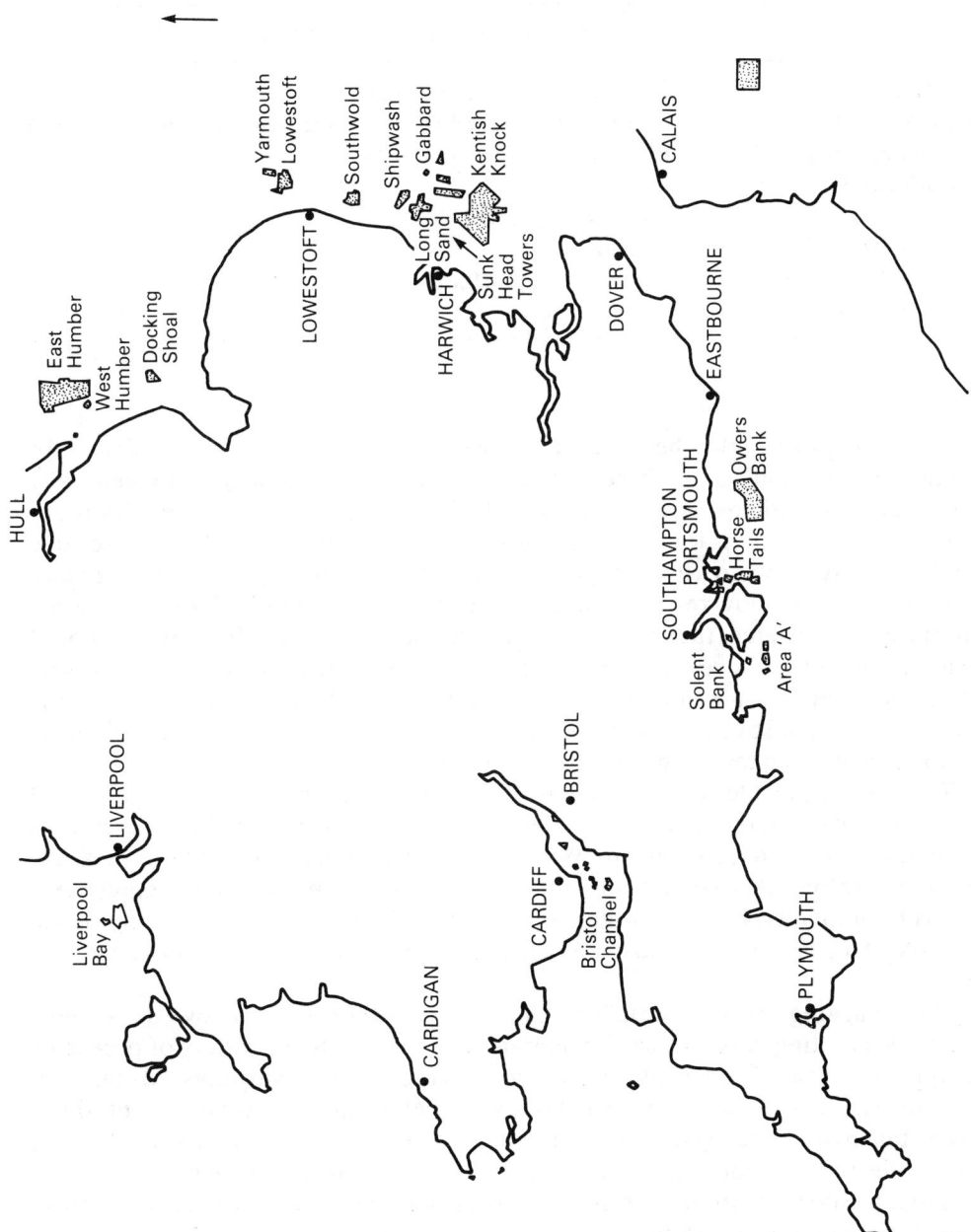

Fig. 3.8 — Marine-dredged sand and gravel licensed areas.

Table 3.5 — Principal petrographic composition of United Kingdom dredged aggregates

Area	Principal petrographic composition
East coast	Granites, basalts, and quartzites
Norfolk coast	Flints, quartzites plus some chert and sandstone derived
Thames Estuary and southern North Sea	from southeast England rocks
South coast	Flints and quartzites
Bristol Channel	Quartz
Liverpool Bay and Mersey Estuary	Quartz and quartzite

the wet aggregate can thus be reduced by washing with fresh water and drainage. In practice a combination of these occurs. The effects of drainage are generally supplementary, and process control is usually linked to chlorides in the wash water.

Use of fresh water for washing without any recycling would cause serious problems in relation to cost, supply, and disposal. Accordingly, most aggregate washing processes allow re-use of water following settlement in silt bays or lagoons. For marine aggregates, the chloride concentration in the recirculated water would eventually increase above acceptable levels and on longer result in adequate chloride reductions. Thus, fresh water is regularly introduced into the washing process. This is normally controlled by measuring the chloride content in the wash water and diluting or replacing it when levels approach a pre-set limit.

There is reasonable correlation between chloride content in the aggregates and that in the wash water, and this can be determined from experimental measurements at any particular processing plant. Meters (Fig. 3.9) are now available to measure and record chloride concentrations automatically and can be coupled to equipment that will sound a warning when the limit is approached. These meters work on the principle of measuring conductivity across platinum electrodes fixed a given distance apart.

Coarse aggregates are normally washed via spray bars placed above the screens used for separating different sized materials (Fig. 3.10), while a variety of processes are applied to sand. More sophisticated systems use water flowing upwards through the sand. This causes the lighter particles to rise, and, in the case of clay and silt, these may be taken off at the top with the rising water. With marine aggregates, wash water also dilutes the chloride content in the sand. There are further sophistications, for example, to allow adjustment of the water pressure; in this case, particular gradings of sand can be drawn off in successive columns or tanks.

The need for such refinements to produce differently graded sands is often unnecessary where the source deposit is of a consistent nature. Simpler systems may also be used if the aggregate has a low silt content, but, whatever system is adopted, chlorides are reduced via the water washing through the sand and any subsequent de-watering process.

Fig. 3.9 — Chloride concentration meter.

Any drainage of coarse and fine aggregates in storage bunkers or stockpiles will further reduce chloride levels, as the water containing chloride drains away. After thorough washing and adequate drainage, only small quantities of chloride are retained in the absorbed water within the pores of the aggregate. If the chloride content of the wash water is measured, then the corresponding chloride ion concentration in the wet aggregate can be calculated, provided that its moisture content is known. This is an approximation, as it makes no allowance for any chloride-containing moisture entrapped deep in the pores of the aggregate, but it does provide a realistic first estimate for aggregates of normal porosity. More accurate assessments can be obtained from full analysis of particular materials. The small chloride residue in the pores is in any event generally likely to be below 0.01–0.02 per cent.

An example of this first-order estimation of chloride content is shown in Table 3.6. It should be treated as a rough guide only; slightly higher levels of chloride may result from the entrapment identified above or subsequent evaporation from stockpiles left exposed to wind or sun. These effects of evaporation should not be exaggerated because, in large stockpiles, small areas are left exposed, relative to the overall volume or mass of aggregate. When aggregate is removed from the stockpile, subsequent blending further reduces such effects.

After extensive research, Course (1984) proposed a more sophisticated equation for predicting chloride ion concentrations. For coarse aggregate subjected to normal washing processes he suggests:

Fig. 3.10 — Coarse aggregate washing with spray bars and size-separating screens.

Table 3.6 — First approximation of chloride ion content in relation to the moisture content of aggregate

Chloride ion content of wash water (% by mass of water)	Moisture content of aggregate (%)				
	2	4	6	8	10
	Expected chloride ion content of agggregates (% by mass of aggregate)				
0.4	0.01	0.02	0.02	0.03	0.04
0.6	0.01	0.02	0.04	0.05	0.06
0.8	0.02	0.03	0.05	0.06	0.08

$$P_{cb} = \frac{W_c(M_b - A_b) + KA_b}{100} \tag{3.1}$$

where P_{cb} = predicted chloride ion concentration, per cent, in the coarse aggregate; W_c = chloride ion concentration, per cent, in the wash water; M_b = moisture content, per cent, of the coarse aggregate; A_b = apparent absorption of the coarse aggregate; K = chloride ion concentration, per cent, in sea-water.

The apparent absorption is a mathematically determined value and is lower than the water absorption value (for example, 1.2. per cent *in lieu* of 1.8 per cent). When Course examined the accuracy of the prediction at various stages of processing, equation (3.1) seemed to work reasonably well for material sampled during processing and samples taken from stockpiles at the end of the first day of processing, and four and eight days after processing. However, predictions at one and fifteen days after processing significantly overestimated the chloride ion concentrations. Course suggested that the chloride level could vary, depending on the degree of washing during processing, subsquent rainfall affecting the apparent absorption and also the efficiency of extraction in the test procedure. The use of this equation may lead to greater accuracy in predicting the chloride ion concentration, but only if an accurate value can be determined for the apparent absorption of the aggregate.

3.1.7 Differences in characteristics between marine and land-based aggregates

The petrology of marine and land-based aggregates is similar, but a number of physical characteristics differ. Most obvious are the surface shape and texture, together with the presence of chloride salts from the sea-water and shells. These and other differences are considered below.

(a) *Shape and texture*

Marine aggregates, with the exception of their shell content, are typically smooth and rounded. These characteristics are also present in some natural aggregates drawn from land-based deposits, and, consequently, the performance of the sea- and land-derived materials could be expected to be similar. The properties of roundness and smoothness result in low inter-particle friction and interaction which is beneficial for some uses, but may require particular attention in other applications. For example, coated macadams and sub-base aggregates require particle interlock and thus, for these applications, it may be necessary to introduce a percentage of crushed aggregate to achieve the required performance. On the other hand, the smooth and rounded characteristics of marine aggregates have been shown to give benefits to some properties of concrete. Even with quite significant quantities of shell present, for instance, a marine aggregate dredged from the North Sea gave increased workability compared with a typical Thames Valley gravel (Chapman & Roeder 1970).

The improved workability is attributable to the lower particle interference associated with rounded aggregates. Advantage of this can be taken in a variety of ways to improve the economics of quality of the concrete. These include:

(i) Increasing workability (slump) to allow improved placing and compaction without loss of strength for the same cement content.
(ii) Reducing the water/cement ratio and improving strength whilst maintaining the same cement content without loss of workability for placing.
(iii) Reducing the cement content whilst maintaining the same water/cement ratio, strength, and workability for placing.

Options (i) and (iii) improve the economics of mix design, whilst option (ii) provides the opportunity to improve significantly the durability characteristic arising from lower permeability and improved strength.

Comparing a marine to a land-based coarse aggregate, Chapman & Roeder reported reductions in water/cement ratio of the order of 0.10 for concrete made to the same aggregate/cement ratio and workability. In this case, they looked at dredged coarse aggregate from the North Sea compared with a typical Thames Valley coarse aggregate, both combined with a Thames Valley sand. This order of reduction in the water/cement ratio was apparent at the three levels of agrregate/cement ratio investigated and is shown in Fig. 3.11.

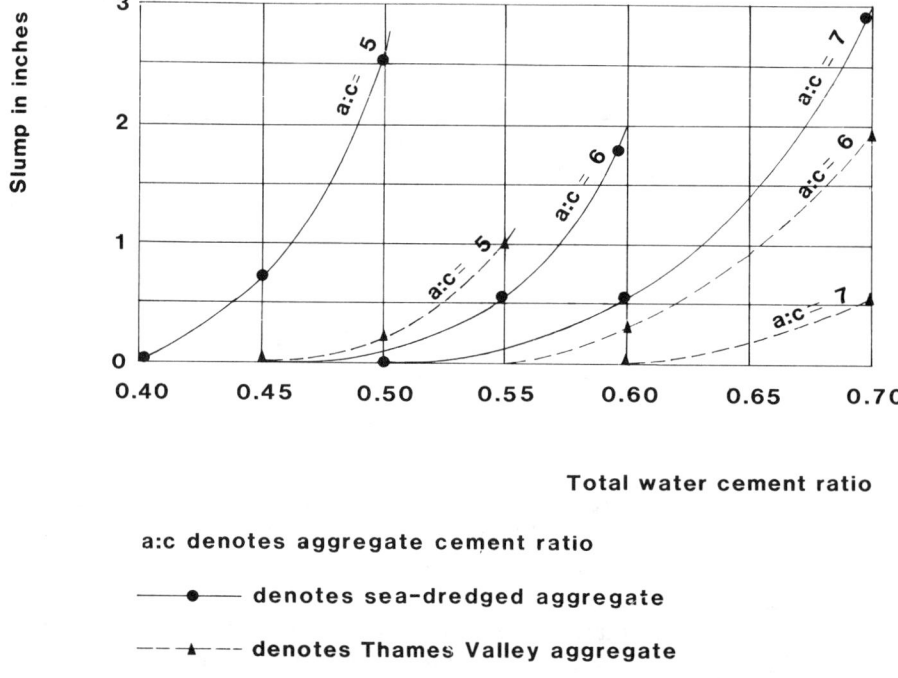

Fig. 3.11 — An example of difference in water/cement ratio for a given slump when a Thames Valley coarse aggregate is replaced with a sea-dredged material.

A similar exercise has been done by the present author, this time comparing combined dredged coarse and fine aggregate fom the southern North Sea against land-based aggregate drawn from a mid-Essex pit. The latter material was typically a mixture of irregular and rounded coarse aggregates having a smooth texture. The shape and texture of the land-won aggregate was thus much closer to the sea-dredged aggregate than that used by Chapman & Roeder, and the reduction in water/cement ratio, though apparent, was less in magnitude and tended to diminish at higher cement contents. Results are shown in Table 3.7; these confirm that the reduction in water/cement ratio or improved workability are directly related to the shape and surface texture characteristics of the marine aggregate. Examples using more rounded land-based gravels could further reduce or eliminate the difference entirely, but in general it may be concluded that the nature of marine-dredged aggregates is such as to allow lower water/cement ratios in concrete for a mixed cement content than land based gravels.

The reductions in water content are apparent at all normal strength levels and are likely to improve the strength performance of marine-dredged aggregates used in concrete. However, at very high strengths (say, characteristic strengths of 45 N/mm^2 and above) some caution may be necessary. Trial mixes should be made to identify any ceiling strengths caused by failure of bond between the cement matrix and aggregate.

Table 3.7 — Comparison of water/cement ratios for similar concretes made with land-based and marine aggregates (at slumps of 50 mm)

Aggregate/cement ratio	Free water/cement ratio	
	Mid Essex pit	Marine-dredged
5	0.47	0.46
6	0.55	0.53
7	0.65	0.61

(b) Silt
Marine aggregates may yield sands with low silt contents. Experts differ, however, as to whether this is beneficial or not. It is sufficient to say that this is the only aspect of grading in which marine aggregates can differ significantly from land-based sands and gravels, and it can easily be accommodated in normal mix design.

(c) Chloride
The presence of excessive chloride and their associated cations (sodium and potassium) in sea-water can cause concern in the following respects:

(i) increased risk of corrosion of reinforcement and embedded metals;
(ii) effects arising with cements containing low levels of tricalcium aluminate such as sulphate resisting cement (BSI 1980);
(ii) provision of cations which could contribute to alkali–silica reactions.

These concerns can be allayed if the chloride concentration is reduced by adequate washing, and limits have been imposed to ensure adequate protection. Most suppliers are prepared to issue certificated identifying chloride ion levels in their aggregates.

3.1.8 Effects of chloride on the corrosion of reinforcement

Under most conditions, Portland cement concrete provides protection against corrosion of reinforcement because of its high alkalinity. This protection can be reduced by a number of factors including carbonation, permeability of the concrete, the presence of cracks, concrete cover to metal and excessive chloride levels.

Corrosion is an electrochemical reaction, and both oxygen and moisture must be present if it is to occur. Chloride above a certain concentration can accelerate corrosion by:

(a) increasing the conductivity of the concrete pore fluid and thus assisting the flow of electrical corrosion currents,
(b) in excessive quantities, overcoming the passivating effects of hydroxyl ions.

It is important to keep the presence of chloride in aggregates in perspective, because it is present in small amounts from other ingredients in concrete; chloride-free concrete is simply not practicable. Chloride may be present in several states:

(i) strongly bound by tricalcium aluminate hydrates and, to a lesser degree, tetracalcium alumino ferrite hydrates;
(ii) loosely bound by calcium silicate hydrates;
(iii) as free ions in the pore solution.

This gives a clue to the means of dealing with chloride for the purposes of specifications by determining threshold values, below which corrosion risk is negligible. Surveys of reinforced concrete structures in Britain by the Building Research Establishment (Everett & Treadaway 1980) led to proposals for three categories for the chloride content of concrete by weight of cement arising from the initial constitutents of concrete:

> low — less than 0.4 per cent
> medium — 0.4 to 1.0 per cent
> high — greater than 1.0 per cent

That research indicated that dense, well-compacted concrete having chloride contents in the 'low' category provided good protection to steel over long periods.

Limits on chloride content for concreting aggregates given in the British Standard, BS 882 for concrete aggregates (BSI 1983), are shown in Table 3.8. These limits allow for combinations of sea-dredged and land-based fractions. For example, in west coast areas of the UK, sea-dredged sand has traditionally been combined with land-based coarse aggregate and, in some cases, blended with other land-based fine aggregate. It should also be remembered in this context that chloride in aggregate is not solely linked to sea-dredged or beach aggregates; several sources laid down by the sea but now forming part of the land mass also contain significant levels of chloride.

Table 3.8 — Maximum recommended chloride content of aggregate — BS 882:1983 Appendix C

Type or use of concrete	Maximum total chloride content expressed as percentage of chloride ion by mass of combined aggregates
Prestressed concrete Steam-cured structural concrete	0.02
Concrete made with cement complying with BS 4027 or BS 4248	0.04
Concrete containing embedded metal and made with cement complying with BS 12	0.06 for 95% of test results, with no result greater than 0.08

It is total chloride ion content by weight of cement in the concrete that is important in avoiding corrosion. Appropriate limits are set down in the British Standard code for structural concrete, BS 8110, (BSI 1985) and are reproduced in Table 3.9. For normal reinforced concrete made with ordinary Portland cement, the

Table 3.9 — Maximum chloride ion contents in concrete by mass of cement — BS 8110:1985

Type of use of concrete	Maximum total chloride content expressed as a percentage of chloride ion by mass of cement (inclusive of pfa or ggbs when used*)
Prestressed concrete Heat-cured concrete containing embedded metal	0.1
Concrete made with cement complying with BS 4027 or BS 4248	0.2
Concrete containing embedded metal and made with cement complying with BS 12, BS 146, BS 1370, BS 4246 or combinations with ggbs of pfa	0.4

* NOTE: pfa = pulverised fuel ash;
ggbs = ground granulated blast furnace slag.

maximum total chloride ion content is expressed as 0.4 per cent by mass of cement. In concrete made with sulphate-resisting Portland cement, the lower tricalcium aluminate content limits its ability to bind chloride ions. Accordingly, lower chloride limits are applied, that is, 0.2 per cent by mass of cement. This limit also applies to BS 4248 supersulphated cement (BSI 1974).

For prestressed and heat-cured concrete the limits are reduced to 0.1 per cent because of the accelerating effects of heat-curing on chemical rections and also because the effects of corrosion on metal components used for prestressing could lead to more serious problems then those associated with ordinary reinforced concrete.

For all but the last case, aggregate suppliers are generally able to produce marine aggregates within the specified tolerances by using normal washing procedures. For prestressed or heat-cured concrete, however, meticulous procedures are necessary to ensure that the required levels are achieved. Most marine aggregate producers operate quality control procedures and are able to certify the chloride levels being achieved in their processing. Table 3.10 shows a typical calculation for the chloride content of a concrete mix arising from the use of sea-dredged aggregate.

Table 3.10 — Typical calculation for chloride content by mass of cement

Cement content	300 kg/m^3 (say)
Chloride ion content — fine aggregate	0.03% by mass of aggregate
— coarse aggregate	0.02% by mass of aggregate
Fine aggregate content	700 kg/m^3 (say)
Coarse aggregate content	1200 kg/m^3 (say)

$$\text{Total chloride ion concrete} = \frac{(700 \times 0.03) + (1200 \times 0.02)}{300} \text{ \% by mass of cement}$$

$$= \frac{21 + 24}{300}$$

$$= 0.15\% \text{ by mass of cement}$$

Problems of corrosion arising from chloride in properly processed sea-dredged aggregates used in the UK are virtually non-existent and should not be confused with the more significant effects of calcium chloride historically used as an accelerating admixture in concrete. Calcium chloride was often included at dosages of chloride ten times higher than the chloride present in processed marine aggregate, or more, and this addition was often in flake form, which tended to lead to inadequate dispersion and local concentrations that were even higher. Calcium chloride also has another effect not attributable to sodium and potassium chlorides in that it can reduce the alkalinity of the concrete. It is thus important to differentiate in consideration between the chloride present in concrete arising from the aggregates and cements, and that derived from deliberate addition of calcium chloride.

A distinction should also be drawn between chloride present in the concrete ingredients and those introduced after the concrete is placed. These may come from exposure to de-icing salts or marine environments, and can be serious in their effects.

Firstly, they may lead to a build-up of chloride ion concentrations large enough to lead to rapid corrosion. Secondly, there is a lower probability that these chloride ions will be incorporated into solid cement hydrates. This can lead to a greater free chloride concentration in the pore water, and it is more significant than the introduction of chlorides through the use of sea-dredged aggregate. When concrete is likely to be exposed to these conditions, attention must be given to reducing its permeability. Consideration may be given to incorporating materials such as ground granulated blast furnace slag (ggbs) or pulverized fuel ash (pfa) as a means of reducing chloride diffusivity.

3.1.9 Methods of testing for chloride content

The best-known test method used for determining chloride content is the Volhard test based on a silver nitrate titration (BSI 1988). Two other proprietary rapid test methods are also widely used. These are the Quantab test strip chloride titrator and the Hach test.

The Quantab test strip (Fig. 3.12) is made of thin plastic in which a capillary

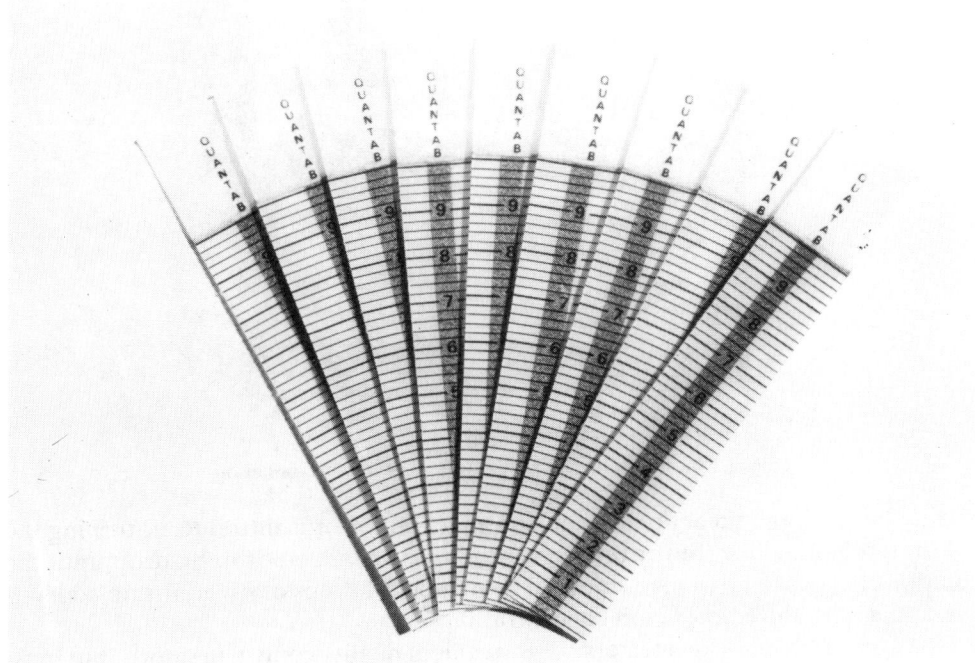

Fig. 3.12 — Quantab test strips for chloride content determinants.

column is impregnated with silver dichromate. Liquid rises in the column by capillary action when the strip is placed in a chloride solution; the reaction between the silver dichromate and the chloride produces a white colour change. The length of the column over which the colour change occurs can be related to the indicated chloride content, using a calibration chart supplied with the test strip.

In the Hach test (Fig. 3.13) a given quantity of chloride solutions is poured into a

Fig. 3.13 — Hach test kit for chloride content determination.

mixing bottle and a sachet of sodium chromate indicator is introduced, turning the solution bright yellow. Silver nitrate is progressively added drop by drop until the solution changes to an orange colour. The number of drops of silver nitrate added is then related to the chloride ion concentration.

The Quantab and Hach testers are convenient quality control methods, but must be calibrated at intervals against the Volhard laboratory method. A comparison of these test methods showed that there was no significant statistical differences at the 95 per cent confidence level between the results (Course 1984).

The Volhard test requires a skilled operator working in an analytical laboratory. The Quantab and Hach tests, in contrast, are relatively simple to carry out and can be performed by less-skilled personnel in the field; they are also cheaper. Although both the simple tests are satisfactory, the Quantab method has been found to be less sensitive to operator error, and Course adopted this for his own investigations into chloride variability in stockpiles. This method has also been widely adopted for quality control in the ready-mixed concrete and aggregate industries.

The importance of methodical and proper sampling and test procedures is stressed if the determination of chloride content is to be meaningful. Additionally, with the so-called rapid tests, it is often necessary to soak the aggregate for at least 24 hours before taking the sample solution for testing. These precautions are necessary if significant amounts of chlorides are entrapped in the pores of the aggregate. Experience and trials may allow shorter extraction times if rapid quality control indications are required.

A more recent test method with considerable appeal employs a hand-held conductivity meter. The meter measures the conductivity arising from the presence of salts in sea-water. Obviously, there are salts apart from sodium chloride present, but it is likely that these are of relatively minor effect and can be at least partly allowed for in the calibration process.

The equipment is also available in forms suitable for use in fixed positions on production plant, or on the laboratory bench. As for all tests, the importance of proper sampling procedures must be stressed. These meters are in the early stage of development (Pike 1987), and calibration against the Volhard method must also be considered. The same precautions for sampling and testing mentioned above will also apply.

Historically, there have generally been three methods of expressing chloride content in concrete, i.e. as percentages of:

(a) chloride ion (Cl^-) — this is the term used in British Standards,
(b) sodium chloride or salt (NaCl),
(c) equivalent anhydrous calcium chloride ($CaCl_2$) — this was used when such admixtures were incorporated in concrete.

The first two expressions are still frequently seen and can be related to each other as follows:

$$\text{Concentration of sodium chloride} = \frac{1.648 \times \text{concentration}}{\text{of chloride ion}}.$$

Table 3.11 — Example of different methods of expressing chloride ion contents

Material	Mass of aggregate (per m^3)	Percentage equivalent sodium chloride (by mass of aggregate)	Percentage chloride ion	
			(by mass of aggregate)	(by mass of aggregate)
Fine aggregate	700	0.05	0.030	0.070
Coarse aggregate	1200	0.03	0.018	0.072
Total combined aggregate	—	—	0.022	0.142

(Note: 300 kg/m^3 cement has been assumed for calculations linked to cement content)

Sea-water contains not only sodium and chloride ions but also many relatively minor constituents including magnesium, potassium, and sulphate ions. For simplicity, however, it is generally assumed that all the chloride ions are present as sodium chloride. The approximation has little effect and is acceptable for most practical purposes. (However, when converting chloride ion content to sodium chloride, it should more correctly be described as the percentage equivalent of sodium chloride.)

Chloride contents are often further expressed in relation to the mass of aggregate or mass of cement in a cubic metre of concrete. An example is shown in Table 3.11 based on a concrete having $300 \, kg/m^3$ cement content and containing chloride at typical values for washed marine aggregates.

3.1.10 Alkali–silica reaction

This is a relatively limited problem in the UK in relation to the quantities of concrete produced (see section 3.2, below), and it is likely to cause damage only when all the following three factors occur together:

(a) sufficient moisture;
(b) sufficient alkalis from all ingredients of concrete, but generally predominantly cement;
(c) a critical amount of reactive silica is present in the aggregate.

Many UK aggregates, and particularly sands from both land-based and sea-dredged sources, contain proportions of flint or chert within which small quantities of reactive silica are present. Some of these can react with excessive alkali very occasionally to cause abnormal expansion in concrete. The principal method of controlling the problem adopted in the UK is to limit the alkali content of the concrete. Sodium and potassium derived from sea-water can contribute to such alkalis and may thus need to be taken into account when calculating total alkalis in the concrete.

A Concrete Society report (Technical Report No 30) recommended that the following equation be used for calculating the equivalent alkali contributed by the ions from any sea-water present in the aggregate:

$$H = \frac{0.76 \times [(NF \times MF) + (NC \times MC)]}{100} \tag{3.2}$$

where H = equivalent alkali contrinbution made to the concrete by the chloride-containing aggregate (kg/m^3), NF = chloride ion content of the fine aggregate as a percentage by mass of dry aggregate, MF = fine aggregate content (kg/m^3, NC = chloride ion content of the coarse aggregate as a percentage by mass of dry aggregate, MC = coarse aggregate content (kg/m^3).

The constant 0.76 is derived from consideration of the composition of typical sea-water and the relation of the sodium, potassium, and chloride ion contents. A typical calculation is shown below, assuming contents in the concrete of fine and coarse aggregate of 700 and 1200 kg/m^3 with corresponding chloride ion contents of 0.03 and

0.02 per cent, respectively. In this case, the equivalent alkali content contribution (H) from the aggregate is:

$$H = \frac{0.76 \times [(0.03 \times 700) + (0.02 \times 1200)]}{100}$$

$$= 0.34 \, \text{kg/m}^3 \,.$$

This contribution is relatively small in comparison with the maximum levels generally specified. (A limit on total content from all ingredients of $3 \, \text{kg/m}^3$ for general concrete construction is a typical specification.) Nevertheless, it again emphasizes the need for the effective processing of aggregate.

It is perhaps worth noting that large quantities of sea-dredged aggregate have been used in the UK and especially in the southeast of England. However, in this area there is no recorded incident relating to abnormal alkali–silica reaction from marine aggregate used alone or in combination with land-based gravels.

A more comprehensive consideration of alkali–silica reaction is given in section 3.2 and the reader is referred to that section.

3.1.11 Shell

Shell is essentially calcium carbonate with a typical particle density of 2.70. Chemically, it is no different from limestone, and only its physical characteristics need separate consideration, in particular the possible effects of flaky and hollow shells.

Concern has been expressed that broken and flaky shells could adversely affect the workability of concrete, and that hollow shells could remain unfilled with mortar, reducing both the strength and elastic modulus of concrete. These concerns were extensively investigated in the late 1960s (Chapman & Roeder 1970). They found that, even with large shell contents, the structural properties of concrete were unimpaired.

Chapman & Roeder conducted a wide range of experiments in which proportions of shell were added to Thames Valley gravel, both as flat and as hollow shells. Comparative trials were undertaken on concretes made at constant water/cement ratio, and a summary of their results is shown in Table 3.12. It can be seen that the addition of shell can cause a modest reduction in workability, but that this is the only negative effect. This is no great problem because, in practice, it is more than counteracted by the benefits deriving from the smooth texture and rounded shape of the marine gravel. Chapman & Roeder confirmed this to be the case by comparing performance between Thames Valley gravel and gravel dredged off the Thames Estuary, and obtained greatly increased workability with the dredged aggregate. They also, confirmed, by sawing through specimens, that hollow shells were filled with concrete mortar.

Other tests, conducted by Messrs Sandberg, are also reported in Chapman & Roeder's paper. These investigated the water absorption of concrete containing mostly whole, hollow shells. Messrs Sandberg, in a separate series of trials, also determined the drying skrinkage of concrete having various shell contents. These

Table 3.12 — Effect of shell on the properties of concrete (after Chapman & Roeder)

Property	Effect of increased shell content
Workability	Modest reduction (counteracted by smoothness and roundness of sea-dredged gravel)
Density after compaction	No difference
Entrapped air	No difference
Cement requirement	No difference
Cube strength:	
at 3 and 8 days	Significant gain
at 28 and 91 days	No difference
Tensile and compressive strength	No difference
Freeze thaw/resistance	No difference

Note: The comparative trials were carried out at a fixed water/cement ratio. In normal use concrete is supplied at a fixed workability, and properties of such concrete containing marine-dredged aggregate would be expected to have a lower water/cement ratio and thereby an enhancement of some of these properties

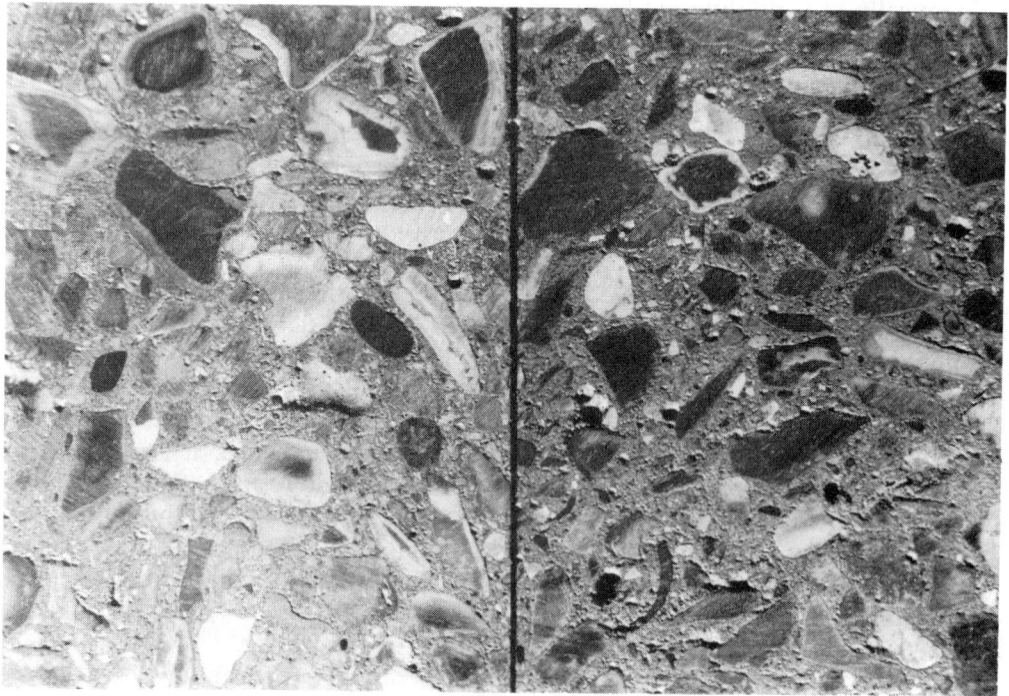

Fig. 3.14.

trials showed that increasing shell content has no significant effect on either the water absorption or drying shrinkage of concrete produced to a fixed water/cement ratio.

Because concrete is normally produced to a given workability suitable for placing and finishing, the water/cement ratio of the concrete produced with marine aggregate is often lower in practice than its land-based counterpart at a fixed cement content. Typical differences are quoted earlier in this chapter, and these reductions could be expected to enhance many of the properties reported above for concrete produced to a given cement content.

As a result of these tests, limits were introduced for shell content into the British Standard BS 882 for concrete aggregates; these are reproduced in Table 3.13. They

Table 3.13 — Limits on shell content after BS 882:1983

Size of aggregate	Limits on shell content (%)
Fractions of 10 mm single size, or of graded or all-in aggregate that are finer than 10 mm and coarser than 5 mm	20
Fractions of single sizes of graded or all-in aggregate that are coarser than 10 mm	8
Aggregates finer than 5 mm	No requirement

appear very conservative in the light of the available evidence on the properties of concrete. Nevertheless, the marine aggregate industry has no apparent problems in complying with them, and most suppliers are prepared to issue certificates identifying the shell content in their coarse aggregates.

The two principal methods for determining shell content are hand sorting and chemical tests. The hand sorting method of separating shell from aggregate and weighing each sorted fraction is suitable only for coarse aggregate (BSI 1985). Chemical tests rely upon acid dissolution of the shell; and any other acid-soluble material, such as limestone, present in the aggregate, may also be dissolved. This could result in an indication of a level of the shell content above that actually present.

3.1.12 Wear and skid resistance

A limited test was carried out by Chapman & Roeder on a sea-dredged aggregate from the Humber to assess its wearing performance in comparison to land-derived aggregates from Surrey and Essex. Tests were conducted, using an early version of the Cement and Concrete Association (now the British Cement Association) machine for measuring accelerated wear. The skid resistance was assessed after various periods of wear and confirmed that there was no significant difference between the aggregates in this respect. Since that work, much satisfactory practical experience of the use of marine-won aggregates in paving is available. Examples include sections of the M3 and M20 motorways.

3.1.13 Organic impurities

The risk of organic impurities such as lignite or coal occurring in marine aggregates is no greater than for land-won aggregates. Occasionally, however, dead fish or crabs are included. Such problems can be overcome by adopting appropriate dredging and/or processing practices. Good management practice should be exercised in these operations. Visual inspection of aggregate remains an important quality control method in determining obvious contamination for both land-won and sea-dredged aggregate.

A warning is necessary for anyone considering pH tests for organic impurities, because the presence of even small quantities of chloride may lead to reduced and anomalous values. Other methods of testing are necessary, including measurements of setting time and/or the strength of concrete test specimens. (Note: reduction of the pH value is related to the aggregate test only, and should not be confused with the alkalinity of the concrete produced which remains sensibly the same.)

3.1.14 Efflorescence

Some suggestions have been made that concrete made with marine aggregates may be more susceptible to producing efflorescence, but studies show that such crystallization of soluble salts is relatively rare, and, because they remain soluble, any such efflorescence would rapidly disappear (Higgins 1982).

The major cause of disfiguring efflorescence is the liberation of free lime during the hydration of cement. This lime can carbonate on exposure to the air, to produce calcium carbonate sometimes referred to as 'lime bloom'. It occurs only where water is able to migrate through the concrete and can remain on the surface in an unsightly manner. This is quite different from the presence of minor quantities of salts from marine aggregates, which would quickly wash away.

References

Advisory Committee on Aggregates (1976) *Aggregates: the way ahead*. HMSO.
British Standards Institution
> BS 812:1985. *Testing Aggregates*. Part 106. *Method for determination of shell content in coarse aggregate*.
> BS 812:1988. *Testing Aggregates*. Part 117. *Method for determination of water-soluble chloride salts*.
> BS 882:1983. *Specification for aggregates from natural sources for concrete*.
> BS 4027:1980. *Specification for sulphate resisting cement*.
> BS 4248:1974. *Specification for supersulphated cement*.
> BS 8110:1985. *Structural use of concrete*. Part 1. *Code of practice for design and construction*.

Chapman, G. P. & Roeder, A. R. (1970) The effects of sea-shells in concrete aggregates. *Concrete*. February.
Collis, L. & Fox, R. A. (1985) *Sand, gravel and crushed rock aggregates for construction purposes*. Geological Society engineering geology special publication No. 1.
Concrete Society Technical Report No. 30. *Alkali-silica reaction — minimising the risk of damage to concrete. Guidance notes and model specification*.
Course, J. L. (1984) *Chlorides in aggregates. An investigation into within batch*

variability. Project for the City and Guilds Advanced Concrete Technology Diploma.

Department of the Environment (1982) *Guidelines for aggregate provision in England and Wales*. Circular 21/82. August.

Everett, L. H. & Treadaway, K. W. J. (1980) *Deterioration due to corrosion in reinforced concrete*. Building Research Establishment Information Paper IP 12/80. August.

Higgins, D. D. (1982) *Appearance matters. Efflorescence on concrete*. Cement and Concrete Association.

Nunny, R. S. & Chillingworth, P. C. H. (1986) *Marine dredging for sand and gravel*. Department of the Environment Minerals Division. HMSO.

Pike, D. C. (1987). *Digital meter for chlorides. The Agtest Salcon* (unpublished). August.

Spruell, W. J. & Uren, J. M. L. (1986) *Marine aggregates and aspects of their use especially in the South East of England*. Sand and Gravel Association 1986.

3.2 ALKALI–SILICA REACTION

3.2.1 Introduction

The alkali–silica reaction, ASR, occasionally induces expansion and visible cracking in concrete subject to external moisture. The affected concrete is normally of high quality and high alkali content. Typical map cracking or oriented cracking induced by ASR is shown in Figs 3.15 and 3.16. The cracking shown is not necessarily

Fig. 3.15 — Map or pattern cracking, probably induced by alkali–silica reaction.

indicative of ASR as abnormal expansion or contraction from other causes can lead to similar visual damage to concrete.

The occurrence of cracking in concrete structures caused by ASR is not new; its

Fig. 3.16 — Oriented cracking probably induced by alkali–silica reaction.

effects were first observed in the USA some 70 years ago. It was not, however, until 1940 that ASR was recognized as a problem in the USA (Stanton 1940). Deterioration due to ASR has been observed in many other countries including Canada, South Africa, Japan, Australia, Germany, and Denmark. It was in 1976 that ASR was confirmed as the probable cause of cracking in a number of structures in the UK (Palmer 1978, Allen 1981). The majority of the affected structures are in the southwest and the Trent Valley. Some of the affected structures date back to the 1930s, but a high proportion were built in the years 1969, 1970, and 1971. It has been suggested that between 50 and 300 structures built in the UK between 1931 and 1975 may have cracked as a consequence of ASR. The structures affected include bridges, foundation blocks at a number of electricity stations, sewage treatment works, reservoirs, a jetty, a hospital, a multi-storey car park, a racecourse stand, and a ventilation shaft. The total number, or even an approximate total, of the affected structures is unknown as few of them have been publicised. Also the reasons as to why the judgement has been made that ASR is the cause of the deterioration have only rarely been given.

The sources of aggregate used in concretes believed to have cracked as a consequence of ASR include:

(i) sea-dredged sand from off the Isle of Wight,
(ii) sea-dredged sand from the Bristol Channel,
(iii) sand from the Wareham area of Dorset, and
(iv) aggregate from a number of sources in the Trent Valley.

3.2.2 The reaction

ASR is a chemical reaction between disordered forms of silica, which occasionally occur in significant quantities in an aggregate, and the hydroxyl ions in the pore water of a concrete. In Portland cement concrete the hydroxyl ion concentration is normally a function of the alkali sodium and potassium hydroxides resulting from the release of alkali compounds from the Portland cement. The reaction forms a gelatinous product that contains silica, calcium, sodium, potassium and water. Because the gel product is much greater in volume than the silica consumed, its formation and growth can occasionally induce internal stresses in the concrete of sufficient magnitude to cause expansion and visually severe cracking (Hobbs 1988).

The rate of reaction is greatest when the reactants first come into contact and declines thereafter. The reaction ceases when either of the reactants is exhausted. Expansion, if induced, ceases when either of the reactants or the free water in the pores is depleted, or if physical equilibrium is established, for example if the rate of growth of the reaction product is so slow that it is able to migrate without inducing additional expansion.

For cracking and expansion to result from the reaction the following combination is normally required:

(i) a significant quantity of reactive silica — if the reactive silica content is low, then the reaction will be of insufficient intensity to induce expansion;
(ii) in the case of fast-reacting forms of silica, a reactive silica content within a critical range (see Fig. 3.17);
(iii) available alkalis above a critical level (see Fig. 3.18); and
(iv) water from an external source.

In small laboratory specimens of concretes stored at 20°C under wet conditions, cracking due to ASR has been observed as early as 14 days (Hobbs 1987a). In concrete structures, perhaps partly owing to the slower ingress of moisture, cracking may take several years to manifest itself. It may not be until an age of five to ten years that the cracking causes concern. Tests on cores taken from affected structures exhibiting severe ASR cracking have shown that the compressive strength may be reduced by up to 35 per cent and the elastic modulus by up to 80 per cent (Koyanagi et al. 1987). It has been concluded from loading tests on concrete structures and concrete beams that ASR expansion may not adversely affect the stiffness or load-carrying capacity of affected concrete members (Koyanagi et al.1987, Imai et al. 1987, Fuji et al. 1987). Thus a core taken from a structure is not necessarily representative of the concrete within the structure because once it has been removed the equilibrium is changed. The core may then expand, establishing a new equilibrium state and lower strength and elastic modulus.

3.2.3 Soures of alkali

Fig. 3.19 shows the effect various sodium and potassium compounds can have upon expansion (Hobbs 1987b). The main source of sodium and potassium compounds in a concrete is normally the Portland cement, but the available alkali content of a

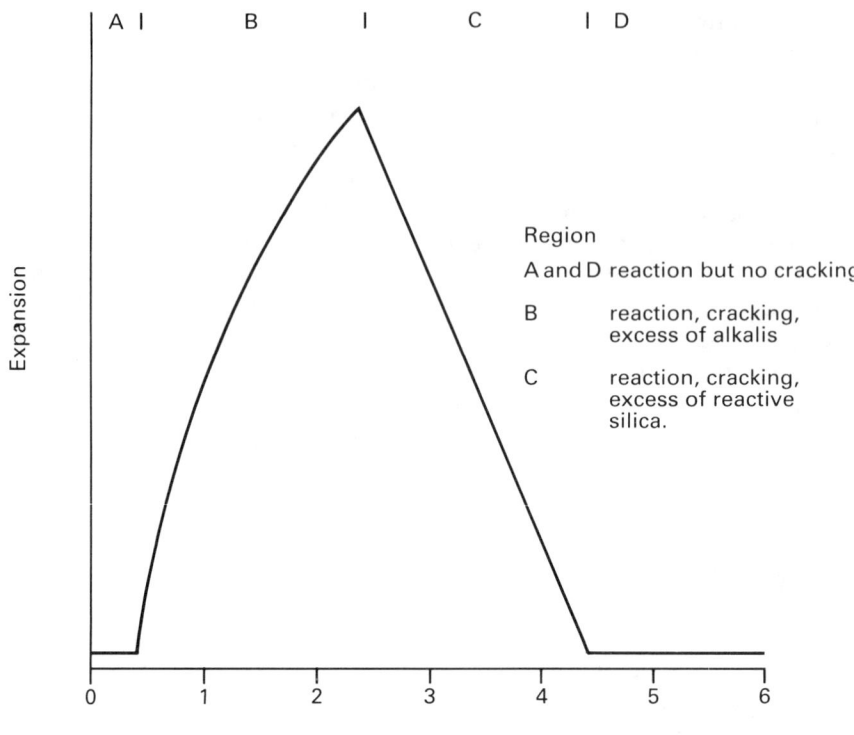

Fig. 3.17 — Pessimum behaviour obtained with OPAL-A, a cryptocrystalline form of silica (W/C 0.4, a/c 2.75, alkali content 6 kg/m³) (by permission of Thomas Telford Publications).

concrete can be enhanced by salt present in sea-dredged aggregate (even after it has been processed), by alkalis entering the concrete from an external source, and by the alkalis released from pulverized fuel-ash and ground granulated blast furnace slag.

(a) Portland cement
The total or acid-soluble alkali content of a Portland cement (BSI 1978) is conventionally calculated as equivalent sodium oxide, using the formula

$$(Na_2O)_e = (Na_2O) + 0.658(K_2O) \tag{3.3}$$

where $(Na_2O)_e$ is the equivalent sodium oxide content, Na_2O is the actual sodium oxide content, and K_2O is the actual potassium oxide content, the assumption being made that equivalent concentrations of sodium oxide and potassium oxide are equal in their effect. The sodium and potassium oxide contents of a Portland cement (and a ground granulated blastfurnace slag) are determined after extracting Na_2O and K_2O from a known mass of Portland cement by boiling in nitric acid (BSI 1970).

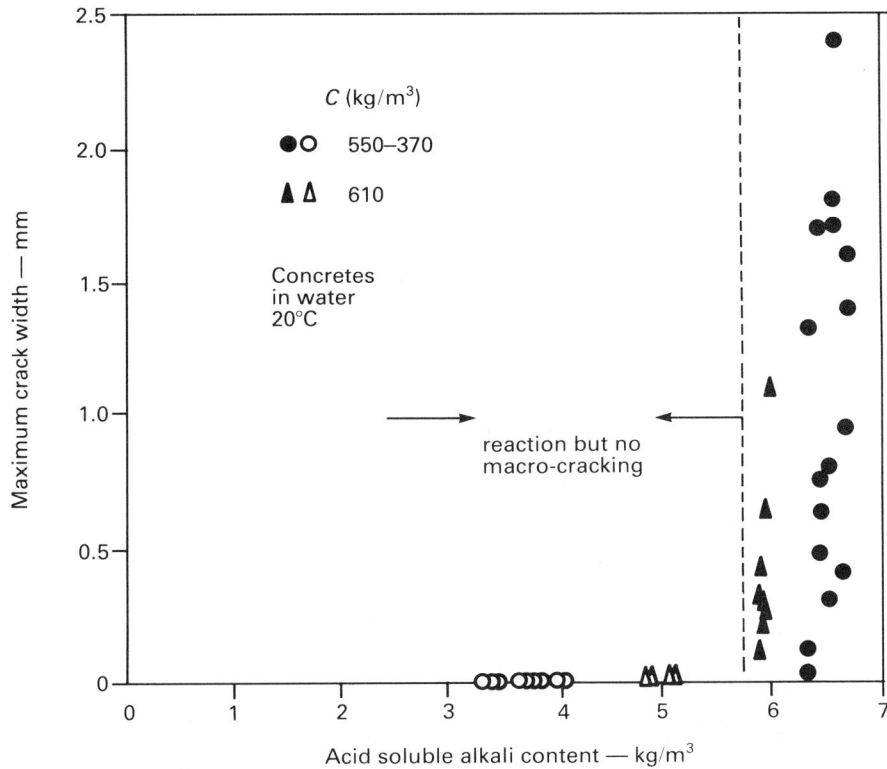

Fig. 3.18 — Variation of maximum crack width at an age of 14 years with acid-soluble alkali content. Thames Valley sand, Mendip limestone coarse (expansion of cracked specimens 0.15 to 0.30 per cent (after Hobbs 1987a) (by permission of Noyes Publications).

In the years 1969, 1970, and 1971, when many of the affected UK concretes were placed, the monthly mean alkali content of the Portland cements in production ranged from 0.3 to 1.3 per cent by mass. Today, rationalization and in some instances changes in raw feed to the cement kiln, have limited the monthly mean alkali content of Portland cements to the range 0.3 to about 0.9 per cent by mass (Hobbs 1988). The hydroxyl ion concentration in a high-alkali content concrete can be up to 10 times as high as in a concrete of low-alkali content. It is only at the higher hydroxyl ion concentrations, or more correctly the higher total hydroxyl ion contents, that the reaction may be of sufficient intensity to induce cracking. This is illustrated in Fig. 3.18 where the maximum crack width at an age of 14 years induced by ASR is shown plotted against the original alkali content in kg/m^3. Note that none of the concretes has cracked at alkali levels below 5 kg/m^3. To the author's knowledge, cracking due to ASR in UK concretes has been observed in controlled tests only when the original alkali level, expressed as equivalent Na_2O, exceeded 5 kg/m^3.

(b) Salt
The alkali content of a concrete can be increased by the salt that is present in sea-dredged aggregate even after washing. This salt, in the pore water of a concrete,

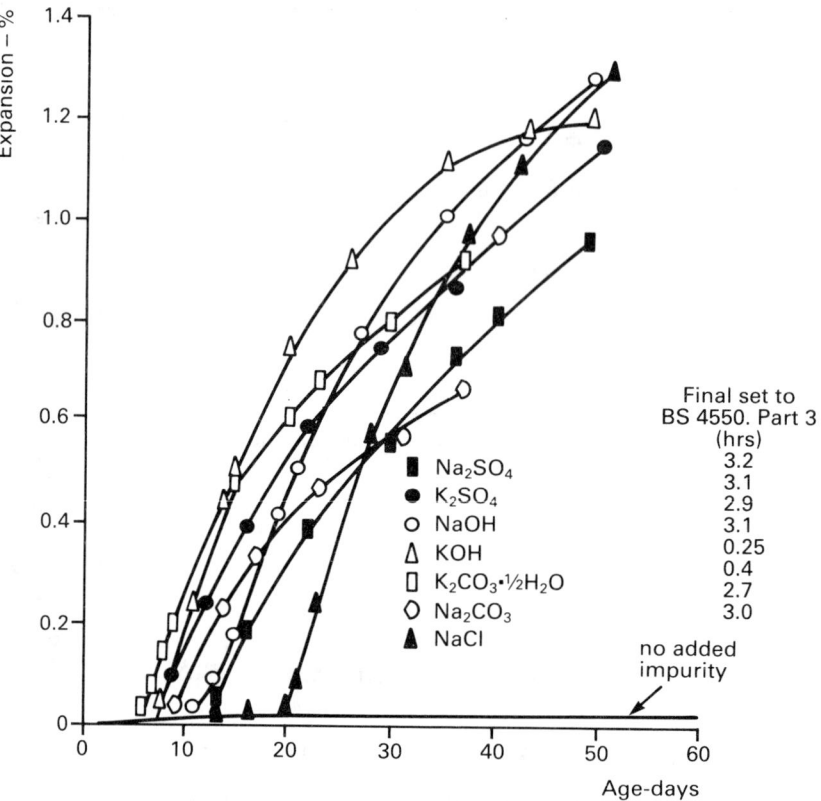

Fig. 3.19 — Influence of some inorganic impurities upon the expansion of mortars due to ASR, w/c 0.41, a/c 2.75, 20°C. (Alkali from Portland cement expressed is Na_2O equivalent 3.26 kg/m^3; alkali from inorganic impurity expressed as Na_2O equivalent 2.0 kg/m^3) (by permission of Conseil International de la Langue Française).

converts to sodium hydroxide which increases the hydroxyl ion concentration and to calcium chloride which reduces the setting time. Both of these factors can increase the risk of cracking due to ASR.

Table 3.14 gives analyses of concrete cores taken from several piers supporting

Table 3.14 — Alkali and chloride contents in cores taken from piers supporting the Hanshin Expresssway

Pier condition	Equivalent Na_2O	Cl$^-$	Equivalent Na_2O kg/m^3	
	(% by mass of concrete)		Total	From salt
Cracked	0.29–0.33	0.072–0.095	6.5–7.4	1.4–1.9
Uncracked	0.19–0.28	0.045–0.060	4.4–6.1	0.9–1.2

the Hanshin Expressway in Japan (Mizuomoto *et al.* 1987). The sand used in these piers was an unwashed sea-dredged sand. The piers affected by ASR had alkali contributions from the salt, expressed as equivalent sodium oxide, of between 1.4 and 1.9 kg/m^3. It is clear from the data in this table that it is unwise to use unwashed sea-dredged aggregate in concrete if it is considered that the aggregate is reactive.

The alkali contribution, as equivalent sodium oxide, made by salt present in aggregate is calculated from:

$$(Na_2O)_e = (0.76 \times Cl \times Ma)/100 \tag{3.4}$$

where Cl is the chloride ion content of the aggregate expressed as a percentage by mass of dry aggregate (BSI 1976a). Ma is the aggregate content in kg/m^3.

(c) Pulverized fuel-ash (pfa) (BSI 1982, 1985a)

The total alkali contents of UK pfas lie in the range 0.7 to 4.6 per cent by mass (Hobbs 1986). This alkali content is determined by evaporation of a known mass with a hydrofluoric/perchloric acid mixture (BSI 1970). Most of the ashes in use in the UK as partial replacements for Portland cements have alkali contents in the range 3.0 to 3.8 per cent by mass. The alkalis in pfa are tied up in its glassy phases and are released slowly as the pfa reacts. There is disagreement over the role played by the alkalis released by pfa in the alkali-silica reaction. Expansion data obtained to date on concretes and mortars made with ordinary Portland cements (OPC) having alkali contents in the range 2.5 to 5.5 kg/m^3, and containing non-UK reactive aggregates, indicate that the effective alkali contribution from a pfa at a cement replacement level above 25 per cent by mass is approximately one sixth of its total alkali content (Hobbs 1986, 1988; Kolleck *et al.* 1986).

Expansion data obtained on concretes of very high alkali contents, 6 to 8 kg/m^3, using a UK aggregate, have indicated no alkali contribution from pfa (Nixon *et al.* 1986). The applicability of this observation to concretes of more normal alkali content is regarded as questionable as, at high hydroxyl ion concentration, alkali release by pfa into the pore solution will be surpressed.

(d) Ground granulated blast-furnace slag (ggbs) (BSI 1974, 1984)

The acid-soluble alkali contents of the ggbs currently available in the UK range from about 0.4 to 1.1 per cent by mass. These alkalis are tied up in the glassy phases of a ggbs and are released at an intermediate rate to those from OPC and pfa. Expansion data obtained on concretes and mortars, with OPC alkali contents in the range 2.5. to 5.5 kg/m^3 and containing non-UK reactive aggregates, indicate that the effective alkali contribution from a ggbs at a cement replacement level above 25 per cent may be approximately one half of its acid-soluble alkali content (Hobbs 1986, 1988; Kolleck *et al.*, 1986)

(e) Aggregates

Certain aggregates, when attacked by hydroxyl ions, can release alkalis into the pore water. Examples of this are artificial glasses and volcanic glass present in rhyolites and andesites. Currently accepted advice (Concrete Society 1987) is that no significant amounts of alkali will be derived from natural aggregates in the UK (chlorides excepted).

(f) External alkalis
The alkali content of a concrete can be enhanced by alkalis entering the concrete from external sources. Alkali salts can migrate into hardened concrete, from seawater, some groundwaters, and some other materials such as de-icing salts. Their influence will depend upon the proportion of reactive silica present in the concrete and the quality of the concrete at the time of exposure. Additional alkalis migrating into mature concrete may not induce significant cracking and expansion if, firstly, the concrete already contains an excess of alkalis (this would be the situation if the reactive silica content is below the pessimum — see next section), and, secondly, the concrete is uncracked at the time of exposure (Hobbs 1988).

3.2.4 Reactive silica
Aggregates are termed reactive if they contain sufficient quantities of reactive silica to induce cracking in concrete. Many UK aggregates are considered not to contain sufficient reactive silica to induce cracking. Table 3.15 gives guidance on the rock types in the UK in which it is currently believed that the potentially reactive constituents present may occasionally induce or possibly induce cracking due to ASR. This information is taken from a recent working party report on diagnosis (British Cement Association 1988).

The alkali–silica reactive constituents which may be present in aggregates include opal, cristobalite, tridymite, siliceous and intermediate volcanic glass, chert, glassy to crypto-crystalline volcanic rocks, synthetic siliceous glasses, some argillites, phyllites, schists, gneisses, gneissic granites, vein quartz, quartzite, and sandstone (ASTM 1987a). The materials in this list, if not finely dispersed and present in low proportions, can be identified by optical examination of X-ray diffraction and by petrographic examination of thin sections. Finding alkali-reactive constitutents in an aggregate does not in itself allow a prediction to be made of the likely performance of that aggregate in concrete. To do this it is necessary to make a judgement based on field performance or to carry out expansion tests on mortars or concretes.

Some researchers (McConnell *et al.* 1947; Gaskin *et al.* 1955) have measured the expansive reactivity of a number of forms of reactive silica in mortar bars of very high alkali content (6.2 and 8.0 kg/m^3) stored above water at 20 or 38°C. The silica minerals tested can be listed in decreasing order of deleteriously expansive reactivity as follows:

> opal,
> chalcedony, volcanic glass,
> cristobalite,
> trydimite,
> crypto-crystalline quartz.

Although micro-crystalline quartz was found to give a reaction it was not

Table 3.15 — Rock types in the UK which may induce cracking due to ASR (after British Cement Association 1988)

Rock (R) or mineral (M) type	Definition	Potentially alkali-silica reactive components that may sometimes be present	Reported occurrences of ASR-related cracking (Note 1)
Chalcedony or chalcedonic silica M	Very fine-grained silica with a distinctive fibrous microstructure	Chalcedony	Possible
Chert R	Micro-crystalline or crypto-crystalline silica	see Flint	Many (some)*
Conglomerate R	Coarse detrital rock containing rounded fragments	see Sandstone	None
Cristobalite M	Very high temperature form of quartz, occurring as a metastable constituent in some acid volcanic rocks	Cristobalite	Possible
Diorite and microdiorite R	Intermediate intrusive igneous rock (differs mineralogically from syenite)	Opaline or chalcedonic veins or vugh-fillings (Note 7)	Some (several)*
Flint R	Micro-crystalline or crypto-crystalline silica Strictly, chert occurring in Cretaceous chalk	Chalcedonic silica and micro-crystalline or crypto-crystalline quartz. Some varieties may contain opaline silica.	Many (some)*
Greywacke R	Detrital sedimentary rock containing rock fragments and mineral grains in a fine-grained matrix	Some greywackes may be alkali-reactive. See Sandstone. (Notes 3 and 4)	Possible
Gritstone R	Sandstone with coarse, angular grains	see Sandstone	
Opal or opaline silica M	Essentially amorphous or disordered forms of hydrous silica	Opaline silica is highly reactive (Note 7)	Some (several)*
Quartz M	Crystalline forms of silica, often occurring as discrete mineral grains especially in fine aggregates	Highly-strained quartz. Some micro-crystalline or crypto-crystalline forms are reactive (Note 2)	Possible (except chert and flint) see
Quartzite R	Sedimentary or metamorphic rock consisting predominantly of quartz: (i) sedimentary or ortho-quartzite (ii) metamorphic or meta-quartzite	see Sandstone Highly-strained quartz and/or poorly crystalline line boundaries between quartz grains. Micro-crystalline of crypto-crystalline quartz (Note 2)	Possible Some (possible)*
Rhyolite R	Fine-grained to glassy acid volcanic igneous rock	Glass or devitrified glass. Tridymite. Crysitobalite. Opaline or chalcedonic veins or vugh-filling	Possible
Sandstone R	Detrital sedimentary rock. The grains are most commonly quartz, but fragments or grains of almost any type of rock or mineral are possible	Highly-strained quartz. Some types of rock cement notably opaline silica, chalcedonic silica, micro-crystalline or crypto-crystalline quartz. (Note 2, 3 and 4)	Possible
Volcanic glass M	Non-crystalline phase within some volcanic rocks; may be devitrified to very fine or incipient crystals	Some volcanic glass or devitrified volcanic glass is reactive	Possible

* The present author's opinion in brackets.

Notes to Table 3.15

(1) This guidance is based upon published information available in 1987 and could be superseded by later reports. In the case of certain rock or mineral types listed for which no occurrences of ASR have been reported in the UK, similar rock or mineral types have been associated with examples of damaging ASR in other parts of the World.

(2) Rocks containing highly-strained quartz, and/or exhibiting poorly crystalline boundaries or crypto-crystalline quartz between contiguous quartz grains, are sometimes found to be alkali–silica reactive. The identification of high undulatory extinction angles for quartz grains (measured in thin section under a petrological microscope) may indicate potentially reactive rocks. At present, no definitive data are available for UK rocks. Overseas experience suggests that an average undulatory extinction angle exceeding 25° would indicate highly strained quartz.

(3) Phyllosilicates may be present; these are sheet silicate minerals including the chlorite, vermiculite, mica, and clay mineral groups. Expansive 'alkali–silicate' reactions involving phyllosilicates have been reported in Canada and South Africa.

(4) Within the UK a few cases of possible alkali-reactions have been reported in coarse aggregates containing greywacke and related rocks. The matrix in such rocks is very finely divided and consists of phyllosilicates, quartz, and other minerals. The nature of this reaction is the subject of continuing research.

(5) Certain types of dolomitic limestone (or calcite dolomite) have been reported to be expansively reactive in North America. Some dolomites have been reported to be non-expansively reactive in the Middle East and North Africa. These are collectively termed 'alkali–carbonate' reactions. No reports of alkali–carbonate reactivity have been published for UK aggregates.

(6) In some other countries, highly altered or exceptionally weathered feldspars in aggregates have been considered to release additional alkalis within concrete. No reports of such alkali content enhancement have been published for UK concretes.

(7) Some concretes in Jersey, Channel Islands, exhibit ASR as the result of opaline silica and chalcedony veining of a dioritic rock aggregate source.

deleterious at an alkali level of 6.2 kg/m^3. Quartz, which is the commonest and most ordered form of silica, was found to exhibit no measurable reactivity at an alkali level of 6.2 kg/m^3.

Reactive forms of silica can sometimes be present as constituents of chert, flint, shale, sandstone, limestone, and other rock types. Chert is a common constituent of affected concretes. In the UK, the reactive forms of silica are generally present in some chert and flint particles, but there is no general agreement on which forms cause deleterious reactions. This lack of agreement arises, in part, because judgements have been based on examinations of affected concretes within which the reacting siliceous constituents may have been depleted. The reactive constituents in the cases of ASR so far identified in the UK, according to a recent Concrete Society Report, are micro-crystalline and crypto-crystalline silica and chalcedony found in flints and cherts and possibly also strained quartz in some quartzites (Concrete Society 1987). However, to the author's knowledge it has not been established that these constituents when present in UK aggregates are deleteriously reactive in UK field concrete.

The expansion of concrete if induced by ASR is influenced by a number of aggregate parameters including the reactivity of the siliceous constitutents, the proportion of reactive silica (see Figs 3.20 and 3.21), the porosity of the aggregate, and the particle size fractions within which the reactive silica is present (McConnel *et al.* 1947, Gaskin *et al.* 1955, Hobbs & Gutteridge 1979, Collins & Bareham 1987).

In affected concrete structures overseas, opaline silica is often found to be a constituent of the aggregate. This is not surprising as it is believed that opaline silica can give rise to cracking at lower alkali levels than other forms of reactive silica. Opaline silica is frequently used in laboratory tests, and many data show that concrete expansion increases with increasing opaline silica content until a maximum is reached and then decreases as the content of opaline silica is increased further. High opaline silica contents eliminate expansion entirely. Fig. 3.20 shows the expansion behaviour for reactive aggregates containing various proportions of opaline silica. The term 'pessimum' was introduced to describe the weight of reactive aggregate which corresponds with the peak expansion (Stanton 1940).

Not all reactive aggregates produce curves of expansion versus reactive-aggregate content that show a sharp pessimum. Such a pessimum will be observed only if the reactive aggregate contains a substantial proportion of a highly reactive form of silica. Many reactive aggregates produce relatively flat curves, whilst others produce no pessimum, the expansion being a maximum when a 100 per cent reactive aggregate is used — see Figs 3.20 and 3.21 — McConnell *et al.* 1947, Kennerley & St Johns 1969, Hobbs 1987a).

A simple explanation for pessimum behaviour is as follows. At high contents of reactive silica the reaction is so rapid that the hydroxyl ion concentration in the pore water is reduced to a threshold value during setting and early hardening. As a consequence the reaction induces negligible expansion. At intermediate contents of reactive silica the reaction continues after the concrete has hardened, inducing expansion, and the reaction ceases when either the reactive silica is consumed or when the hydroxyl ion concentration in the pore water is reduced to a threshold value. At low contents of rective silica the intensity of the reaction is insufficient to induce expansion.

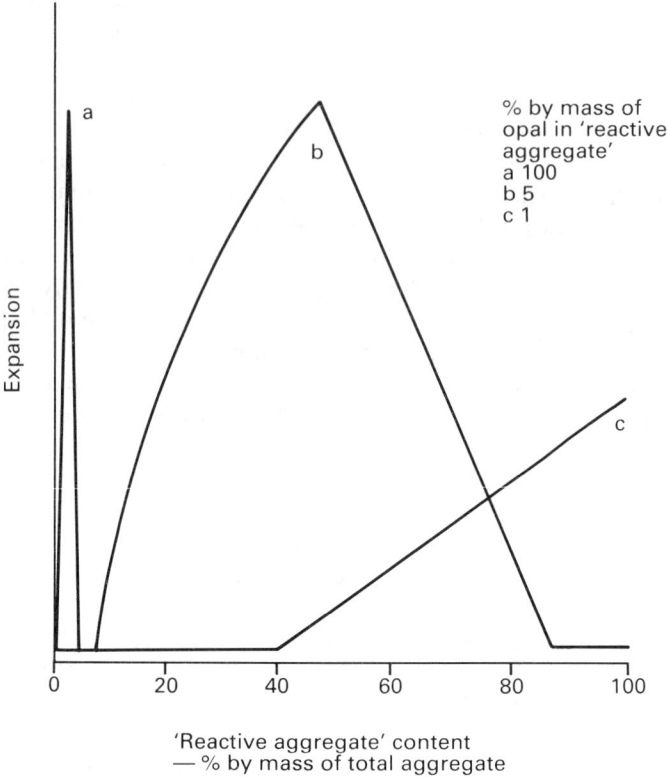

Fig. 3.20 — Expansion behaviour of 'reactive aggregate' containing various proportions of opaline silica (w/c 0.4, a/c 2.75, alkali content 6 kg/m^3).

The expansion induced by reactive silica present in the fine aggregate fraction is greater in the presence of a coarse aggregate having low porosity than in the presence of a relatively porous coarse aggregate. This effect may result from the availability of extra voids in the more porous aggregates into which gel may be exuded without causing deleterious expansion. In the author's view this explains the apparent pessimum behaviour which has been observed with some UK aggregate combinations. For example, replacing Thames Valley flint gravel in a mix containing 100 per cent Thames Valley flint aggregates by a relatively low-porosity limestone can lead to expansion and cracking in abnormally high alkali content cements (see Fig. 3.22) (Nixon & Bollinghaus 1983, Hobbs 1988).

3.2.5 Testing aggregates and cement–aggregate combinations for their reactivity

In this section a number of tests currently being used in the UK for assessing the reactivity of aggregates and cement–aggregate combinations are discussed. The results obtained in these tests have yet to be correlated with field performance in the UK, so none gives definitive values for the deleterious reactivity of aggregates and

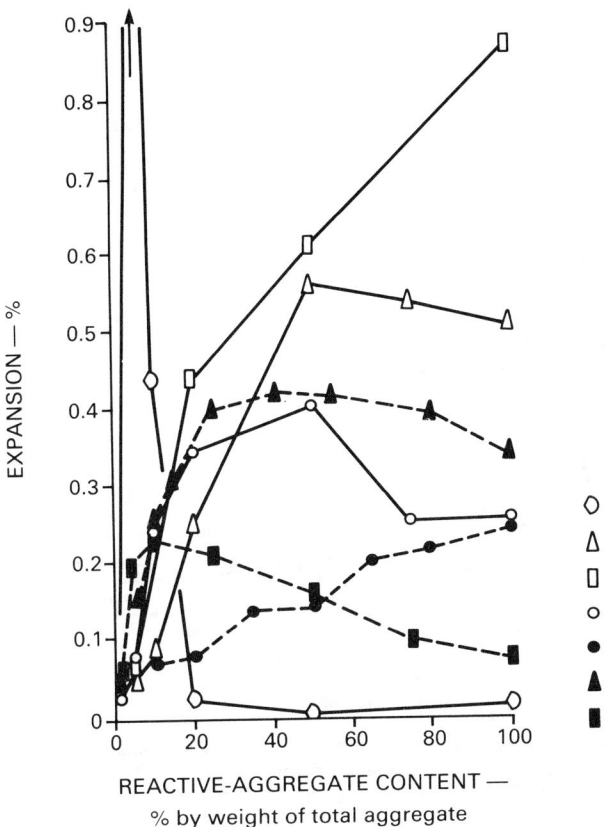

Fig. 3.21 — Influence of reactive-aggregate content upon expansion. High-alkali cement, temperature 38°C. (After McConnell, Mielenz, Holland, & Green 1947, Kennerley & St Johns 1969) (by permission of Palladian Publications Ltd.)

given aggregate–cement combinations. A procedure for petrographical examination of aggregates is being considered for inclusion in BS 812; this method requires the visual recognition and quantification of rock and mineral constituents of an aggregate sample using both macroscopic and microscopic techniques. This examination must be carried out by a suitably qualified person, preferably experienced in aggregate petrography, having access to a laboratory equipped with a good quality, polarizing microscope and the facilities for the manufacture of thin sections. Potentially reactive siliceous minerals, if not finely dispersed, can be identified by this method.

(a) ASTM quick chemical test
In the ASTM C289 quick test (ASTM 1987b) a representative sample of crushed aggregate, 150 to 300 μm in size, weighing 25 g is placed in 25 ml of a 1-molar sodium hydroxide solution and maintained at 80°C for 24 hours. The solution is then analysed for dissolved silica and reduction in alkalinity. The test is repeated three

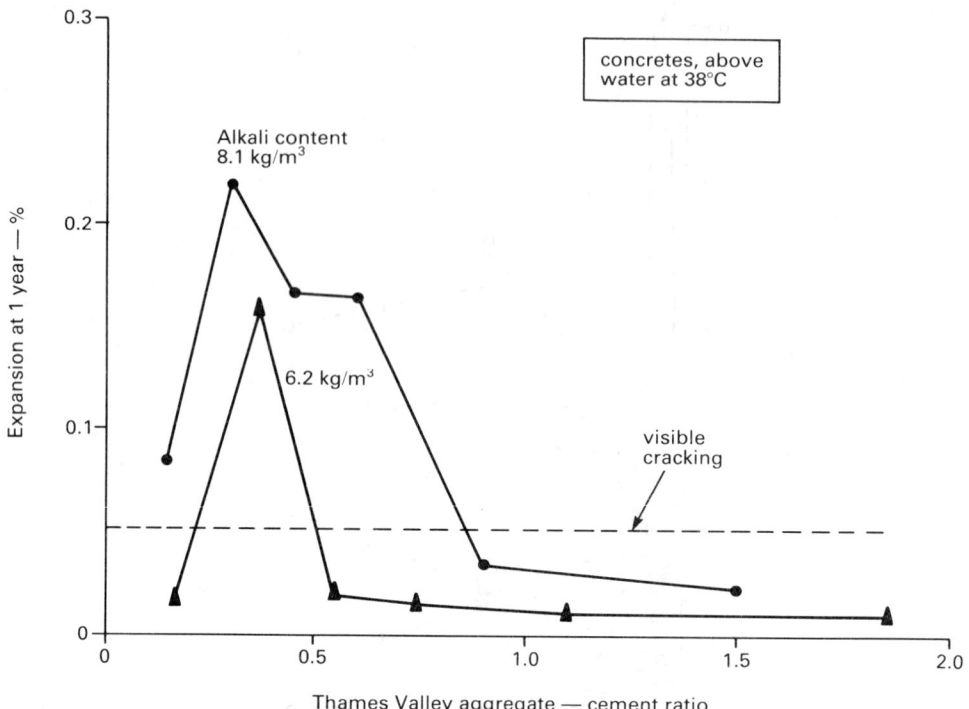

Fig. 3.22 — Variation in expansion at 1 year with Thames Valley aggregate–cement ratio. The Thames Valley aggregate used in combination with a high-quality limestone. (Hobbs 1988, Nixon & Bollinghaus 1983) (by permission of Thomas Publications.)

times. The results are then checked against the curve shown in Fig. 3.23. If any result lies to the right of the curve, then the aggregate may give rise to delerious expansion in high-alkali content concrete. If the aggregate fails this test, then it should be considered to be reactive unless demonstrated to be unreactive by service records or by the ASTM mortar expansion test. If any result falls in the potentially deleterious aggregate zone represented by points above the dotted line shown in Fig. 3.23, it is recommended that expansion tests should be carried out, using a series of proportions of test aggregate to inert aggregate ranging from about 5:95 to 50:50 by mass.

The ASTM quick chemical test will often classify UK aggregates containing chert or flint (and sometimes quartzite) as being 'deleterious' or 'potentially deleterious', even if these are known to have good service records. As a consequence the test is regarded as being unduly pessimistic (Concrete Society 1987).

(b) The gel pat test

In the gel pat test (Jones & Tarleton 1958) a representative sample of aggregate is embedded in a pat of cement paste. After setting, the face of the pat is ground flat to expose the aggregate particles. The pat is then immersed in an alkaline solution and the ground face is examined periodically for the presence of gel. In a method being prepared for possible adoption by BSI, gel pats are prepared with exposed aggregate

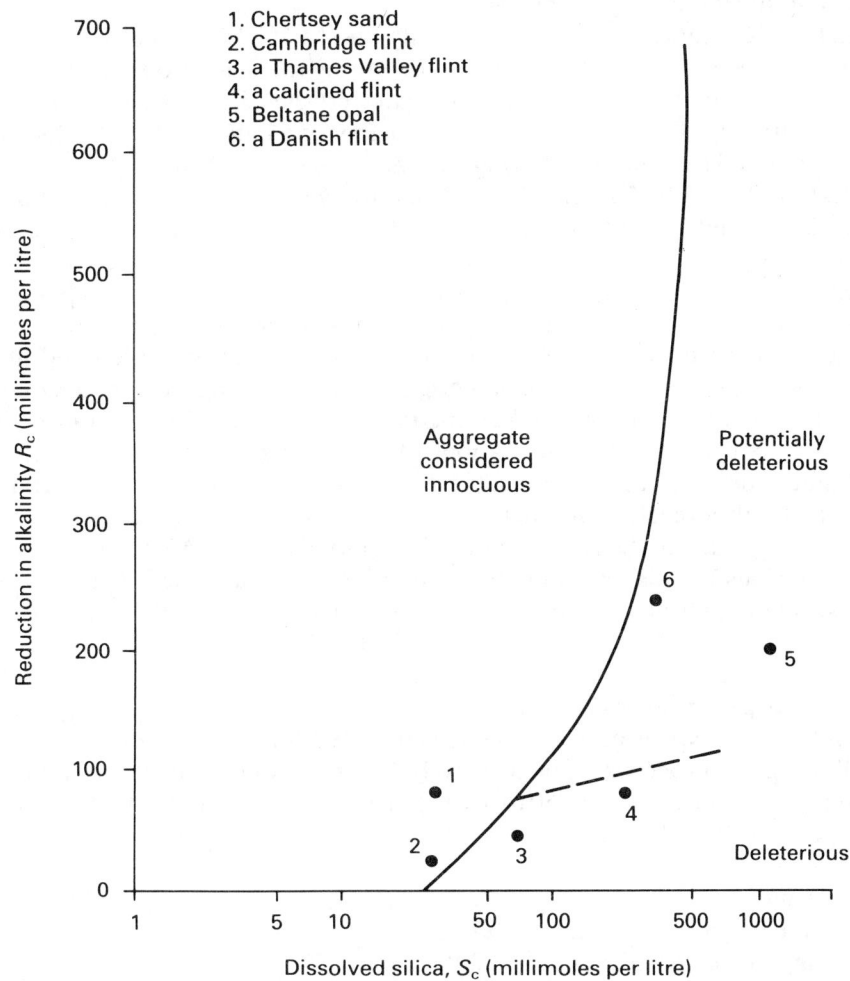

Fig. 3.23 — Division between innocuous and deleterious aggregate on basis of reduction in alkalinity test. ASTM-C289 (ASTM 1987b) (by permission of Thomas Telford Publications).

particles, 1.18 to 5 mm in size, taken from the aggregate in the condition in which the sample was received and also with exposed aggregate particles 2.36 to 5 mm in size obtained by crushing aggregate particles retained on a 5 mm sieve. The pats are immersed in a solution consisting of distilled water containing 20 g/l of sodium hydroxide and 28 g/l of potassium hydroxide. According to this method the sample can be classified as non-reactive if no signs of gel growth are visually apparent after 28 days storage at 20°C. The gel pat test can readily detect reactive silica which is too finely dispersed to be visible under the petrographic microscope. If little is known about the field performance of an aggregate that fails this test, then it is necessary to carry out an expansion test on mortar prisms or concrete prisms to check whether the reaction can give rise to deleterious expansion.

(c) ASTM mortar bar expansion test

The ASTM C227 mortar bar expansion test (ASTM 1987c) is considered internationally to be the most reliable method of measuring the deleterious reactivity of an aggregate. For this test the suspect aggregate, crushed if need be, is made to a prescribed grading. Four prisms, two from each of two batches, $25 \times 25 \times 250$ mm in size, are prepared from mortars having an aggregate–cement ratio of 2.25 and a flow of 105 to 120 (ASTM 1987d). Either a high alkali cement, or the cement to be used in practice, is adopted for the test. The prisms, after demoulding at 24 hours, are stored vertically above water in sealed containers maintained at 38°C, and their length changes are monitored. Mortars that show expansions greater than 0.10 per cent at six months (or more than 0.05 per cent at three months) are usually considered to be deleterious. When expansions in excess of 0.10 per cent are observed and little is known about the field performance of the aggregate, it is strongly recommended that tests be carried out to confirm that ASR has induced expansion (ASTM 1987c). Such tests include petrographical examinations of the aggregate for reactive silica, the ASTM quick chemical test, the gel pat test, and the examination of the mortar bars for the alkali–silica reaction product.

No UK aggregate of the types involved in reported cases of ASR has been shown to be deleterious by this test. Consequently it has been suggested that the ASTM C227 test is not applicable to UK aggregates (Concrete Society 1987).

(d) Concrete prism expansion test

A concrete prism expansion test is being considered for possible adoption by BSI. The following test procedure is being used by a number of investigations in the UK. Four concrete prisms, $75 \times 75 \times 250$ to 300 mm in size, are prepared with the following mix proportions by volume:

cement	22.2
free water	22.8
aggregates 20–10 mm	22.0
10–5 mm	16.5
5 mm–150 μm	16.5

The cement content is 700 kg/m^3 and the cement alkali level is enhanced when necessary to 1.0% by mass by adding potassium sulphate to the mixing water. The aggregate is normally tested in the combination to be used on the job. However, when the aggregate combination to be used is not decided, the fine aggregate is tested in combination with a coarse aggregate from the same or a mineralogically similar source, and separately in combination with a non-reactive coarse aggregate. A similar procedure is followed for the coarse aggregate. It is considered that this will cover most eventualities.

The concrete prisms are demoulded at 24 hours, wrapped in wet towels, and placed in closely-fitting polythene sheaths. The wrapped prisms are stored in closed containers maintained at 20°±2°C for 7 days after demoulding. The length and mass of each prism are then measured, and the wrapped prisms, within their closed containers, are stored at 38±2°C. At the ends of periods of 2, 4, 12, 24, 36, and 48 weeks after demoulding, the length and mass of each prism, after equilibrating at $20\pm2°C$, are measured. Alternatively, the prisms may be stored at $20°\pm2°C$ and the measurements continued for 96 weeks.

A major question associated with the test, which has yet to be resolved, is the choice of expansion level for distinguishing between deleterious and innocuous cement–aggregate combinations; some UK aggregates with good service records give expansions well in excess of 0.10 per cent.

(e) Concrete cube test

The concrete cube test is a simpler and cheaper test than the concrete prism test. This test is being investigated at the British Cement Association. Three concrete cubes are prepared with a water–cement ratio of 0.4, and an aggregate–cement ratio of 3, using the suspect aggregate combination with a cement having an alkali level of about 1 per cent by mass. If the aggregate combination for the job is unknown, then the test is repeated to ensure that a complete range of possible aggregate gradings is covered. After compaction of the concrete, the moulds are covered with a polythene sheet until they are demoulded at 24 hours. The cubes are then examined, and any visible plastic shrinkage cracking noted. One cube is then immersed in water at 40°C for 6 months, one at 30°C for 6 months, and one at 20°C for 12 months. Periodically the cubes are examined, and any cracking and reaction spots noted. If cracking is observed and little is known about the field performance of the aggregate, then it is recommended that tests be carried out to confirm that ASR is the cause of the observed cracking. Such tests include the ASTM quick chemical test, the gel pat test, and examination of the concrete prisms for the alkali–silica reaction product.

In the cube test the reactivity of a particular cement–aggregate combination is given by the time elapsed from casting to cracking at each temperature, or by the maximum crack width at the end of the test. As is the case with the prism expansion method, criteria for distinguishing between deleterious and innocuous cement–aggregate combinations have yet to be established.

3.2.6 Specifying and minimizing the risk of deleterious expansion in new construction

(a) Minimizing the risk by using selected aggregates

(i) General

There are no accepted UK tests for measuring the expansive alkali-reactivity of UK aggregates and UK cement–aggregate combinations. The British Cement Association has proposed that one way of minimizing the risk of cracking due to ASR is to comply with the following draft clause:

> Both the coarse and fine aggregate, separately and in the combination used, must have a good performance record for at least ten years when used in high cement content concretes made using cements of similar or higher alkali levels than the chosen cement.

A similar recommendation is made in BS 8110 (BSI 1985b). The Concrete Society

Working Party on ASR has suggested that there may be many cases where, although there may be no guarantee, it is not considered necessary to introduce clauses into the specification to minimize the risk of cracking from ASR, for example when the engineer is satisfied that the sources of aggregate have a long history of satisfactory use in concrete (Concrete Society 1987).

To comply with the above draft clause an engineering judgement must be made. If a new source of aggregate and/or a new source of cement is used, judgement based on performance is not possible.

(ii) Concrete Society Working Party on ASR

The Concrete Society Working Party (1987) has proposed the following model specification clauses for minimizing the risk by using selected aggregates.

(1) Fine and coarse aggregate material shall comply with the requirement of BS 882 (BSI 1983a), BS 1047 (BSI 1983b) or BS 3797 (BSI 1976b).
(2) The aggregate shall be classed as non-reactive if the engineer is satisfied that the source does not contain opaline silica and one of the following sub-clauses is satisfied.
   (a) The fine and coarse aggregate each consist of at least 95% of one or more of the rock types or artificial aggregates listed in Table 3.16 and provided that the engineer is satisfied that the source does not contain a quantity of flint, chert, or chalcodony that could cause damage from alkali–silica reaction.
   (b) Fine aggregate shall be obtained from the following source: (Specify source) ...
   Coarse aggregate shall be obtained from the following source: (Specify source) ...
   (c) The proportion of chert and flint in the sources of aggregate is such that the proportion of chert and flint in the total aggregate is greater than 60% by mass when the fine and coarse fractions are combined.

Table 3.16

Aircooled blast-furnace slag	Expanded clay/shale/slate	Microgranite
Andesite	Feldspar[†]	Quartz[†,‡]
Basalt	Gabbro	Schist
Diorite	Gneiss	Sintered pfa
Dolerite	Granite	Slate
Dolomite	Limestone	Syenite
	Marble	Trachyte
		Tuff

[†] Feldspar and quartz are not rock types but are discrete mineral grains occurring principally in fine aggregate.
[‡] Not highly-strained quartz and not quartzite.

(iii) Department of Transport

To control cracking due to ASR the Department of Transport (DTp 1986a,b) has proposed that aggregates for concrete structures should be examined in a specified manner by an independent qualified geologist approved by the engineer (DTp 1986c) and can be taken to be unreactive if:

(1) both the coarse and fine aggregate contain at least 95% of one or more of the rock types given in Table 3.17 or of the following artificial types, air cooled blastfurnace slag, expanded blastfurnace slag/clay/shale/slate or sintered pfa provided they are not contaminated with any opal, trydimite, or cristobalite or contain a total of more than 2 per cent of chert, flint, or chalcedony taken together;

Table 3.17 — Non-reactive aggregate types listed in the DTp specification

Andesite	Feldspar	Marble	Syenite
Basalt	Gabbo	Microgranite	Trachyte
Diorite	Gneiss	Quartz	Tuff
Dolerite	Granite	Schist	
Dolomite	Limestone	Slate	

(2) in the case of quartz in the 95% stated above it should not be in the form of quartzite or contain more than 30% by weight of highly-strained quartz;
(3) the proportion of chert or flint is greater than 60% by weight of the total aggregate provided it contains no opal, trydimite, or cristobalite.

(b) Minimizing the risk by controlling the alkali content

If it is considered that the aggregate to be used is deleteriously reactive or may be deleteriously reactive in high-alkali content concrete then, to minimize the risk of cracking, it is necessary either to place a limit on the available alkali content of the cement or the binder or to place a limit on the available alkali content of the concrete from all sources. The limit commonly being specified for a cement or other binder is 0.60 per cent by mass and for concrete, $3\,kg/m^3$ (Concrete Society 1987, DTp 1986a). There are, however, differences in view regarding the alkali limit deemed to be safe and of the ways in which the alkalis from OPC, ggbs, pfa, sodium chloride, and admixtures should be combined (Hobbs 1988).

References

Allen, R. T. L. (1981) Alkali–silica reaction in Great Britain — a review, *Proceedings of the Fifth International Conference on Alkali–aggregate Reaction in Concrete, Cape Town, 30 March–3 April, NBRI, Pretoria*, paper S252/18.

American Society for Testing Materials. ASTM C295-85 (1987a) Standard practice for petrographic examination of aggregate for concrete. Section 4, *Construction*, Vol. **0402**, *Concrete and Aggregates*, p 225.

American Society for Testing and Materials, ASTM-C289-81, (1987b) Standard test method for potential reactivity of aggregates (Chemical method). *Annual Book of ASTM Standards*, Section 4, *Construction*, Vol. **04.02**, *Concrete Aggregates*, p 204.

American Society for Testing and Materials, ASTM C227-81 (1987c) Standard test method for potential alkali reactivity of cement aggregate combinations. (Mortar-bar method). *Annual Book of ASTM Standards*, Section 14, *Construction*, Volume **04.02**, *Aggregates*, p. 159.

American Society for Testing and Materials, ASTM C109-86 (1987d) Standard test method for compressive strength of hydraulic cement mortars (using 2 in or 50 mm cube specimens). *Annual Book of ASTM Standards*, Section 4, *Construction*, Vol. **04.01**, *Cement; lime, gypsum*, p. 74.

British Cement Association (1988) *The diagnosis of alkali–silica reaction.* Report of a Working Party. British Cement Association, Wexham Springs, Slough.

British Standards Institution (1970) *Methods of testing cements*, London, BS 4550:Part 2:1970.

British Standards Institution (1973) *Specification for Portland–blastfurnace cement*, London, BS 146:Part 2:1973: *Metric units.* AMD 2615, June 1978, AMD 4419, March 1984.

British Standards Institution (1974) *Low heat Portland–blastfurnace cement*, London, BS 4246:Part 2:1974:*Metric units.*

British Standards Institution (1976a) *Methods for sampling and testing of mineral aggregates, sands and fillers*, London, BS 812:Part 4:1976. *Chemical Properties*, AMD 4295, June 1983, AMD 4617, August 1984.

British Standards Institution (1976b) *Specification for lightweight aggregates for concrete*, London BS 3797:Part 2:1976. *Metric units*, AMD 3518, January 1981.

British Standards Institution (1978) *Specification for ordinary and rapid-hardening Portland cement*, London, BS 12:1978, AMD 4259, May 1983.

British Standards Institution (1982) *Pulverized fuel ash*, London, BS 3892:Part 1:1982. *Specification for pulverized fuel ash for use as a cementitious component in structural concrete.*

British Standards Institution (1983a) *Specification for aggregates from natural sources for concrete*, London, BS 882:1983.

British Standards Institution (1983b) *Specification for air-cooled blastfurnace slag aggregate for use in construction*, London, BS 1047:1983.

British Standards Institution (1985a) *Specification for Portland pulverized fuel ash cement*, London, BS 6588:1985.

British Standards Institution (1985b) *Structural use of concrete.* Part 1: Code of Practice for design and construction, London, BS 8110:Part 1:1985.

British Standards Institution (1986) *Specification for ground granulated blastfurnace slag for use with Portland cement.* London, BS 6699.

Collins, R. J. & Bareham, P. D. (1987) Alkali–silica reaction: suppression of expansion using porous aggregate. *Cement and Concrete Research*, **17**, No. 1, p 89.

Concrete Society (1987) *Alkali–silica reaction: Minimizing the risk of damage. Guidance notes and model specification clauses*, Report of a Working Party, Concrete Society Technical Report, No. 30.

Department of Transport (1986a) *Specification for highway works*, Part 5, Her Majesty's Stationery Office, London.

Department of Transport (1988b) *Specification for highway works*, Part 7(ii), Her Majesty's Stationery Office, London.

Department of Transport (1986c) *Specification for highway works*, Part 7 Appendix E, Her Majesty's Stationery Office, London.

Fujii, M., Kobayashi, K., Kojima, T., & Maehara, H. (1987) The static and dynamic behaviour of reinforced concrete with cracking due to alkali–silica reaction. *Proceedings of the 7th International Conference on Concrete Alkali–aggregate Reactions, Ottawa, 1986*, Ed. Patrick E. Grattan-Bellew, Noyes Publications, p 141.

Gaskin, A. J., Jones, R. H., & Vivian, H. E. (1955) Studies in cement-aggregate reaction XXI. The reactivity of various forms of silica in relation to the expansion of mortar bars. *Australian Journal of Applied Science*, **6**, p 78.

Hobbs, D. W. (1978) Expansion of concrete due to alkali–silica reaction: an explanation. *Magazine of Concrete Research*, **30**, No. 105, p 215.

Hobbs, D. W. (1986) Deleterious expansion of concrete due to alkali–silica reaction: Influence of pfa and slag. *Magazine of Concrete Research*, **38**, No. 137, pp 191–205.

Hobbs, D. W. (1987a) Some tests on fourteen year old concretes affected by the alkali–silica reaction. *Proceedings of the 7th International Conference on Concrete Alkali–aggregate Reactions, Ottawa, 1986*, Ed. Patrick E. Grattan-Bellew, Noyes Publications, p 342.

Hobbs, D. W. (1987b) Mix design. Quality of mixing water. w/c ratio. Homogeneity, *Proceedings of a Seminar on 'Le Beton et L'eau, College International des Sciences de la Construction, Saint-Remy-les Chevreuse, France, 18–20 June, 1985*. Conseil International de la Langue Française, 103, Rue de Lille, 75007, Paris, p 46.

Hobbs, D. W. (1988) *Alkali–silica reaction in concrete*. Thomas Telford Publications.

Hobbs, D. W. & Gutteridge, W. A. (1979) Particle size of aggregate and its influence upon the expansion caused by the alkali–silica reaction. *Magazine of Concrete Research*, 31, p 235.

Imai, H., Yamasaki, T., & Maehara, H. (1987) The deterioration by alkali–silica reaction in Hanshin Expressway concrete structures — Investigation and repair. *Proceedings of the 7th International Conference on Concrete Alkali–aggregate Reactions, Ottawa, 1986*, Patrick E. Grattan-Bellew, Noyes Publications, p 131.

Jones, F. E. & Tarleton, R. D. (1958) *Part VI Alkali–aggregate interaction. Experience with some forms of rapid and accelerated tests for alkali–aggregate reactivity. Recommended test procedures*. National Building Studies Research Paper No. 25, pp 17, London, Her Majesty's Stationery Office.

Kennerley, R. A. & St Johns, D. A. (1989) Reactivity of aggregates with cement alkalis. *Proceedings of a National Conference on Concrete Aggregates, Hamilton, New Zealand*, p 35.

Kolleck, J. J., Varman, S. P., & Zaris, C., 1986, Measurement of OH^- ion concentration of pore fluids and expansion due to alkali–silica reaction in composite cement mortars. *8th International Congress on the Chemistry of Cement. 22–27 September*, Theme 3, **IV**, pp 183–189.

Koyanagi, W., Rokugo, K., & Ishida, H. (1987) Failure behaviour of reinforced concrete beams deteriorated by alkali–silica reactions. *Proceedings of the 7th International Conference on Alkali–aggregate Reactions, Ottawa, 1986*, Ed. Patrick E. Grattan-Bellew, Noyes Publications, p. 141.

McConnell, D., Mielenz, R. C., Holland, W. Y., & Greene, K. T. (1947 Cement-aggregate reaction in concrete. *Journal of American Concrete Institute, Proceedings*, **44**, No. 2, pp 93–128.

Mizuomoto, Y., Kosa, K., Ono, K., and Nakano,, K. (1987) Study on cracking damaged concrete structures due to alkali–silica reaction. *Proceedings of the 7th International Conference on Concrete Alkali–aggregate Reactions, Ottawa, 1986*, Ed., Patrick E. Grattan-Bellew, Noyes Publications, p 204.

Nixon, P. J. & Bollinghaus, R. (1983) Testing for alkali reactive aggregates in the UK. *Proceedings of the Sixth International Conference on Alkalis in Concrete; Copenhagen, June 1983*, Eds G. M. Idorn & S. Rostan, Danish Concrete Association, p 329.

Nixon, P. J., Page, C. L., Bollinghaus, R., & Canham, I. (1986) The effect of pfa with a high total alkali content on pore solution composition and alkali–silica reaction. *Magazine of Concrete Research*, **38**, No. 134, pp 30–35.

Palmer, D. (1978) Alkali–aggregate reaction — recent occurrences in the British Isles. *Proceedings of the Fourth International Conference on the Effect of Alkalis in Cement and Concrete, Purdue University, West Lafayette*, Publication No. Ce-MAT-1-78, p 285.

Smith, M. A. & Halliwell, F. (1979) The application of the BS 4550 test for pozzolanic cements to cements containing pulverized fuel ashes. *Magazine of Concrete Research*, **32**, No. 108, p 159.

Stanton, D. E. (1940) The expansion of concrete through reaction between cement and aggregate. *Proceedings of the American Society for Civil Engineers*, **66** p 1781.

3.3 THE DRYING SHRINKAGE OF AGGREGATES

3.3.1 Introduction

Over fifty years ago it was recognized that certain aggregates change volume when changes in moisture content occur (Davis 1930). Some aggregates, commonly referred to as shrinkable aggregates, can exhibit significant changes in volume from the wet to the dry state and vice versa. However, it was some twenty years later that this phenomenon was first researched as a possible source of deterioration of concrete. Even today it is a subject that is not fully understood by many practising in the construction industry, and only recently has a British Standard method of test been developed.

Shrinkable aggregates are found in many parts of the World and have been the subject of reports from, notably, South Africa (Stutterheim 1954) and the USA (Hansen & Nielson 1965). Within the United Kingdom the use of shrinkable aggregates is predominant in Scotland although it is known that they are also quarried in Northern Ireland.

At present attention is concentrated on the problems arising from the use of shrinkable aggregates in concrete. It seems likely that there might be fewer problems where such aggregates are used in bituminous mixtures because the inherent flexibility of the binder might absorb the volume changes without disruption. Certainly there appears to be no published case in which a failure of a bituminous material has been attributed to aggregate shrinkage.

3.3.2 The mechanism of aggregate shrinkage in concrete
It is thought that some aggregates suffer volume changes because they contain clay minerals which absorb water and swell or dry out and shrink. The cement paste in a concrete also changes in volume during setting and afterwards. This exerts forces on the aggregate during a change in moisture content and these forces can cause a change in volume of the aggregate — the magnitude of which will be related to the modulus of elasticity of the aggregate and the volume proportion of the cement paste and aggregate. Consequently, in considering the drying shrinkage of aggregate in concrete it is reasonable to suppose that no aggregate is non-shrinkable and that all aggregates will, to a greater or lesser extent, change in volume. One can then consider a classification system for aggregates from low to high shrinkage types.

3.3.3 Experience in Scotland
In the early 1950s the Scottish Laboratory of the Building Research Station (SLBRE) compared concretes made from flint aggregate from southern England with concretes made from local crushed whinstone aggregates, in an attempt to gain information about aggregates that would be more representative of use in Scotland. The concrete specimens made with whinstone (dolerite) aggregate developed micro-cracking within a year and subsequently disintegrated after the first severe frost in 1952. Concrete specimens made with flint aggregate suffered no defects and continued to gain in strength over a long period. At the time it was believed that dolerite was the only rock type that led to such poor performance, but later research and evidence from South Africa (Stutterheim 1954) showed quite conclusively that a variety of types were suspect.

The significance of these findings was recognized when excessive deflections were found in reinforced concrete floor slabs in Scottish Special Housing Association domestic housing, and the warping and cracking were found to be associated with the shrinkage level of the aggregate used. Over the next few years a number of other cases were found. In some instances total disintegration of concrete exposed to the weather was reported. All these failures were found to be attributable to the use of shrinkable aggregate.

3.3.4 Shrinkable rock types
Over the ten years up to 1963, SLBRE carried out a survey of virtually all working sources of aggregate in Scotland. In general, dolerites, basalts, greywackes, and mudstones, and gravels containing these rock types, tended to have excessive shinkage. Granite, limestone, flint, quartzite, and fresh felsite were found to give low shrinkage. Such broad petrological descriptions are not sufficiently precise to prescribe a reliable classification of shrinkage based on rock type. It is probable that other factors, such as degree of weathering, should also be considered. Work carried out at Glasgow University (Moore & Gribble 1980) demonstrated that the drying

shrinkage of concrete could change dramatically with the formation of hydrous clay minerals.

It is interesting to note that fresh felsite has been reported both as a non-shrinkable rock type (BRE 1968) and a a high-shrinkage rock type (Snowdon & Edwards 1962). This anomaly may arise from the term felsite being used to cover a range of rock types, or perhaps a difference can be caused by weathering. Whatever the explanation, this example does illustrate the difficulty and danger of classifications based on loose descriptions.

3.3.5 The effects of shrinkage

Use of aggregates having higher than normal shrinkage can lead to unacceptable movements in concrete. Provided that this is allowed for in design — by additional reinforcement or allowance for loss of prestress, for example — it should not be a problem in itself. In essence the solution lies in an estimate of the shrinkage of the concrete to be expected with a particular mix, aggregate source, and climatic conditions (particularly relative humidity).

Another effect of shrinkage is a reduction in durability of the concrete. This can occur when the shrinkage of the aggregate leads to cracking of the concrete, making it vulnerable to attack from weathering, and to corrosion of the reinforcement. This risk can be reduced by using concrete mixes of low water/cement ratio (<0.5) and including entrained air. Clearly it would be difficult to measure the effectiveness of prudent mix design in maintaining durability, and therefore it is necessary to restrict the use of shrinkage aggregates for some types of concrete.

The stress within concrete caused by aggregate volume change should be at its greatest when the difference between the shrinkage of the coarse aggregate and that of the mortar is at a maximum. A combination of low-shrinkage coarse aggregate with high-shrinkage sand has been shown (Snowdon & Edwards 1962) to produce concrete with the greatest internal stress and consequently most inferior weathering properties. These researchers also showed that the wet and dry length of specimens increased with the number of cycles, and that the increases were greater the higher the shrinkage of the aggregate.

The research carried out by SLBRE led to the publication of *BRE Digest* 35 in 1963, followed by an amended version in 1968 — see section 3.3.7. This *Digest* incorporated a method of measurement for drying shrinkage and gave recommendations for the use of aggregates of different shrinkage — these are reproduced in Table 3.18.

3.3.6 The measurements of drying shrinkage

Attempts have been made to relate aggregate shrinkage to parameters such as water absorption and relative density which are easily measured. Although some measure of correlation has been established within particular rock types, there is no reliable correlation across a range of different rock types. More sophisticated methods such as measurement of specific surface by gas absorption (Fulton 1961) may be more promising but have not so far been developed sufficiently for use as standard methods.

At the time of the research by SLBRE, a method of measuring concrete

Table 3.18 — Recommendations for using concretes of various shrinkages. (After *BRE Digest* 35, 1968)

Drying shrinkage of concrete (%)	Remarks	Use	Number of Scottish sources examined
Not exceeding 0.045	Possible with several types of very low shrinkage aggregates such as granite, limestone, unaltered felsite, blastfurnace slag and a few dolerites and gabbros. Quartz, flint gravel and marble which are very hard and non-shrinking will produce concrete having a shrinkage less than 0.025%.	Suitable for all concreting purposes. If very low values of drying shrinkage for precast products are required, it may be necessary to cure with high pressure steam but this would not apply to concrete made with quartz, flint gravel and marble.	37
0.046–0.065	Possible with hard dense aggregates having low shrinkage and with softer non-shrinkable aggregates. Most of the aggregates in Central Scotland are in this range.	Suitable for most applications. Special care should be taken in the design of such units as cladding panels and cast *in-situ* floors, particularly if heated. Span/depth ratios should be reduced below the values given in Table 13 of CP 114: 1957 or Table 12 of CP 116 or alternatively, additional reinforcement should be provided in the compression zone of the section. For prestressed concrete an extra allowance should be made for loss of prestress.	49
0.066–0.085	Possible with aggregates covering a wide geological range. At this and higher levels of drying shrinkage the durability of the concrete is likely to become affected.	Suitable for all general structural purposes but for reinforced or thin reinforced members, exposed to the weather, air-entrained concrete should be used.	31
Exceeding 0.085	These are gravels consisting mainly of sedimentary rocks such as greywacke, shale and mudstone. They have produced concrete which has resulted in severe warping and widespread deterioration	Suitable for all general positions where complete drying out never occurs, for mass concrete surfaces with air-entrained concrete and for members that are symmetrically and heavily reinforced and are not exposed to the weather.	

Note:
1. Different batches of ordinary Portland cement may be expected to give rise to variations of the order of 0.005% in the drying shrinkage of the concrete.
2. In general, although the coarse aggregate plays the major part in determining the shrinkage of the concrete, the type and grading of the fine aggregate also affect the result. When a low shrinkage fine aggregate is used with a high shrinkage coarse aggregate (in place of the fine aggregate from the same source) the shrinkage of the resulting concrete may be reduced by as much as 0.03%. Also, changing the grading of the fine aggregate from Zone 4 to Zone 1, BS 882: 1965, may in extreme cases decrease the shrinkage by as much as 0.02%.
3. It is assumed that the normal recommendations for quality control given by the appropriate Code of Practice for the particular job will be observed.
4. Attention is drawn to the effect of changing the fine aggregate, as discussed at the end of the Appendix.

shrinkage was available in the British Standard for tests on concrete (BSI 1970), and this was adapted and given as a method for determination of drying shrinkage in the Appendix to *BRE Digest* 35. This is described below.

3.3.7 The BRE test method

Aggregates to be assessed are mixed with ordinary Portland cement and water and the resulting concrete is cast into three prisms, each measuring $50 \times 50 \times 200$ mm. The mix has a total water:cement ratio of 0.6 and an aggregate:cement ratio of 6.0. One day after casting, 6 mm diameter stainless steel balls are fixed into preformed indentations at the ends of the prisms to act as references for length measurements. The prisms are cured in air for a total of 28 days and then immersed in water for four days. After the water curing, the prisms are measured for length and then placed in an oven at 50°C and 17 per cent relative humidity for a further 28 days. On completion, the prisms are measured for length for the second time. The difference between the two length measurements, expressed as a percentage of the prism length, is the drying shrinkage.

Drawings of a typical measuring apparatus and moulds are shown in Figs 3.24 and 3.25. The use of an Invar steel rod enables the dial gauge to be set at zero before each reading, or lengths can be compared as differences between prism measurements and Invar rod measurements.

This method of determination of drying shrinkage has been the basis for assessing aggregates in Scotland for twenty five years up to 1988.

3.3.8 The use of *BRE Digest* 35

Since its publication, *Digest* 35 has been used by specifiers as the basis for applying specification limits to the shrinkage of aggregates in concrete. Unfortunately many specifiers in Scotland adopted a value of 0.050 per cent as their maximum shrinkage limit despite the recommendations of *Digest* 35 (see Table 3.18). Such a restrictive value could not be justified and was wasteful of aggregate resources. Many aggregates within the central industrial belt of Scotland have shrinkage values close to 0.050 per cent. It is likely that specifiers chose that limit because they didn't want to be seen to be restricting aggregates to the lowest range of Table 3.18 (up to 0.045 per cent). They wanted to exclude aggregates tending towards the higher ranges without increasing costs, but there have been severe commercial problems, because the poor precision of the test has brought the effective limit even lower in practice.

With such a long test cycle (over 2 months) some suppliers could not risk waiting for re-rests when a sample of their material failed. So they adopted the habit of having their sources tested by more than one laboratory, and selecting the most favourable result for certification purposes.

Although the *Digest* proved successful in preventing problems from aggregate shrinkage over many years, dissatisfaction with the test method and application of the *Digest* led the Sand and Gravel Association to carry out an experiment in 1981 to assess the precision of the test method. A medium-shrinkage aggregate was chosen for the experiment and randomly distributed samples were tested by nine participating laboratories. In this experiment the reproducibility was found to be 0.012 per cent (20 per cent of the average of 0.060 per cent). This confirmed the belief that the precision of the method was poor.

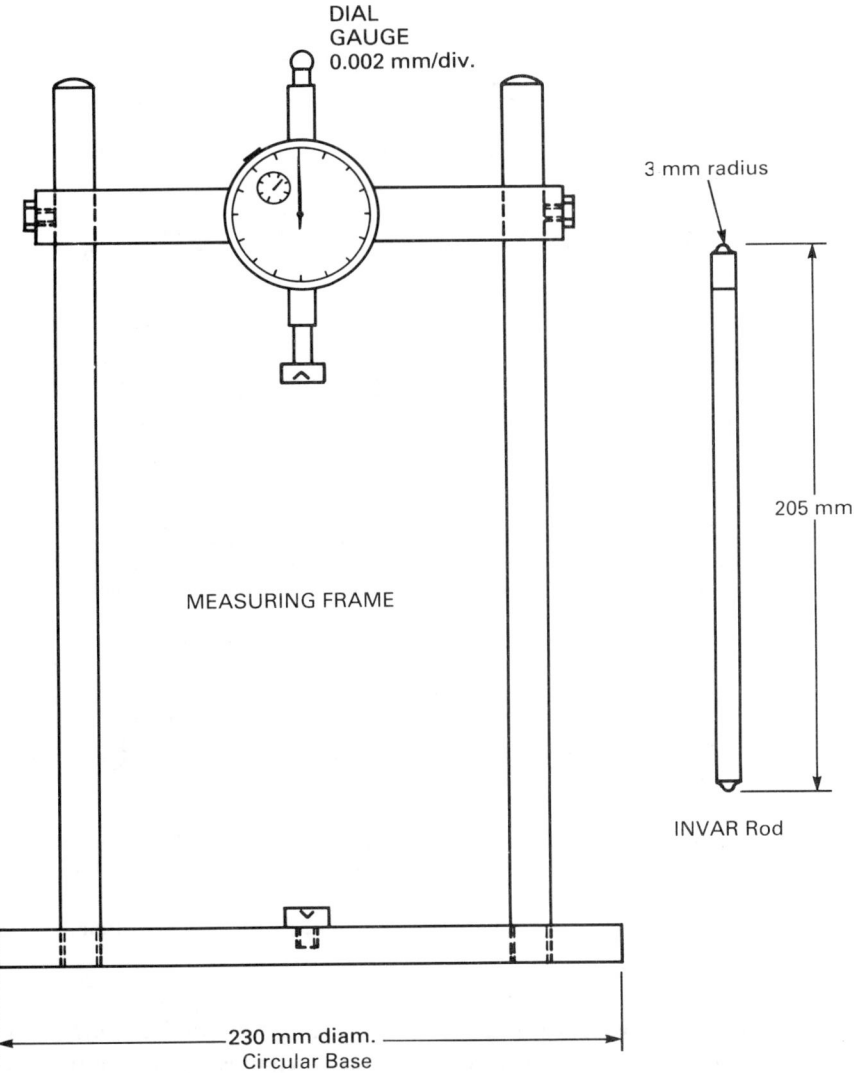

Fig. 3.24 — Conventional apparatus for length measurement of prisms.

A number of other features of the test method were considered unacceptable in what was, in all but name, a standard method. For instance: variations in humidity were known to occur but were uncontrollable; the range of temperature at measurement was too wide; and there was a lack of information on precision. So a BSI working group (CAB 2/5/WG6) was established in 1982 with the task of developing a new method as Part 120 of BS 812.

3.3.9 Introducing a test to British Standard BS 812
As part of the revision of the British Standard for test methods for aggregates (BSI 1975), consideration has been given to the introduction of a shrinkage test.

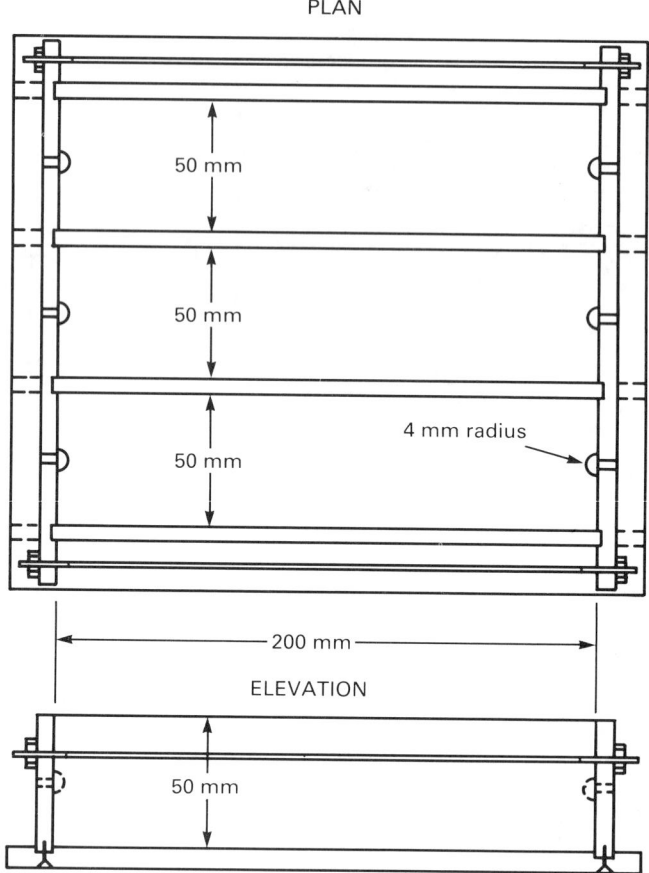

Fig. 3.25 — Typical BRE pattern mould for three prisms 200×50×50 mm.

Three methods have been considered: these are a modified *Digest* 35 method incorporating more precise procedures and minor changes in method; a RILEM method modified for curing at 20°C and 65 per cent relative humidity with a minimum 98 days test cycle; and a ten-day method developed by Hobbs at the Cement and Concrete Association, based on drying at 105°C.

Both of the methods that control humidity and temperature are lengthy. The ten-day method makes no attempt to control humidity; it was offered only as a means of sorting aggregates into low- and high-shrinkage classes. Its great advantage is the short test cycle time.

The ten-day test incorporates many of the features of the *Digest* 35 method — for the first two days of the test the procedures are virtually identical. After that prisms are cured in water at 20°C for five days; measured to obtain wet length; and then measured again after three days drying in a conventional drying oven at 105°C.

Apart from the curing and drying regime, changes were made to tighten up the

3.3.10 Precision of the methods

A precision trial was arranged to extend and complement the results of the 1981 SAGA trial. This time, aggregate having a shrinkage of about 0.10 per cent was used.

The random distribution of bulk samples to ten participating laboratories and subsequent division into test portions was planned so that it would be possible to compute separately the errors caused by sampling; sample reduction; and within- and between-laboratory differences for each method of test. These are shown in Table 3.19, from which it will be evident that the highest error variance, V_L, arises

Table 3.19 — 1985 precision trial results — components of variance

Source of variation	Symbol	Standard deviation (%)	
		10-day test	Mod.D35 test
Prisms	$\sqrt{V_{Prisms}}$	0.0045	0.0033
Test portions	$\sqrt{V_{Portions}}$	0.0028	0.0000
Laboratory samples	$\sqrt{V_S}$	0.0010	0.0033
Laboratories	$\sqrt{V_L}$	0.0081	0.0072
General average	$\bar{x}$	0.115	0.101

from between-laboratory variations. Calculations were extended to examine the improvement in precision that might be possible by either increasing the number of prisms from one test portion, or increasing the number of test portions, as shown in Table 3.20. It is apparent that, although repeatability could be halved by quadrupling the number of test portions, the improvement in reproducibility would be negligible. For this reason the conventional system of three prisms per test portion seems worth retaining.

Using the data from the 1981 SAGA trial and the BSI precision trial it was possible to deduce that the precision estimates were proportional to the level of shrinkage measured. Because the BSI precision trial compared results from the *Digest* 35 method and the ten-day method, it was also possible to estimate the precision for either method of test at any level. For the ten-day method, the repeatability was 9 per cent, and the reproducibility 22 per cent, of the average measured, that is, not significantly worse than the *Digest* 35 method.

Although precision data have been available for a number of tests for many years

Table 3.20 — Effect of replication on precision estimates

Number of test portions per lab. sample	Number of prisms per test portion	10-day test			Mod.D35 test		
		$R_1\%$	$R_1\%$	$R_2\%$	$r_1\%$	$R_1\%$	$R_2\%$
1	1	0.0148	0.0271	0.0272	0.0092	0.0228	0.0240
1	2	0.0119	0.0256	0.0258	0.0065	0.0212	0.0231
1	3	0.0107	0.0251	0.0252	0.0053	0.0209	0.0228
2	3	0.0076	0.0239	0.0241	0.0038	0.0205	0.0225
3	3	0.0062	0.0235	0.0237	0.0031	0.0204	0.0224
4	3	0.0053	0.0233	0.0235	0.0027	0.0203	0.0223
General average		0.115%			0.101%		

there has been very limited practical use of them by specifiers. Possibly this is because there has been a lack of guidance on their application. One way of expressing precision is as confidence limits, that is, limits set around a test result within which it can be expected that the 'true' result should lie. For the ten-day method the confidence limits are ±15.5% relative (at a probability level of 95 per cent).

3.3.11 Correlation between the *Digest* 35 and ten-day methods

The BSI precision trial established a relationship between the *Digest* 35 and ten-day methods at a single level of shrinkage. If the recommendations given in the *Digest* were to be transformed into a new specification based on the ten-day method, it would be essential that a relationship should be established between the methods over a wide range of values, using a variety of rock types. A relationship is shown in Fig. 3.26. This relationship is based on routine tests carried out at the Scottish Laboratories of Harry Stanger Limited from 1985–1987. Some fifty results are plotted covering a wide range of crushed rock and gravel aggregates from mainland Scotland, Skye, Shetland, and Northern Ireland. No anomalies were found, and it is suggested that this relationship is sufficiently reliable to be used for practical purposes for all natural aggregates to be found in the UK.

3.3.12 Factors affecting measured values of shrinkage

Any method of test suffers from unavoidable errors, and the shrinkage test is susceptible in particular to errors in temperature at measurement (see Fig. 3.27). There are, however, other sources of error that have been discussed over a number of years. Some of these arise form the use of different cements and the effect of variations in particle size distribution for a particular combination of aggregates.

BRE Digest 35 suggested that different batches of ordinary Portland cement could cause variations of up to 0.005 per cent in measured values of shrinkage. Other researches (Lerch 1946, Hobbs & Parrott 1979) have suggested that variations may be much higher. The coefficient of variation of 10 per cent given by Hobbs & Parrott would indicate variations four times larger than those suggested by BRE.

The cements used by the participating laboratories in the BSI trial were analysed by the Cement and Concrete Association. No correlation was found between cement properties measured and the shrinkages reported by each laboratory. This is one reason why the use of a standard cement has not been incorporated in the test method as yet, but it is a subject worthy of future research.

With regard to variations in the grading of aggregates, there is again a variety of opinion. *BRE Digest* 35 states that a change in the sand from coarse to fine grading can increase measured shrinkage by as much as 0.02 per cent. This is probably an exceptional case but, in the present author's experience, shrinkage has been reduced by up to 0.01 per cent by the use of coarse instead of fine sand from the same source. Even so, despite the fact that the worst combination for weathering properties of concrete is the use of high-shrinkage sand with low-shrinkage coarse aggregate, in practice it is normally the coarse aggregate which receives the greater attention in assessments.

Commonly, gravel aggregates are tested for shrinkage with the sand from the same source. Because crushed rock fines are not often used for concrete in Scotland,

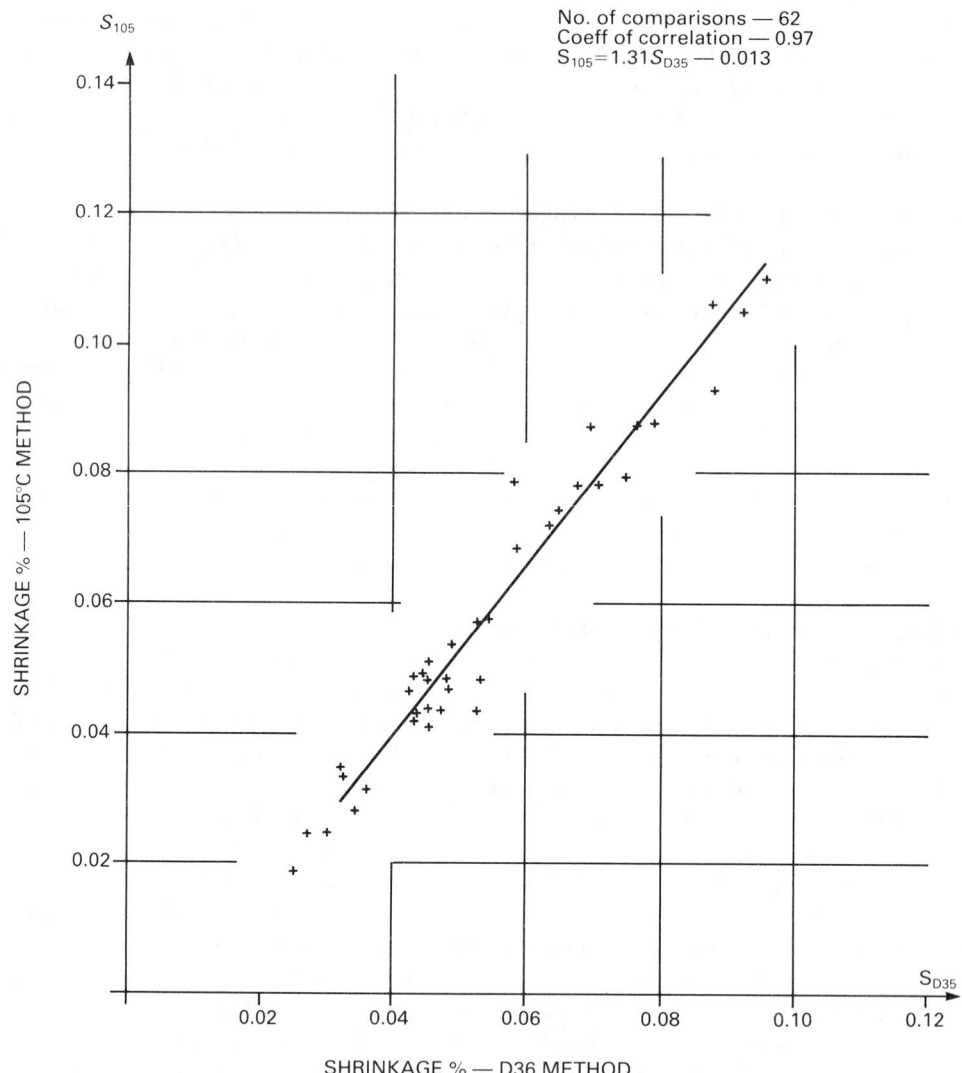

Fig. 3.26 — Comparison between 105°C method and *Digest* 35 method. Routine tests on different aggregates carried out by one laboratory.

it is usual to assess crushed rock aggregate with a low-shrinkage sand for certification purposes. It is suggested that a user would be well advised to check the drying shrinkage of the particular combination of aggregates intended for a concrete.

Both the *Digest* 35 method and the ten-day method use concrete prisms made from a mix having a constant total water content with no allowance for aggregate absorption. In research at SLBRE (Edwards 1956) it was demonstrated that, although different aggregates produced mixes of variable workability (as expected),

Sec. 3.3] The drying shrinkage of aggregates 123

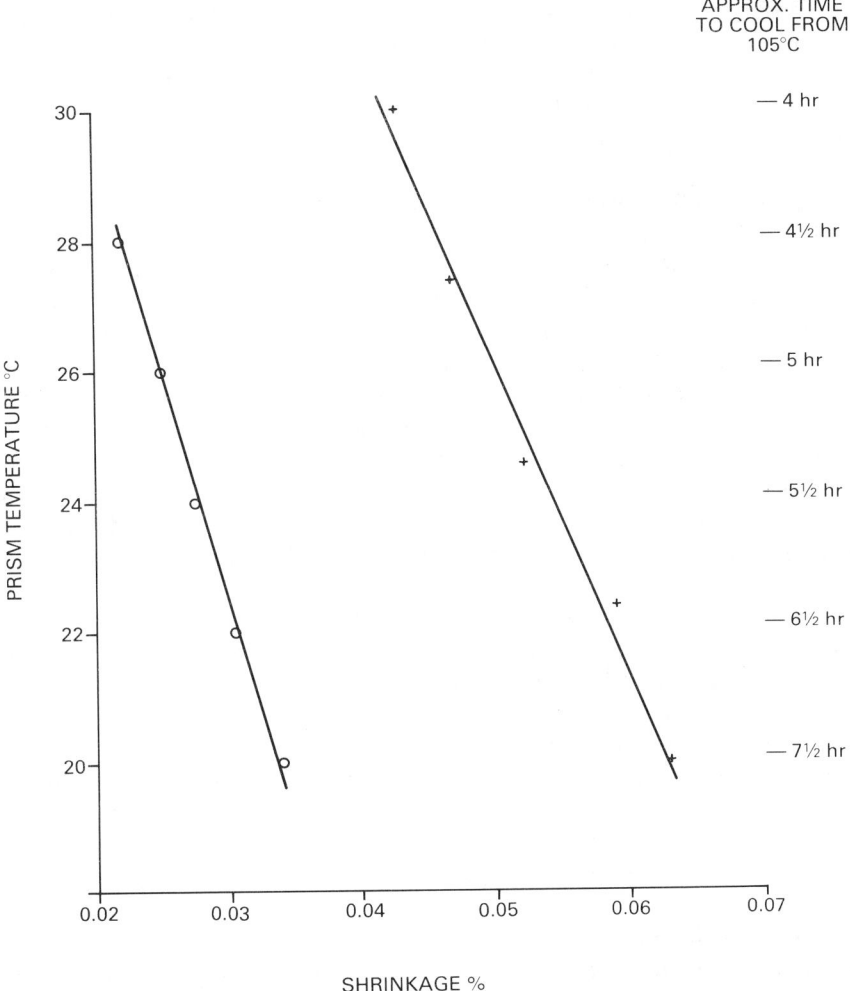

Fig. 3.27 — The effect on shrinkage values of measuring prisms at different temperatures.

there was no significant difference in measured shrinkage, even at high levels of shrinkage, between mixes using dry aggregate and those using saturated surface dry aggregate. In the development of the ten-day method this matter was considered at some length, and it was decided that, though the use of saturated surface dry aggregates would be desirable, this might adversely affect the precision of the method because the determination of absorption is imprecise.

As part of a research progamme on the use of low-grade aggregates for use in concrete, the Building Research Establishment are carrying out tests to assess shrinkage. Naturally-occurring low-grade aggregates tend to have higher water absorption and lower particle density than conventional aggregates for structural concrete, and difficulty has been experienced in making test prisms when using the constant mass proportions given in the ten-day method. For this reason it has been recommended that the ten-day method should be restricted to aggregates with absorptions not greater than 3.5 per cent and particle densities not less than 2450 kg/m^3. Aggregates having properties outside these values should be assessed by using procedures to be published by the Building Research Establishment.

3.3.13 Specification for shrinkage of aggregate in concrete
Ideally, the only variable in a test for the determination of aggregate shrinkage in concrete should be the aggregate shrinkage itself. Although one cannot ignore the possibility of unknown interactions between aggregate and cement, the ten-day test was designed to indicate true differences in aggregate shrinkage. Variations caused by cement and aggregate grading have already been discussed. Also, standard mix proportions, curing conditions, and measurement procedures were selected to provide the basis on which one aggregate could be compared reliably with another. It follows that test results are peculiar to this method and that it would be most unlikely that those results could provide useful indications of concrete shrinkage in practice. Infinite variations in mix proportions and curing are possible; estimates of concrete shrinkage are best made by using methods described in BS 1881: Part 5, 1970, or ASTM C157-80.

The ten-day test, coupled with a single limit at, say, 0.075 per cent, could usefully be employed to sort aggregates into two simple categories, that is, 'low shrinkage' and 'high shrinkage'. This is being considered at BSI.

Concern was expressed by representatives from the aggregate industry that the method of test might be published before the specification for its use. It has therefore been proposed that a classification procedure should be incorporated together with recommendations for frequency of testing. It has been suggested that aggregates having a shrinkage value above 0.05 per cent should be tested annually, and that aggregates up to this level of shrinkage should be tested at intervals of five years.

3.3.14 Future research
Hobbs & Parrott concluded that, for a concrete containing non-shrinking aggregate, the shrinkage of a plain or reinforced concrete could be estimated by using an equation where, for a particular design of member, the shrinkage was related to drying time, relative humidity, and initial water content of the concrete mix. It is possible that data from a modified ten-day test could be used in extending such an equation to arrive at estimates for concrete with shrinkable aggregate.

Work carried out at Dundee University (Tharmabala *et al.* 1976) demonstrated that aggregate resources could be used more effectively by blending aggregates of different shrinkages. It is not thought that there have been practical applications of this work as yet, but, with dwindling resources of aggregates in some parts of the UK, this is a subject that might prove to be of interest in the future. Certainly it would not be economically viable within Scotland at present, but, if aggregates were to be transported over long distances, the additional costs of blending could become less significant. Further work could be required to assess the efficiency of the blending techniques that might be used in practice.

The precision of shrinkage tests is not very good. Several suggestions have been made about ways of reducing the between-laboratory errors that are the main source of variation in results, for example the accreditation of laboratories and the use of standard reference materials. These ought to be examined as part of the process of continuing development.

References

American Society for Testing and Materials (1987) Length change of hardened mortar and concrete. Designation C157-80.

British Standards Institution (1970) *Testing concrete.* BS 1881: Part 5: 1970 (and its predecessors).

British Standards Institution (1975) *Testing aggregates.* BS 812: 1975 et seq.

British Standards Institution (1983) *Specification for aggregates from natural sources for concrete.* BS 882: 1983.

Building Research Establishment (1968) *Digest 35: Shrinkage of natural aggregates in concrete.* HMSO, London.

Collis, L. & Fox, R. A. (1985) *Aggregates: sand, gravel and crushed rock aggregates for construction purposes.* The Geological Society.

Davies, R. E. (1930) A summary of investigation of volume changes in cement, mortars, and concretes produced by causes other than stress. *Proceedings of the American Society for Testing Materials*: Part 1: **30**: pp 668–685. Appendix 13 to Report of Committee C9 on concrete and concrete aggregates.

Edwards, A. G. (1956) *BRE Internal Report* SL/C6. (unpublished).

Fulton, F. S. (1961) A co-ordinated approach to the shrinkage testing of concrete and mortars. *Mag. Conc. Res.* **13**: No. 39: pp 133–140.

Hansen, T. C. & Neilson, K. E. C. (1965) Influence of aggregate properties on concrete shrinkage. *J. Amer. Conc. Inst.* **63**: No. 7: pp 783–794.

Hobbs, D. W. & Parrott, L. J. (1979) Prediction of drying shrinkage concrete: 13: No. 2: pp 19–24.

Lerch, W. (1946) The influence of gypsum on the hydration of Portland cement pastes. *Research Laboratories of the Portland Cement Association, Bulletin* No. 12: pp 41.

Moore, I. C. & Gribble, C. D. (1980) The suitability of aggregate from weathered Peterhead granites. *Q. J. Eng. Geol. London*: **13**: pp 305–313.

Snowdon, L. C. & Edwards, A. G. (1962) The moisture movement of natural aggregate and its effect on concrete. *Magazine of Concrete Research*: **14** No. 41.

Stutterheim, N. (1954) Excessive shrinkage of aggregate as a cause of deterioration of concrete structures in South Africa. *Transactions of the South African Institute of Civil Engineers.* **4**: No. 12: pp 351–367. Discussion Vol. 5: No. 6: June 1955: pp 199–201.

Tharmabala, T., Sangha, C. M., & Dhir, R. K. (1976) The use of moderate to high shrinkage aggregates for making concrete. In *Advances in Ready Mixed Concrete Technology.* Pergamon Press pp 75–88.

4

Sands for building mortars

N. Beningfield and T. P. Lees

4.1 TYPES OF AGGREGATES IN USE, AND PRODUCTION METHODS

Natural sands supplied from quarries relatively close to the point of use predominate in the production of mortars. Aggregates from a distant source are sometimes specified so that a particular effect can be achieved, for example to match masonry or rendering in an existing structure undergoing extension or repair, but these instances are relatively rare. Some of these aggregates can make mortars that are harder to use than those made with local materials. In areas where there is a dearth of suitable natural sands, mortars may be made with crushed rock sands. These give satisfactory properties when they are hardened, but they are more difficult to work with, and are thus unpopular with bricklayers and plasterers.

Natural sands can be found in many different types of deposit. Relatively coarsely-grained, well-graded sands (i.e. having a wide particle size distribution) are found in association with gravels laid down by rivers and glaciers. More single-sized and finer sands are found in deposits formed in lakes and estuaries. Marine deposits yield both coarse and fine sands, generally lacking in fines (i.e. material finer than 75 μm). The particle shape and texture of sands is governed partly by mineralogy and partly by the aggregate production process. In the south and east of England the dominant petrological type in the gravels is flint, whereas in the Midlands it is quartzite.

Siliceous sands are the commonest petrological type throughout the UK, although there are many exceptions. For example, the gravels of the Upper Thames have a high content of limestone. Isolated deposits of uncemented oolitic limestone sand of Jurassic age are also known, as in one case in Oxfordshire; the spherical particle shape of the ooliths promotes good workability of mortars. Other examples of sands from geologically older deposits can be cited, for example sands of Permian age are worked in County Durham.

The extensive Quarternary glaciation led to many superficial deposits of mixed petrology in the UK. These can yield suitable sands, especially if the products are free from lignite. Marine-dredged sands have been used very widely in some coastal areas, e.g. in Cardiff, Bristol, and Liverpool and their environs. Concern is sometimes expressed about contamination with salts, but adequate drainage of the sand when it is landed, or a washing process, can reduce salt contents to acceptable limits. Dredged or washed sands often lack fines, leading to a loss of workability in the mortar, which has to be compensated for by the addition of a filler, e.g. lime, or a chemical admixture.

Many sands are produced, without washing, by a simple process of digging and rough screening to remove oversize particles. Bricklayers often prefer these materials because their high contents of fines lead to very workable mortars. Artificial and other special aggregates are in limited use. Perlite is a lightweight material, consisting of nearly spherical particles. Mortars made from it are easy to use, and are used in special plasters to achieve good thermal insulation. Expanded vermiculite can also be used in this way, but the particle shape is flaky and thus the mortars are difficult to work. Sintered expanded clays are not available in the appropriate size range in the UK. It would be possible to crush the coarser sizes, but most of their beneficial properties would then be lost. Barytes and boron carbide are used in mortars associated with shielding X-ray and nuclear apparatus.

4.2 SPECIFICATIONS FOR MORTAR SANDS

Specifications for mortar sands in the UK have not been securely grounded on evidence of performance. There is some reason to believe that this is the case in other countries too. Most specifications, including the British Standards BS 1199 and 1200 (BSI 1976) lay emphasis on particle size distribution. Some specifiers, understandably, insist on full compliance with the Standards. Bricklayers and plasterers, familiar with local, non-complying materials, sometimes complain that sands produced to comply with the Standards are difficult to use. Sand producers sometimes cite examples of the successful use of sands that lie outside the specified grading limits, with no apparent durability problems. An attempt was made by BSI to bridge the gap between what is specified and what is used by bringing in amendments to BS 1199 and 1200 in 1984. This was started after investigations by the Construction Industry Research and Information Association (Ragsdale & Birt 1976), Edinburgh University (Currie & Sinha 1981), and the Sand and Gravel Association. The grading requirements for sands and mortars for laying bricks and blocks are reproduced in Table 4.1.

It is important to realize that the changes made in the 1984 amendment to BS 1200 were agreed almost entirely because of a change in test method from dry sieving to washing-and-sieving. The original specification had been based on dry sieving, but by the mid 1970s it had become recognized by many people that dry sieving often failed to provide a true result, as fines could bind the coarser particles together. Nevertheless, BS 812 allowed the tester discretion as to whether a preliminary washing should be carried out before sieving, so some testers were using washing-and-sieving, and others were using dry sieving. This led to confusion, and a need for standardization became obvious.

Table 4.1 — Grading requirements for sands for mortars for laying bricks and blocks from BS 1200

BS sieve mm	Before the 1984 amendment	After the 1984 amendment Type S	After the 1984 amendment Type G
6.3	100	100	100
5.0	100	98–100	98–100
2.36	90–100	90–100	90–100
1.18	70–100	70–100	70–100
μm			
600	40–100	40–100	40–100
300	5–70	7–70	20–90
150	0–15	0–15	0–25
75	†	0–5[‡]	0—8[§]

[†]No requirement was made, although 'clay' was specifically excluded by name.
[‡]This limit is increased to 10 per cent for crushed stone sands.
[§]This limit is increased to 12 per cent for crushed stone sands.

In the UK it is now agreed that washing-and-sieving should be used, because that method gives a better indication of the true result, and better precision, than the dry sieving test (Pike & Limbrick 1981) — see section 4.5. The grading limits agreed in 1984 greatly increase the number of building sands that comply with the grading limits of the Standard when they are tested by washing-and-sieving, but, even so, there are still sands in use that are finer than Type G. Grading specifications from several other countries are set out in Table 4.2.

Table 4.2 — Grading specifications for sands for laying bricks and blocks from Belgium, Japan, and the USA

Sieve sizes mm	Belgium	Japan	US
5[†]	100	100	100
2.36	100	90–100	95–100
1.18	100	70–100	70–100
μm			
600	80–100	40–100	40–75
300	40–85	5–75	10–35
150	10–25	0–25	2–35
75	0–7[‡]	0–10	—

[†]Or close alternative used in the national standard.
[‡]Increased to 20 per cent if the fines are of the same material as the rock.

Some countries have grading requirements that are varied according to the different types of sand, and for the different kinds of application. For example, South Africa has six grades, as shown in Table 4.3.

There are also separate grading limits for sands for mortars for plastering and rendering in the UK. These are given in BS 1199, which was amended in 1986, again to take better account of the materials that are used in practice. The grading limits are reproduced in Table 4.4.

There is no separate French national standard for building sands, but reference is sometimes made to the standard for concrete sands. In the Federal Republic of Germany (FRG), Albrecht & Wiszotsky (1968) showed that, as in the UK, a wide range of sand gradings was in use, and recommended that gradings should be specified at sieve sizes between 7 mm and $20 \mu m$, but this recommendation appears not to have been implemented. The only restriction on grading in the current FRG standard for building sands is a maximum limit of 8 per cent 'settleable solids', i.e. fines passing the 63 μm sieve. Some national standards permit the use of sands outside the specified limits if tests on the mortar, or field experience, show satisfactory performance. In general, then, there seems to be little evidence to support restrictive grading limits. Grading tests are a convenient method of specifying and exercising compliance control, but the results have limited meaning in terms of performance. The specifications of other countries sometimes also regulate other properties of building sands, such as the contents of clay and of lightweight particles. Some of the test methods that are used are unrefined, and show poor precision. In the UK, more reliable test methods for these properties are being developed, but are not yet used in practice — see section 4.5.

4.3 THE PROPERTIES OF MORTAR

The annual tonnage of aggregates used in concrete in the UK is much greater than that used in mortars. The technical as well as the commercial aspects of the mortar industry are influenced to a degree by the concrete industry. People entering the field of mortars with a background of concrete technology may regard mortar simply as a concrete without coarse aggregate. Attempts to predict the general properties of mortars from that starting point have been markedly unsuccessful. Some mortars, e.g. those used as screeds for floors and roofs, can be treated to some extent as concretes without coarse aggregate, but this is a wholly inadequate approach to the mortars that are used for laying bricks and blocks, and to those that are used for plastering and rendering. Two of the differences that can be discerned between concrete and mortar are: (a) the water content of concrete does not usually change greatly from the point of placing over the succeeding few hours, if good practice is observed, but mortars can suffer substantial water losses; and (b) it is easy to characterize the main structural contribution of concrete by means of the cube compression test, whereas for mortars compressive strength is of strictly limited significance.

Losses of water in mortars arise from two main causes. If the masonry units with which a mortar is used are dry and exhibit strong suction, this can remove a high proportion of the water in the mortar very quickly. Again, if the ambient conditions of temperature, relative humidity, and windspeed create pronounced drying con-

Table 4.3 — The South African grading requirements for building sands — percentages finer than indicated sieves

Sieve size	Requirements for natural sand			Requirements for sand made by crushing rock for		
mm	internal plaster	external plaster and high strength mortar	general purpose mortar	internal plaster	external plaster and high strength mortar	general purpose mortar
4.7	100	100	100	100	100	100
2.36	90–100	90–100	90–100	90–100	90–100	90–100
1.18	70–100	70–100	70–100	70–100	70–100	70–100
µm						
600	40–100	40–100	40–100	40–100	40–100	40–100
300	5–56	5–56	5–57	5–70	5–56	10–75
150	0–20	0–15	0–24	5–25	5–20	10–25
75	0–10	0–7.5	0–10	0–10	0–10	0–15

Table 4.4 — Grading requirements for sands for mortars for plastering and rendering from BS 1199 after the 1986 amendment

Bs sieve mm	Percentage by mass passing indicated sieve	
	Type A	Type B
6.3	100	100
5.0	95–100	95–100
2.36	60–100	80–100
1.18	30–100	70–100
μm		
600	15–100	55–100
300	5–50	5–75
150	0–15	0–20
75	0–15	0–5

ditions, the mortar may dry out very quickly. It is standard practice to cure concrete carefully, but scant attention is often paid to curing mortar. In extreme cases so much water may be removed from mortar that insufficient remains to allow full hydration of the cement.

Furthermore, mortar is rarely used on its own in a structure; it is almost always part of a composite material, i.e. masonry. Research by Bowler (1988) demonstrates that exposing fresh mortar to different regimes of suction, by varying the substrate, profoundly alters the characteristics of the hardened mortar. Thus a simple compressive strength test on a cube of mortar may provide little useful information on the strength of the masonry. For this, it is necessary to carry out tests on samples of masonry built with the bricks or blocks, and the mortar, in question. Relevant tests are the wallette test given in BS 5628 (BSI, 1978), and the US crossed couplet test (ASTM 1976). Anderson & Held (1983), Held & Anderson (1988) have applied these tests to study the effects of changing the grading and fines content of sands on the properties of mortar and masonry. As might be expected, they showed that sand grading can have a marked effect on the properties of mortar and the tensile bond strength of mortar to bricks. They found that increasing the content of non-plastic fines in the sand increased the compressive, tensile, and flexural strengths of their experimental mortars, but decreased the tensile bond strength. As tensile bond strength is an important determinant of the structural performance of masonry, this disparity in the effects of changing fines content is of considerable significance. Held & Anderson (1988) recommend bond testing as a method of construction control.

Descriptions of the mortar before it is used are not necessarily a useful basis for predicting the properties of the mortar when it is hardened as part of the masonry composite. Perhaps because it is difficult to relate the properties of the hardened mortar to the performance of the masonry, when the properties of mortar are being considered in the field, most attention is given to the plastic properties of fresh mortar. Plasticity is not a term much used in practice, but it can be taken to embrace a number of subsidiary terms that describe particular features of the working properties of mortars as they are perceived by the mason or the plasterer. 'Cohesion' describes the ability of a mortar to remain in a body without breaking into its components.

Excessive cohesion makes a mortar 'puggy' or 'sticky' and difficult to spread; insufficient cohesion means that the mortar does not 'hang together' and again is difficult to spread. 'Adhesion' describes the ability of a mortar to strick to other surfaces, e.g. to a trowel or to masonry units. Adhesion is particularly necessary when a mortar has to stick to a vertical surface, as in plastering and rendering, and for spatterdash and Tyrolean finishes. For these finishes, mortar is thrown at a prepared wall surface by hand, or with a small machine.

Shortcomings in cohesion and adhesion can usually be overcome by adjustments in mix design, by altering the sand grading, changing the content of cement and/or filler, or by the use of chemical admixtures. Care must be exercised here; increasing binder content may cure inadequate adhesion in the plastic state, but it can then lead to excessive shrinkage. Mortars that contain sands with an excessive amount of coarse or poorly shaped particles, or poor gradings (such as a deficiency of a middle size, or of fines), may be difficult to finish; they are then described as 'harsh'. Flaky and elongated particles lead to tearing of the mortar surface during final trowelling of a render, or when brickwork mortar is pointed. Ideally, there should be a smooth gradation of particle size, with particle shapes not deviating much from the spherical, and with the voids in the sand filled with a mixture of the natural fines (but not clay) and the cement (or cement and filler).

'Segregation' and 'bleeding' lead to some problems that are similar to those caused by excessive substrate suction, i.e. loss of workability before the task of bricklaying or rendering is finished. Again modification of the mix can alleviate the problems. For example, unduly single-sized sands can cause bleeding and segregation; blending of two sands can improve the grading.

Recently, more attention has been paid to the durability of mortars. Beard (1986) has linked general fineness of building sands to poor durability of mortars, although this evidence is not absolutely conclusive on this point, as other relevant factors (e.g. poor control over mix proportions, and so inadequate cement contents) can be found in his published data. Harrison (1986) found that, for some mixes, the strength and frost resistance of mortars were reduced when sands finer than Type G were used, but that sand grading had no effect on sulphate resistance. Four grades of mortar for brickwork and blockwork are specified in the British Standard Code of Practice for masonry, BS 5628 (BSI 1985), see Table 4.5.

It will be seen that, for Designations II, III, and IV, a range of sand contents is permitted. The intention is that the higher sand contents are for use with the Type S sands, and the lower sand contents are to be used with the Type G sands.

When the designations were standardized it was intended that, for any given designation, the various options would give mortars of about the same compressive strength. So, for example, the sand contents permitted for the mixes made with masonry cement are lower than those permitted with ordinary Portland cement, because masonry cement gives lower compressive strengths as it includes an inert, or

Table 4.5 — Volume proportions specified for mortars specified in BS 5628

Mortar designation	Nominal volume proportions		
	Cement: lime: sand	Cement: sand	Masonry cement: sand
I (15–29)[†]	1: 0.25: 3	1: 3	—
II (6–10)	1: 0.5: 4 to 4.5	1: 3 to 4	1: 2.5 to 3.5
III (4)	1: 1: 5 to 6	1: 5 to 6	1: 4 to 5
IV (2)	1: 2: 8 to 9	1: 7 to 8	1: 5.5 to 6.5

[†]The values in brackets are indicative of the levels of compressive strength commonly achieved when mortar cubes are tested at 28 days.

low-strength, filler. When the various options are tested in practice, however, they may not give the same compressive strength (see the values in brackets in Table 4.5).

Even lower strengths than those of Designation IV are found with some mortars made with lime and sand but no cement, similar to those used from Roman times until the invention of Portland cement in the 19th century. Such materials are only rarely used today.

Some mortars are made with deliberately retarded setting times. Retarded, ready-to-use mortars usually have their setting times retarded by times in the range of 12 to 72 hours. Retardation is achieved by one of a group of chemical admixtures, and the dosage controls the retardation time. It is sometimes assumed that the relation between dosage and length of retardation is linear, but this is not the case; permanent repression of set is achieved above a limiting dosage.

Mortars in masonry undoubtedly experience creep, but problems have not been reported. This may be because volume changes in mortar can accommodate volume changes in the masonry units, which would otherwise cause cracking. Mortars undergo more shrinkage than concrete because they have higher water contents. Water:cement ratios of around 0.5 are typical of concrete; in mortars the water:cement ratio may be five times higher than that. Much of the shrinkage of mortar may, however, take place immediately after application, particularly when high-suction masonry units are being used.

The amount of air entrained in mortars is important to their freeze/thaw resistance. There are misconceptions about the optimum and maximum amounts of air to be included. The crucial factor is the distance between the adjacent air pores in the hardened mortar, which is controlled mainly by the number and size of the air bubbles. The limits given in BS 4721 (BSI 1981) take this into account.

Mortars may be subjected to attack by acids, e.g. by carbonic acid and stronger acids carried in rain, and by waterborne sulphates from the ground or from masonry units.

An important indication of the acceptability of a sand is its proven, satisfactory performance, including durability in appropriate conditions of exposure over a period of decades. There are, however, limitations to this approach as some sands have been used successfully only in restricted localities where the craftsmen have developed expertise with them. Such sands would not necessarily be regarded as suitable elsewhere; also the importation of masonry units into areas where their use is not familiar to local bricklayers can cause problems.

4.4 SAND PROPERTIES AND THEIR EFFECTS ON MORTAR

The particle size distribution of sand is most important; it affects both the working properties of fresh mortar and the long-term performance of the hardened mortar. However, some of those aspects of the size distribution which are most important have not been satisfactorily defined. It is obvious that particles larger than the eventual thickness of a mortar joint or render coat cannot be tolerated, but terms such as 'well-graded' or 'fineness' are often used too loosely.

Particle size distributions, or 'gradings', are usually tabulated as the cumulative percentage by mass passing a series of sieves having stated aperture sizes, as is illustrated in Tables 4.1 to 4.4, which show specifications in terms of ranges of percentages at each of the specified sieve sizes. Correlating information of this nature with resulting mortar properties is difficult, but various attempts to do so have been made by using single numbers which have been developed to express gradings in a simple way, e.g. equivalent mean diameter (Hughes 1960), sorting coefficient (Bessey & Purton, 1865), and fineness modulus (Abrams, 1925). These have been advocated, and may be satisfactory, for use with the coarser sands that are often used in concrete, but they do not satisfactorily describe the extremes of size distribution that are so important for mortar sands.

Special consideration must be given to the fine end of a grading, say the fraction finer than 75 μm. These fines are important in both workability and long-term durability. It is known that sands with an excessive content of fines require extra water to make a workable mortar, and this can be explained in terms of the extra surface area to be wetted. A small increase in fines content, which would not show a great increase in any of the three descriptors mentioned in the previous paragraph, can increase surface area, and hence water demand, greatly. To illustrate this, consider the grading shown in Fig. 4.1, which complies with sand Type S of BS 1200. First, the surface area of the particles retained on the 75 μm sieve is calculated. If it is assumed that all the sand grains are spherical and that the particles in each size fraction have a diameter at the mid-point between the two sieves containing that fraction (so that all the particles in the 600 to 300 μm fraction have a diameter of 450 μm, etc), then the total surface area of the material retained on the 75 μm sieve (which makes up 96 per cent of the sample) is 5.4 m^2 kg^{-1} if the sand has a particle density of 2650 kg m^{-3}.

The 4 per cent of fines contributes much more surface area than this. Table 4.6 shows calculated values of surface area for the whole sand when the fines have various particle diameters.

Conventionally, fines with an equivalent diameter finer than 2 μm are regarded as clays. Clay minerals tend to be platey, and thus have greater external surface areas than calculations based on spheres would suggest. Also, some clays have the ability

Sec. 4.4] Sand properties and their effects on mortar

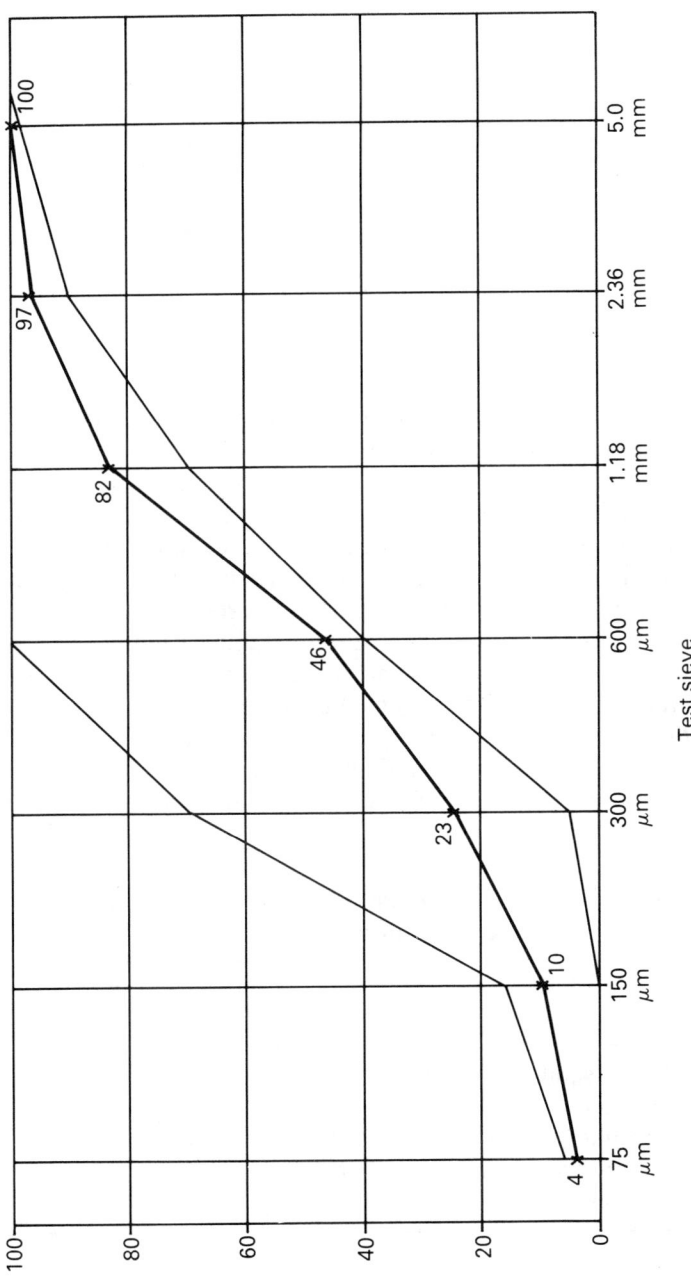

Fig. 4.1 — A grading result for a mortar sand plotted inside the envelope of limits for Type S, BS 1200.

Table 4.6 — Surface area of sand plotted in Fig. 4.1, depending on the particle size of the 4 per cent of fines

Assumed particle size of 4 per cent of fines — μm	Total surface area of the sand — $m^2\,kg^{-1}$
50	7.2
25	9.0
10	14.5
5	23.5
1	96.0

to adsorb water on internal surfaces. Thus, below about 5 μm, the surface area of fines can be even larger than the values shown in Table 4.6.

A fairly high content of fines in mortar sands may be attractive to the bricklayer and plasterer because it will promote workability, but the long-term durability of mortars can be adversely affected if the fines exceed the limits given in BS 1199 and 1200 (see Tables 4.1 and 4.4). Clay minerals in particular can greatly increase water demand because they are very fine and have a large surface area. Reliable and simple diagnostic tests for clay in mortar sands are not standardized in the UK, although attempts to find such methods are being made (see section 4.5). When clay is present as small lumps or aggregations they may not be readily identifiable. Clay lumps make mortar susceptible to frost attack, causing localized expansions which lead to pop-outs and cracks. These are unsightly and impair durability.

Fines also affect air entrainment. Table 4.7 shows the dosage of a commercial air-

Table 4.7 — Dosages of admixture required to yield 15 per cent of air in a 1:6 mortar

Sand description	Dosage — ml per 50 kg of cement
Washed sand to Type S of BS 1200	57
Fine sand to Type S of BS 1200	149
Very fine sand, high clay content	515
Very coarse sand	10

entraining mortar plasticizer required to yield 15 per cent of air in a mortar having a cement:sand ratio of 1:6 (Goodwin, J., personal communication).

It is clear from these data that the finest particles inhibit air entrainment. (It has also been found that prolonged mixing of mortars made with sands having high contents of fines decreases the air content still further.) Conversely, coarser sands may entrain excessive air unless the admixture dosage is cut back.

Sands with a preponderance of fine particles, and lacking coarser particles, have reduced bulk densities. To an extent, this offsets their higher water demand when mortars are batched by volume, as the reduction in bulk density leads to an effective increase in cement content by mass. Even so, mortars made with fine sands will require significantly more water than those made with coarser sands to achieve a given level of workability. This will result in lower compressive and flexural strengths of mortar test specimens. When such mortars are applied to masonry units of high suction, the resulting hardened mortar will not show such a marked difference in strength when compared with a mortar made with a coarser sand. However, when mortars made with fine sands are used with masonry units having low suction, the long-term properties of the mortar will almost certainly be adversely affected. When such mortars are applied as renders, the resulting drying shrinkage may lead to cracking and, in extreme cases, to failure associated with de-bonding.

For sands used in renders it is desirable that a significant proportion of the sand should consist of coarser particles in the range of, say, 1.18 to 5.0 mm so as to reduce shrinkage. Recognizing this, many craftsmen mix a medium (Grade M) concreting sand with a building sand to provide a suitable grading.

At the present time, all that can be done to control gradings in specifications is to impose realistic limits on the maximum content of the coarsest particles (the upper size depending on the type of work) and to place an overall limit on the fines content (i.e. material finer than 75 μm). It must be recognized, however, that this simple approach can lead to the rejection of suitable sands and, possibly, the acceptance of material badly contaminated with deleterious fines.

The effects of particle shape and texture of sands have not been extensively studied, but, as has already been noted, angular, flaky, and elongated particles can reduce workability, and may therefore cause complaints from bricklayers and plasterers.

'Bulking', i.e. an increase in volume associated with an increase in moisture content, is more pronounced with some mortar sands than with most concrete sands. Because mortars are generally batched by bulk volume, richer mixes may result if wet sands are used, but allowance for bulking should not be made unless the effect has been observed and measured.

Organic contaminators that markedly retard the hydration of Portland cement are not often observed in commercial mortar sands in the UK. Organic impurities such as coal and lignite, however, are likely to be found in some areas and can cause problems. Even small quantities can cause unsightly stains on renders, and, at higher levels of contamination, may be visually unacceptable in bedding mortars. Pyrites is another source of unacceptable stains. It is possible to analyse for sulphides, but the sampling problems are considerable because pyrites is often an impersistent contaminant in UK aggregates. The most reliable indicator of suitability is established use with a sand from a particular source, or of similar materials from the same area.

If chloride is present in significant concentrations (say more than 0.1 per cent by mass of the sand) it may promote the corrosion of any metals, e.g. wall ties, in the mortar. At higher levels, chloride can exacerbate efflorescence of brickwork and lime-bloom of mortar; these effects are, however, temporary, and disappear after a few bouts of rain. If there is to be no unprotected metal embedded in the mortar, unrealistically tight limits on chloride should not be imposed.

4.5 TESTING MORTAR SANDS

The sampling procedure of BS 812: Part 102 should be followed. Mortar sands as delivered are usually fairly homogeneous and cohesive. The segregation that is typical of some coarser aggregates is generally not nearly so pronounced with sand. This, and the smaller sample sizes required, makes the sampling of sands easier then the sampling of coarse aggregates. Nevertheless, the number and size of increments specified should be observed scrupulously. The bulk sample should be well mixed before the laboratory sample is taken, and the laboratory sample should be well mixed again before test portions are taken. Unless the sand is dry and free-flowing, the use of a riffle box is not satisfactory. Damp sand is cohesive, so sample reduction by mixing and sub-sampling by increments is usually sufficiently accurate.

The washing and sieving test procedure given in Part 103 of BS 812 should be used to determine particle size distribution. This is especially important for building sands because many of them contain fines that agglomerate the sand particles. The procedure is to take test portions, weighing say 300 to 500 g, and then to dry them overnight at 105°C. The dried test portions are accurately weighed and then transferred to vessels in which they can be mixed with about five times their own volume of water. Mechanical agitation, e.g. by rolling in a metal or plastic cylinder, can loosen any attached fines. If that kind of apparatus is not available, the test portion must be well soaked and agitated.

The soaked and agitated test portion is allowed to settle very briefly, and then the supernatant liquid, with the fines in suspension, is decanted through a 75 μm test sieve which is protected by a 1.18 mm sieve. The test portion is then repeatedly soaked, agitated, and decanted until the wash water runs clear. The water and fines passing the 75 μm sieve are allowed to run to waste. The solids retained on the sieves are washed back into the original container, and then dried to constant mass at 105°C. The amount of material passing the 75 μm sieve is found as the difference in the mass before and after washing. The dried residue is sieved in the normal way, with a 75 μm sieve included in the sieve series. Any particles passing this sieve are weighed and their mass is added to that lost during washing to give the total mass passing 75μm. The sieve grading is then calculated.

Only if it can be demonstrated that simple dry sieving gives the same result as washing-and-sieving can the simple, dry method be used, and then only for the purposes of production control. It is often found that dry sieving is acceptable for the coarse sands used for floor screeds, but this possibility must never be assumed without a preliminary check.

If it is required to characterize the material finer than 75 μm a wide range of techniques is available, but, in BS 812, gravitational sedimentation (Andreasson pipette) is preferred. Unless very careful control is exercised on temperature, it is unlikely that the method can be applied reliably to sizes below 6 μm. Automated sedimentation apparatus is available. Other methods for measuring the particle size distributions of very fine particles, such as the Coulter Counter and devices using laser-scattering, are in use in some related industries, but they have not yet been extensively applied to mortar sands.

It is possible to obtain electroformed sieves having apertures down to 5 μm, although the finest sieves are very difficult to use as their effective open area is so small. It is considered that 20 μm is the smallest practicable aperture size in this application. Woven wire sieves down to 38 μm are made, but are fragile (Pike & Limbrick, 1981).

It is sometimes assumed that all very fine particles in a mortar sand are clay, possibly because of a convention used in soil mechanics and geology that clay is material having a settling velocity equivalent to or less than spherical particles having a diameter of 2 μm. This can be misleading on two counts. Because clay minerals are plate-shaped they settle more slowly than spheres of the same mass; in their larger dimensions the plates may be as large as 40 μm (although some are also extremely small, i.e. sub-micron size). Also, some material finer than 2 μm consists of material other than clay minerals. Furthermore, clay minerals themselves vary widely in their properties; some have much greater effects on mortar than others. There are no reliable British Standard tests for clay at present, although a study of the available methods is underway. In other countries use is made of the Sand Equivalent test, which is based on the tendency of clays to bulk more than other fines, and of the methylene blue dye adsorption test. Clay minerals, especially smectites, have significant ion exchange and adsoprtion capacities; their ability to adsorb methylene blue dye has been shown to correlate with their harmful effects on concrete and mortar.

Lumps of fines, including clay, can be detected by visual examination. This is best done by observing the residue retained on a 1.18 mm sieve after gentle washing and decantation. It is sometimes necessary to measure the bulk density of mortar sands, for example for laboratory experiments.

As has been noted above, British Standards for mortars use volumetric proportions. It is often not clear if the BS 812 procedure for the loose bulk density, or that for the rodded condition, should be measured. In these circumstances it is wise to measure and report both. It has been noted by the authors that more repeatable values of loose bulk density are obtained if the container is filled with a lot of small increments rather than a few large increments. For researchers and others interested in mix design, measurements of bulk density at higher levels of compaction energy, e.g. the compactability test in BS 5835 (BSI 1980) may be useful, particularly if they are combined with measurements of optimum moisture content.

Organic contamination of mortar sands leading to retarded cement hydration is not common. If it is suspected, a sodium hydroxide discoloration test is probably the most convenient to apply. This method has been dropped from BS 812 in recent years because it can give misleading results, but there is now a proposal to reinstate it as a preliminary sorting method. Materials failing that test can then be used to prepare a standard mortar, which would be compared with an identical mortar made with the same sand freed of organic contamination by ignition in air at 600°C. Comparisons of the setting times and rates of strength development of the two mortars give a basis for evaluation. This technique has been published for public comment (BSI 1988), but not yet extensively used in practice. There could be some interference effects from other contaminants such as clay.

Organic contaminators that cause discoloration are not easy to test predictively. It may be possible to make mortar pats and then to observe them for the development of stains, but it is difficult to identify the curing regime that is likely to maximize stain development. Some stains are caused by lightweight materials such as lignite, and such materials can be identified by a flotation test — a draft method has been circulated for public comment (BSI 1988).

No standard tests for pyrites are in use in the UK. It has been suggested (Midgley 1958) that, if a sample of sand is immersed in a saturated solution of calcium hydroxide, pyritic particles will develop a green colour, but this test has not proved to be completely reliable.

The test for chloride content described in Part 117 of BS 812 is appropriate, but, for high productivity, modern instrumental methods offer advantages, and are being examined by BSI. A hand-held conductivity meter specially developed for chloride is commercially available and has been shown to give accurate results.

To achieve rapid results for production control it is unnecessary to insist on full oven drying of the sample before taking an aqueous extract. In these circumstances, the extract can be taken from the sample immediately, and an estimate may be made of the moisture content by rapid methods and used to calculate the chloride content on the basis of the dry mass of the sand. In addition to these standard laboratory tests, makers and users of mortars are well-advised to carry out practical trials with any type of sand with which they are unfamiliar. In particular they should measure the amount of water required to give adequate workability in standard mixes, and the setting times and the compressive strengths of these mortars in accordance with the standard methods for testing mortar given in BS 4551 (BSI 1980).

REFERENCES

Abrams, D. A. (1925) *Design of concrete mixtures*. Bulletin 1. (revised edition) Structural Materials Research Laboratory. Chicago.

Albrecht, W. & Wisotzky, T. (1968) Survey of the particle size composition of sands for masonry and plastering mortars. *Bau und Bauindustrie*. pp. 692–697.

American Society for Testing and Materials (1976) *Standard test method for bond strength of mortar to masonry units*. ASTM C952-76.

Anderson, C. & Held, L. (1983) The effect of sand grading on mortar properties and tensile bond. *8th international symposium on load-bearing brickwork. British Ceramic Society. November 1983*.

Beard, R. (1986) Brick masonry. *Chemistry and Industry*. December, pp. 848–854.

Bessex, G. E. & Purton, M. J. (1965) The effect of the grading of the aggregate upon the strength of calcium silicate bricks and blocks. *International Symposium on Autoclaved Calcium Silicate Building Products*. Society of Chemical Industry.

Bowler, G. K. (1982) Physical and chemical interactions of brick masonry materials. *Proc. 6th international masonry conference. Rome*.

British Standards Institution (1975 *et seq.*) *Testing aggregates*. BS 812.

British Standards Institution (1976 *et seq.*) *Specifications for building sands from natural sources*. BS 1199 and 1200.

British Standards Institution (1980) *Methods of testing mortars, screeds and plasters*. BS 4551.

References

British Standards Institution (1981) Specification for ready-mixed mortars. BS 4721.
British Standards Institution (1985) *Code of practice for use of masonry*. BS 5628. Part 1.
British Standards Institution (1980) *Compactibility test for graded aggregates*. BS 5835.
British Standards Institution (1988) Draft Part 122.1 of BS 812. *Organic contaminators which influence the setting and hardening of Portland cement mortars*. BSI Document 88/13116.
British Standards Institution (1978) Draft Part 122.2 of BS 812. *Lightweight contaminators which may disfigure concrete and mortar*. BSI Document 88/13117.
Currie, D. & Sinha, B. Survey of Scottish sands and their characteristics which affect mortar strength. *Chemistry and Industry*. pp. 639–645.
Harrison, W. H. (1986) Durability tests on building mortars — effect of sand grading. *Magazine of Concrete Research*. **38**, No. 135.
Held, L. & Anderson, C. (1988) Effect of fines content of sand on tensile bond. *Proc. 7th International conference on brick masonry. Melbourne University. February 1988*. pp. 959–968.
Hughes, B. P. (1960) Rational concrete mix design. *Proc. Inst. of Civil Engineers*. **17**. London, pp. 315–332.
Midgley, H. G. (1958) The staining of concrete by pyrite. *Magazine of Concrete Research*, **22**, No. 20, pp. 42–44.
Pike, D. C. & Limbrick, A. J. (1981) A study of sieve tests for building sands. *Chemistry and Industry*, pp. 626–630.
Ragsdale, L. & Birt, J. C. (1976) *Building sands: avcailability, usage and compliance with specification requirements*. Construction Industry Research and Information Association, Report 59.

5

Aggregates for bituminous materials

G. E. Broadhead and **J. F. Hills**

5.1 PROPERTIES AND SPECIFICATIONS OF BITUMINOUS MATERIALS

5.1.1 Introduction

Bituminous materials used in roadworks are associations of bituminous binder with mineral matter. They may occur naturally as, for example, lake asphalt or rock asphalt, or be manufactured by using various combinations of mineral aggregate and refined bitumen or road tar.

Nowadays, the predominant binder is refined bitumen, not only because of the steep decline in the production of coal gas, of which crude tar was the major by-product, but also because of the superior durability, versatility, and resistance to deformation of the petroleum product.

The aggregates may be one or more of crushed rock, slag, gravel, sand, or similar materials selected and graded in accordance with the technical and economic requirements. They form the bulk of a manufactured bituminous mixture, occupying between 75 and 90% of its compacted volume.

Surface dressing, coated macadam, rolled asphalt, asphaltic concrete, and mastic asphalt are all examples of bituminous materials used in road construction. Their performance depends on the individual properties of the aggregate and the bitumen, and on the way in which these components are combined and placed in the road. So, to appreciate the demands made on aggregates it is appropriate to have some acquaintance with bituminous materials.

This major section (5.1) begins with a brief outline of their technology which includes reference to the intrinsic properties of the materials, the manufacture and laying of asphalt mixes, surface treatments, and some of the factors that influence the modes of failure of an asphalt pavement. This is followed by a review of the mechanical properties of bituminous mixtures and how they are controlled by the relative proportions and physical properties of the bitumen and aggregate and the compacted density of the mixture.

Sec. 5.1]   **Properties and specifications of bituminous materials**   143

The final section gives an account of the properties currently specified by British Standards BS 594 and BS 4987, the Department of Transport *Specification for Highway Works*, and the Property Services Agency *Airfields Branch Specification* Part 4, for aggregates in bituminous mixtures at all levels of construction in flexible pavements for roads and airfields. Reference is made to the modified dense bitumen macadam base and basecourse material DBM 50, which has a 50 pen† bitumen and HDM, which has, in addition, a high filler content of 7 to 11%. A brief description is given of the Marshall test, used to determine the optimum bitumen content of a bituminous mixture for a given aggregate and grading.

5.1.2 Bituminous materials: applications, processing, and performance

The most extensive applications of bituminous materials are for the surfacing of roads, airfield runways and taxiways, and other paved areas. Asphalt and coated macadam can provide a smooth running surface for vehicles, coupled with skid-resistance, and the dense graded mixes act as important load spreading layers in a pavement construction.

In addition to its traffic-carrying abilities, asphalt is valued for its impermeability. This is an important characteristic for roads and airfields, where the underlying soil must be protected from the ingress of water, and it is an essential property when asphalt is used in hydraulic engineering applications such as canal linings, reservoir revetments, and the upstream facings or cores of dams.

Impermeability is also the chief property of mastic asphalt when used for the waterproofing of bridge decks, flat roofs, and basements of buildings.

Coated macadam may be dense- or open-graded, depending on requirements; and for the maintenance of a road, a surface dressing or slurry seal may be used for both sealing the surface and improving skid-resistance.

The physical and mechanical properties of a mix depend on those of its constituents, and also on the composition — that is to say, on the relative proportions of bitumen, aggregate, and air voids, and on the grading of the aggregate. In asphaltic concrete, for example, the grading must be chosen to allow dense packing of the aggregate. If only the largest size of particles that are to be used were to be packed together, the volume proportion of air voids would be about 35%. One can imagine the voids then being filled with smaller particles, the smaller voids in turn with still smaller particles, and so on. In this way it can be seen that certain proportions of each size of aggregate are required. In practice, the particles are continuously graded in size.

When experimenting with concrete, Fuller & Thompson (1907) found that the densest packing could be obtained when the cumulative percentage passing plotted against sieve size gave a curve that could be described by an empirical equation. The equation used today for dense packing is:

$$p = 100(d/D)^{0.5}$$

where p is the total percentage passing a sieve of size d, and D is the largest size sieve

† Penetration.

on which particles may be retained, i.e. the maximum (nominal) size of the aggregate grading. This equation expresses Fuller & Thompson's result in a simplified form, and is usually referred to as describing a Fuller curve. The US Federal Highway Administration (FHA) employs a similar relationship for the design of dense asphalt mixes but with an exponent of 0.45 instead of 0.5.

The nominal maximum particle size should not exceed about one third of the thickness of the layer in order to promote good compaction of the mix. Particles finer than 75 μm are usually called filler.

When designing a continuously graded asphaltic concrete mix (that is, choosing the materials and their proportions) the aggregate grading curve is selected to approximate to a Fuller or FHA maximum density curve. The densest packing obtainable with particular materials is not necessarily the best, however. When an uncoated aggregate is compacted, the voids in the mineral aggregate (VMA) should not be less than a certain minimum amount which depends, inversely, on the nominal size. For example, the VMA for a 10 mm nominal size continuous grading should not be less than about 16% by volume, and, for a 20 mm size, 14%. These amounts of voids are needed to leave room in the mix for the binder (approximately 10% by volume and 5% by mass) and the air voids (approximately 3 to 7% by volume). Such asphaltic concrete owes its properties to the development of inter-particle friction when the material is compacted.

In a gap-graded material, such as hot rolled asphalt wearing course with a fine aggregate of natural sand, one or more of the aggregate size fractions is missing. Consequently the coarsest particles are embedded in a mortar made with fine aggregate, filler, and binder, like the nuts in nougat, and the stability of the compacted mix depends to a large extent on that of the mortar. At high road temperatures there is stronger dependence on the consistency of the binder than is the case with asphaltic concrete.

The binder is usually bitumen specified in accordance with BS 3690: Part 1: 1989. It is supplied by the petroleum industry in a range of grades designated by the penetration, or 'pen'. The penetration is a measure of the consistency of the bitumen and is expressed as the depth of penetration of a standard needle in tenths of a millimetre under specified conditions of load, time, and temperature (usually 25°C). The bitumen properties may be modified by the addition of polymers or by mixing with lake asphalt or pitch (BS 3690: Part 3: 1990). For open graded mixtures, cutback bitumen or bitumen emulsion (BS 434: Parts 1 & 2: 1984) may be used. Road tar, which is prepared from crude tar as a by-product of the carbonization of coal, is also used as a bituminous binder — high-viscosity grades for dense mixes, and medium or low-viscosity grades for open graded mixes and surface dressing.

For the manufacture of bituminous materials, rock or gravel is quarried and then crushed and screened to give the range of sizes of coarse aggregate required. Crushed and screened slag or screened natural gravel may also be used. The fine aggregate may be sand, crushed rock, or slag fines. Added filler, if required, is usually ground limestone, but, depending on the circumstances, Portland cement, hydrated lime, crushed rock, or slag are also used.

At the mixing plant, the different sizes of aggregate are stored in separate stockpiles. The filler is normally delivered in a road tanker and conveyed pneumatically into a storage silo. The binder is supplied hot in a road tanker and pumped into an insulated storage tank that is maintained at a suitable temperature by electric immersion heaters or by the circulation of hot oil.

The mixing of the components may be carried out either as a continuous process in a drum mix plant, or in a batch plant. In a batch plant used for high-temperature mixes such as asphaltic concrete or rolled asphalt, aggregate in the required proportions is conveyed to a continuous, oil-fired, rotary drier where its moisture is removed and the temperature is raised to between about 150 and 190°C. The aggregate is then screened into size fractions and kept in buffer storage in hot bins. To make a batch, the requisite quantities of the various sizes of aggregate are fed sequentially to a weigh-hopper and released into the mixer. The filler and bitumen are then added, the bitumen being at a temperature at which its viscosity is about 2 poise (0.2 Pa.s). When mixing is complete, the material is discharged either into hot storage or direct into an insulated wagon for transport to the laying site. The load is covered with a tarpaulin to reduce heat loss.

Bituminous materials are usually laid continuously by a paving machine or 'paver' fitted with a towed 'floating' screed and tampers, by which means the initial thickness and density of the layer are controlled. After laying, the voids content of the asphalt is some 15 to 20%.

The asphalt layer is then compacted by rollers to reduce its voids content to about 5% and produce a smooth and even surface. It is important, of course, that the aggregate should be strong enough not to be crushed during this operation. Rolling is effective so long as the viscosity of the binder does not rise above about 200 poise (20 Pa.s). The temperature corresponding to this viscosity depends on the grade of bitumen and ranges from about 75 to 90°C.

The laying and compaction of asphalt are facilitated if the mix is 'workable'. A mix is more workable if its fine aggregate is rounded sand with a smooth surface rather than crushed rock with a rough surface texture and angular shape, and also if there is sufficient binder to provide lubrication. However the characteristics of a mix that make it workable also tend to make it less stiff in service, and so there has to be a compromise.

Rolled asphalt wearing course is provided with a roughened skid-resistant surface by rolling in coated chippings that have been spread over the surface after the asphalt has been laid by the paver. The chippings are single-sized (14 or 20 mm) and are coated with bitumen so that they are wetted by the binder at the surface of the layer of rolled asphalt to give good adhesion. An analysis of data on the loss of chippings suggests that, under the climatic conditions in the United Kingdom, this loss is minimized by having a laying temperature greater than 140°C. The chippings give a knobbly 'macrotexture' that aids skid-resistance at high speeds by providing channels for the dispersal of water under the tyres of vehicles and so preventing aquaplaning. The chippings themselves must have a surface with a 'microtexture' that gives a sandpaper finish which is not polished or seriously abraded by the action of traffic, thereby providing skid-resistance at low speeds and conservation of macrotexture.

A treatment that may be used for maintenance purposes and to restore skid-resistance is surface dressing. The existing road surface is sprayed with a thin layer of hot bitumen, bitumen emulsion (fine bitumen particles dispersed in an aqueous solution), or cutback bitumen (a solution of bitumen in a hydrocarbon solvent).

Chippings are then deposited on the layer of binder shoulder-to-shoulder to form a stone mosaic, and are rolled in. The chippings should be single-sized, cubic in shape, and free of dust — because dust impairs the adhesion of the binder.

Another restorative treatment is slurry seal which consists of bitumen emulsion, water, fine aggregate, and, usually, an additive such as Portland cement or hydrated lime. It is mixed immediately before being poured into a screed box, fitted with an adjustable rubber screed, which is towed by the mixing machine along the surface of the road and fed by it during laying. Most slurry seals benefit by being rolled with a pneumatic-tyred roller after setting.

In service, the asphalt pavement is required to carry traffic without unacceptable deterioration during its design life under the prevailing conditions of traffic density and climate. The sub-grade is protected from undue vertical stress from the wheel-loading by the load-spreading action of the asphalt layer. This depends on the thickness of the asphalt and its stiffness modulus.

There are various defects which may appear in the course of the life of an asphalt pavement whose serviceability gradually falls and eventually reaches a level at which either maintenance or reconstruction is required.

In common with other materials, asphalt is subject to fatigue. Fatigue cracks are initiated on the underside of an asphalt layer, where the tensile strain is greatest, when the cumulative number of strain pulses caused by wheel-loadings exceeds the fatigue life.

Rutting in an asphalt road may occur when the sub-base deforms because the load-spreading capability of the pavement is insufficient. On the other hand, the asphalt layer itself may rut if the resistance of the asphalt to permanent deformation is inadequate. This resistance depends on the mix design — that is, the choice of the constituents of the mix and their proportioning — and on good compaction.

Another important characteristic of an asphalt mix is its resistance to age-hardening. The hardening occurs when air has access to the bitumen and oxidises it, causing it to become brittle. This may lead to fretting of the asphalt and the initiation of surface cracks that propagate downwards. This age-hardening process is more rapid in hot climates, and may be retarded by increasing the thickness of the binder films coating the aggregate particles and by having a well-compacted, dense mix whose permeability to air is as low as possible.

In an open-graded mix, rainwater can penetrate and may cause the loss of adhesion, or 'stripping', between the binder and the aggregate. This situation is strongly affected by the density of the traffic loading. High and low temperatures may also have a significant effect.

It is necessary, of course, for the aggregate itself to be sound. Rock that is susceptible to either chemical or physical degradation is unsuitable for use either in an asphalt mix or in unbound layers of the pavement construction.

5.1.3 Mechanical properties of bituminous mixtures

The role of the aggregate in a bituminous mix is to provide a stony skeleton of rigid particles. The binder coats the particles in a thin film whose thickness may be reduced to zero at the points of contact between them, thus giving dry frictional contact. The finer the aggregate the greater the specific surface. It follows that how well the bitumen adheres to the aggregate is particularly important for the fine aggregate fractions in a mix. The finest aggregate, of course, is the filler; it is essential that this has good adhesion to the bitumen, and, for this reason, added fillers are usually ground limestone or Portland cement which both perform well in this respect. (Adhesion is considered further in section 5.3.)

The contribution of the binder to a mix depends on the properties of bitumen, which is a mixture of high molecular weight hydrocarbons that is a solid at low temperatures and a fluid at high temperatures. For 50 pen bitumen, for example, the viscosity is low enough at around 160°C for the bitumen to wet and coat dry aggregate readily. When the bitumen cools its viscosity increases rapidly until at 90°C it has become greater by a factor of about one hundred. At high road temperatures its viscosity is greater still by a further factor of one hundred, and it then behaves as a fluid that is highly viscous.

At very low temperatures, and at intermediate temperatures with short loading pulses such as those generated by moving traffic, bitumen behaves virtually as an elastic solid — that is, the strain resulting from the application of a stress is fully recovered when the stress is removed. If the loading time is long, bitumen behaves viscously, i.e. the strain is proportional to the time of loading and is not recoverable, although the viscous resistance is very high.

To provide a convenient measure for the whole range of rheological properties of bitumen, Van der Poel (1954) defined a stiffness modulus as the ratio of tensile stress to total strain. The stiffness is analogous to Young's modulus for elastic materials, but encompasses viscous behaviour too and so may be used to describe the effect of stress on bitumen for both low and high temperatures and short and long times of loading.

Van der Poel developed a nomograph for predicting the stiffness of the bitumen. This nomograph is entered by using the values of two properties that characterize the bitumen (softening point and penetration index) and two quantities that define the loading conditions (the time of loading and the temperature). While the form of the nomograph was guided by rheological theory, it was drawn by making direct use of the results of extensive laboratory tests on a wide range of bitumens. Because it is empirically based, the nomograph is in no danger of becoming outmoded by advances in theory. A computer version of the nomograph is available in a program called PONOS (de Bats 1973).

Two British Standard tests for the quality control of bitumen are those for penetration (at 25°C) and softening point. The results of these two tests can be used to calculate the penetration index, which is a measure of the temperature susceptibility of a bitumen and indicates its type. Outside the United Kingdom, a viscosity test for bitumen is sometimes preferred to the softening point test. Ductility and the susceptibility to cracking at low temperatures, using the Fraass test, may also be determined.

Other tests are the solubility in trichloroethylene, which is specified as a minimum of 99.5% and so amounts to a definition of bitumen, and the loss of mass on heating. Both the maximum acceptable loss of mass after heating, caused by the evaporation of low boiling point components, and the reduction in penetration, caused by both evaporation and oxidation, are specified.

There are chemical characteristics of a bitumen that may influence its emulsifiability, for example, or its ability to peptize the filler in a mastic so that the mastic is easy to spread. For bituminous materials in general, however, the most important aspect of the binder is its rheology — and the rheology of a bitumen is known (by using Van der Poel's nomograph) when its penetration and softening point have been determined.

Apart from its grade, the effect of a binder in a mix will depend on the binder content. This refers to the proportion of bitumen in the mix as a whole. It is often more pertinent, however, to consider the average thickness of the binder films, and this depends on both the binder content and the specific surface of the aggregate. Because of their large specific surfaces, the proportions of fine aggregate, and particularly the filler, are important. The latter may be taken into account by specifying the ratio of filler to bitumen by mass, and values in the range 1/1 to 1.3/1 are regarded as satisfactory.

The properties of a mix also depend generally on the grading of the aggregate and the voids content. While the voids content is a measurable characteristic of a mix that is important, it is not always as informative as might be wished; it is not a direct measure of permeability, or of the state of compaction that determines the mechanical properties — which depend also on the mode of compaction (Hills, 1973).

When describing the mechanical properties of mixes, it is convenient to consider separately the behaviour at short and long times of loading. The aggregate properties themselves are largely independent of the time of loading and the temperature, but, as outlined above, the bitumen properties do depend on these factors.

For short loading times and low temperatures the bitumen binder has a high stiffness modulus and is elastic. As, then, both the binder and aggregate components are elastic, the mix behaves elastically. Its stiffness modulus is high and is a function of the stiffness of the binder and the volume concentration of the mineral aggregate. The voids content has a modifying effect on high mix stiffnesses (Bonnaure *et al.* 1977), but the grading of the aggregate is immaterial. The mode of failure is by fatigue cracking of the bitumen. Failure occurs when the cumulative number of tensile strain repetitions reaches a critical value — the fatigue life; this depends on the type of bitumen (which may be characterized by its softening point), the bitumen content by volume, and the amplitude of the strain pulses. In the design of pavements to carry traffic on roads and airfields, the thickness of the asphalt layer is chosen so that the maximum horizontal tensile strain (which is at the lower surface of the layer) gives a fatigue life that is equal, to the design life.

At long times of loading, or when there is an accumulation of short times of loading, asphalt undergoes permanent deformation called creep (Hills *et al.* 1974). In the very act of deforming, however, the internal structure of the mix changes in such a way as to make it more resistant to further deformation — an example of strain hardening. The result is that, under a constant load, the rate of strain in a satisfactory mix gradually diminishes to virtually zero. The precise shape of the creep curve (strain against time of loading) is determined by the aggregate and the binder content. To obtain a mix with good resistance to creep, the aggregate should have a continuous grading curve, its surface texture should be rough rather than smooth, and its shape angular rather than rounded. Further, the binder content should not be too high, so as not to impair the effectiveness of the aggregate grading. It is also beneficial for the bitumen to have a high viscosity — in practice this means a high softening point.

In principle, mix design procedures are aimed at achieving a mix with satisfactory creep properties. Creep tests can be used for this purpose, and they have the advantage that their results are given in fundamental units and so can be used for the prediction of rut depths (Hills 1975, Shell 1978).

Wheel tracking tests have been correlated empirically with rut depth by Szatkowski & Jacobs (1977). More conventional tests for mix design are also empirical ones whose results correspond to one point on a creep curve. A popular test of this type is the Marshall test (ASTM D 1559, BS 598: Part 3) which, at a constant rate of compression, measures a breaking load termed 'stability' and a deformation at break termed 'flow'. A brief description of the BS 598 test procedure is given later in the chapter.

5.1.4 The requirements for aggregates in asphalt and coated macadam specifications

The structure of a modern highway pavement in flexible construction typically comprises three principal layers — sub-base; roadbase, which may be subdivided into a lower and upper roadbase; and surfacing, which may consist of a base course and a wearing course. When an upper roadbase is employed, the surfacing may be represented by the wearing course alone, as the functions of the base course are expected to be incorporated in the upper roadbase.

A capping layer of granular or stabilized material should be included in the design whenever the California bearing ratio (CBR) of the subgrade is likely to be less than 5%. Its purpose is to improve the effective strength of the subgrade economically so as to provide a platform for the compaction of the sub-base, but, notwithstanding that it effectively reduces the required thickness of the sub-base, it is included in the earthworks as a sub-formation material and not as part of the pavement structure.

Each of the layers is intended to perform specific functions in the pavement, and almost without exception their materials, manufacture, and placement are specified in varying detail in the project specification. The specifications for all major highway projects in England and Wales, and for many of the minor ones, have their roots in the Department of Transport (DTp) *Specification for Highway Works* (SHW) (1986) and the complementary *Notes for Guidance* (NG) (1986) which, in turn, refer extensively to the mandatory requirements and recommendations of appropriate British Standards, DTp Publications, and Transport and Road Research Laboratory reports. Nevertheless, it is neither obligatory nor always desirable for the engineer to adopt all the applicable clauses of the national specification for flexible pavement construction without giving due consideration to the specific requirements of his scheme and local sources of materials, such as aggregates, provided that any variation that is introduced is economically viable and that the engineering properties of the alternative materials can be assured. It is not unknown, in fact, for scheme specifications to expect more demanding standards than the national specification in response to local experience of the performance of materials customarily used in the region.

The primary objective of the pavement structure is to attenuate the repetitive or cyclic compressive stresses, generated at the surface by the traffic, to a value that can be imposed on the subgrade without causing significant deformation during the design life of the road. It therefore follows that the strength of the various layers of a flexible pavement may be reduced with depth. Similarly the specified properties of the aggregates used in the wearing course should be more exacting than those required in the lowest layer of the sub-base.

(i) Sub-base

In the UK, sub-bases are usually constructed using unbound aggregates such as Granular Sub-base Materials Type 1 and Type 2 which are described in the DTp *Specification for Highway Works* Part 3, Clauses 803 and 804 respectively. Cement-stabilized and lime-stabilized sub-bases are occasionally laid on moisture-susceptible sub-grades in periods of wet weather, and the process is also used to upgrade unbound materials that would not, otherwise, be acceptable for sub-bases. Granular sub-base materials are dealt with in Chapter 6, and, in view of this, they are not discussed here. It must be borne in mind, nevertheless, that the function of the sub-base is not only that of distributing relatively low-level in-service stresses over the subgrade, because, if so, aggregates of relatively poor quality would probably suffice. It has, in addition, to act as a levelling course, as a working platform for the haulage, laying, and compaction of the materials in the superimposed layers, and, if within 450 mm of the surface (in the UK), as a frost-resistant layer which also insulates the subgrade. Furthermore, for some site conditions it may be desirable for it to be permeable, particularly when there is no permeable capping layer, so as to allow water that might otherwise accumulate and weaken the pavement and subgrade to drain away. If it is to fulfil all these requirements the quality and grading of the sub-base aggregates must not be neglected, and appropriate minimum standards should be set and maintained.

In flexible construction, it is possible to replace part or all of the granular sub-base with rolled asphalt or dense bitumen macadam, the latter then being referred to as a full depth asphalt pavement. The advantages of full depth asphalt are quicker and thinner construction, stronger pavements, reduced maintenance, and less damage/disruption to statutory undertakers' plant during reconstruction (Pooley & Clark 1982). Bituminous mixtures are more expensive than unbound aggregates, but the financial savings inherent in these advantages, together with those in reduced excavation and haulage, can lead to a reduction in the overall cost of the pavement.

The savings in thickness can be substantial. Pooley & Clark (1982) using the Asphalt Institute of America method of design, substituted rolled asphalt for the granular layer in the equivalence ratio of between 0.2 and 0.3. At TRRL, Nunn & Leech (1986) carried out trials on test pavements having subgrade CBRs of 2 to 3% and 12%. The investigation demonstrated that even at a low subgrade CBR of 3% the bituminous bases, particularly rolled asphalt, were properly compacted. Below a CBR of 5%, however, it was concluded that without a capping layer special consideration would be necessary to ensure that the subgrade would support the paver and asphalt supply lorries without undue deformation. Tests were made to measure stress, strain, and deflection under a rolling wheel load and elastic theory used to adjust the measured stresses and strains for the effect of temperature, and hence modulus, and for thickness. Comparing the results for the conventional and full-depth asphalt pavements, the ratio of thickness of rolled asphalt equivalent to a given thickness of granular sub-base was estimated to be about 0.3, irrespective of the thickness of bituminous material over the range 100 to 250 mm (Powell 1987a).

Sec. 5.1] **Properties and specifications of bituminous materials** 151

Although this value could be used with caution, however, more authoritative designs would be established only by extensive road trials.

For partial asphalt substitution and full-depth asphalt the bituminous material is integrated into the roadbase, and the quality and properties of the aggregates must then, understandably, comply with the requirements of that specification.

(ii) Roadbase

The mixes used for roadbases in flexible construction are wet-mix macadam, cement-bound materials, dense coated macadam, and rolled asphalt. The use of dry-bound macadam for new road pavements carrying other than light traffic is rare in the UK, and there has been a shift from wet-mix macadam roadbases into bituminous roadbases since the publication of the reports on the investigation into the slippage of some rolled asphalt wearing courses, that occurred under traffic shortly after the roads had been constructed (Kennedy 1978, TRRL 1979). These reports attributed the slippage to a plane of incipient failure at or near the wearing course/base course interface, formed when compacting the wearing course. They concluded that the slip plane had occurred primarily because the roadbase had a low structural stiffness and the wearing course had been laid and compacted at a high, but not unduly high, temperature on a cold basecourse. The majority of the failed roads had roadbases of dry-bound or wet-mix macadam.

The gradings of the aggregates used for coated macadam dense roadbase are continuous, such as those of the 40 mm and 28 mm materials of Group 1 in BS 4987: Part 1: 1988. The calculated midpoints of these gradings are plotted in Fig. 5.1, and, apart from the requirement for ±2.5% and ±3.5% at the coarse and fine ends of the curves, the tolerances specified in the Standard range from ±12% to ±7%. The quantities of the various size fractions are well distributed throughout a continuously graded aggregate, and it is mainly this property, together with angularity, which gives the stability and good load-bearing capacity of these materials when they are well compacted.

The coarse aggregate, which is the material substantially retained on the 3.35 mm test sieve, may be one of the following:

- crushed rock from one or more of the relevant groups inferred to be in the BS 812: Part 1: 1975 classification (withdrawn), except schist;
- blast furnace slag;
- steel slag;
- gravel from the same groups as crushed rock;
- flint.

Blast furnace slag must comply with BS 1047: 1983, and steel slag must be weathered until it is no longer susceptible to falling and have a bulk density of between 1700 and 1900 kg/m^3. Particle shape is defined by the flakiness index which must not exceed the rather tolerant limits of 45% for crushed rock and crushed gravel or 50% for uncrushed gravel. The quantity of material finer than 75 μm in gravel must not exceed 1% by mass of the coarse aggregate when determined either by the sedimentation or the decantation method in BS 812. The current procedure of the sedimentation test, however, is designed to determine the quantity of minus 20 μm

152 Aggregates for bituminous materials [Ch. 5

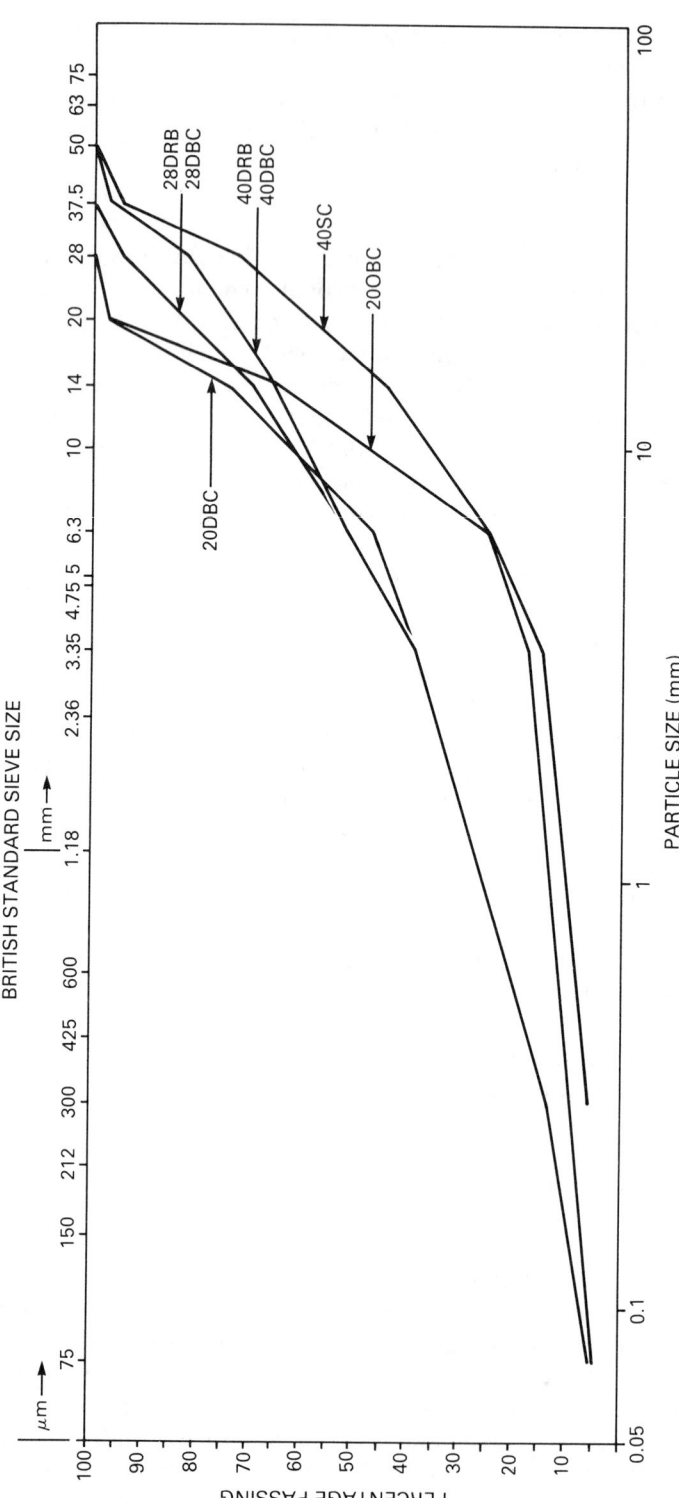

Fig. 5.1 — Midpoint gradings of BS 4987: Part 1: 1988 Coated macadam dense roadbase and basecourse mixtures.

material (medium and fine silt and clay sizes) rather than that of the minus 75 μm. Nevertheless, in the event of a dispute the sedimentation test has to be used, and a rather controversial note is added to the effect that the test gives some indication of the clay content.

The Standard points out that if an aggregate is prone to stripping then it will be beneficial to add an adhesion agent or 2% of hydrated lime or Portland cement to the mix. When the coarse aggregate is gravel, other than limestone, however, it gives no option, and 2% of hydrated lime or Portland cement, by mass of total aggregate, must be added. The DTp *Specification for Highway Works* (1986) Part 3 has a similar requirement, but the percentage is calculated on the total mix and the percentage of fine aggregate reduced accordingly. Moreover, with tar binders only Portland cement is allowed because, the *Notes for Guidance* point out, there is evidence to suggest that hydrated lime can cause hardening and brittleness.

The fine aggregate, which must substantially pass the 3.35 mm test sieve, may be fines crushed from a suitable coarse aggregate, sand, or a mixture of the two. Sand must not contain more than 8% by mass of the fine aggregate finer than 75 μm when determined — with the same provisos as for the coarse aggregate — by the sedimentation or decantation method. There is no limit prescribed for crushed rock or slag fines. If filler is added, then, subject to the special requirements for gravel coarse aggregate, it may be crushed rock, crushed slag, hydrated lime, Portland cement, or other approved material, and at least 75% must pass a 75 μm test sieve.

Various target binder contents are prescribed in the Standard according to the type and grade of binder (bitumen or tar) and the type of aggregate, but they may be adjusted by agreement if the evidence from trials or previous satisfactory performance with the particular aggregates and binder concerned justifies a change. Examples of the prescribed binder contents for 40 mm size dense roadbase are 3.5±0.6% using bitumen with crushed rock, and 4.5±0.6% using bitumen with gravel. Slightly higher percentages are specified for tar.

Dense roadbase macadams in the UK have conventionally used relatively soft 100 and 200 pen bitumens and a mean filler content (minus 75 μm) of about 5%. A study of French experience with grave-bitume (Hingley *et al.* 1976), however, followed by a collaborative laboratory investigation and site trials conducted by the Transport and Road Research Laboratory (TRRL), the British Aggregate Construction Materials Industries (BACMI), and the Refined Bitumen Association (RBA), established that the elastic stiffness was 1.7 times that of the conventional material when a 50 pen bitumen was substituted for a 100 pen bitumen, and 2.2 times the conventional value when, in addition, the filler content was raised from 5% to 8% (Leech 1982, Nunn *et al.* 1987). Neither of these modifications had a significant effect on the fatigue resistance of the material. The relationship between elastic stiffness, E, the penetration of the recovered binder, P, and the percentage filler content, F, was found to be:

$$\log E = 0.81 - 0.0066P + 0.039F$$

This is illustrated in the TRRL diagram reproduced in Fig. 5.2 (Powell 1987b).

The consequent improvement in load-spreading ability and increased resistance

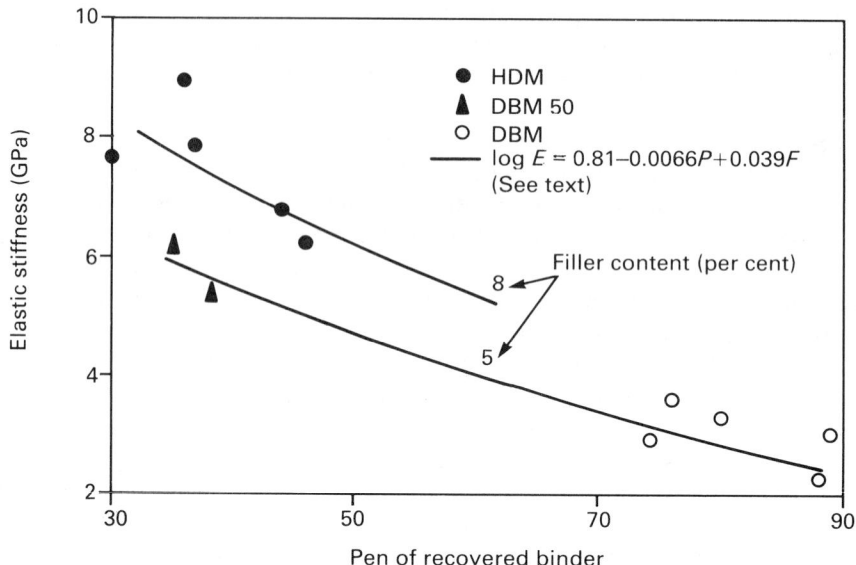

Fig 5.2 — Elastic stiffness of trial materials at 20°C and 5 Hz.

to deformation permit a reduction in the design thickness of a flexible pavement of 20 to 40 mm when using the bitumen-modified dense bitumen macadam DBM50, and 30 to 60 mm when using the bitumen- and filler-modified material heavy duty macadam or HDM, both over the range of 100 to 10 000 commercial vehicles per day, cv/d, at opening (Nutt 1987). These reductions are derived from the design charts in the DTp Departmental Standard HD 14/87 *Structural design of new road pavements*, amended to take account of the new materials. Cautiously, however, modular ratios of 1.5 and 2.0 were assumed for DBM50 and HDM respectively, instead of the 1.7 and 2.2 obtained as a result of the full-scale trials. Both HDM and DBM50 are now included in the DTp *Specification for Highway Works* Part 3 — roadbase in clauses 930 and 932, and basecourse in clauses 933 and 934. For HDM the filler content is specified to be in the range of 7 to 11% by mass of aggregate, and some restrictions are imposed for both materials on the use of gravel coarse aggregate.

Rolled asphalt is also used as a roadbase. Its composition is specified in Table 2 of BS 594: Part 1: 1985 together with that of basecourse and regulating course, all of which comprise the materials in Group 1. The mix designation of roadbase is 60/40 or 60/28, which signifies that the mixture contains 60% of coarse aggregate having a nominal size of 40 mm or 28 mm respectively. The size of the aggregate is related to the nominal thickness of the layer, 60/40 being laid at 75 to 150 mm and 60/28 at 60 to 120 mm. The calculated ratios of thickness to aggregate size for these values are 1.9 to 3.8 and 2.1 to 4.3 respectively, and it might be inferred that the permitted minimum nominal thicknesses of these mixes are rather low for achieving an entirely satisfactory compacted layer.

Sec. 5.1] **Properties and specifications of bituminous materials** 155

The combined aggregates in rolled asphalt roadbase are, or tend to be, gap-graded (Fig. 5.3). They may lack the intrinsic stability of a continuously graded aggregate, but their greater specific surface enables them to retain a higher bitumen content than dense bitumen macadam, which enhances the durability and fatigue resistance of the mix. Furthermore, satisfactory resistance to deformation is achieved by the normal use of a relatively hard grade bitumen of 50 penetration. As a consequence, and because the material can be compacted more easily than coated macadam roadbase on a granular sub-base, rolled asphalt roadbase with crushed rock or slag coarse aggregate is specified in DTp Departmental Standard HD 14/87 as the only material for the 125 mm thick lower roadbase in the construction of heavily trafficked new roads (3000 to 10 000 cv/d). It is also specified as an alternative roadbase material when no lower roadbase is required.

The specified requirements for the coarse aggregate of rolled asphalt roadbase are, with some exceptions, similar to those for coated macadam roadbase. The exceptions are that it is substantially retained on the 2.36 mm test sieve (instead of the 3.35 mm), and that the quantity of minus 75 μm material must not exceed 5% by mass of the coarse aggregate for crushed rock and slag and 3.5% by mass for gravel, when determined by the decantation method of BS 812: Part 1: 1975. There is no mandatory requirement for adding hydrated lime or Portland cement to gravel mixes, although in Part 1 Appendix A2 of the Standard attention is drawn to the beneficial effects of adding an adhesion agent or 1 to 2% of hydrated lime or Portland cement to all of the rolled asphalt mixes if the aggregate is prone to stripping. With regard to the DTp Specification (clause 904), the addition of 2% of Portland cement (not hydrated lime) by mass of total mix is mandatory for rolled asphalt roadbase when the coarse aggregate is gravel other than limestone gravel.

The requirements in the Standard for fine aggregate are also similar to those for coated macadam roadbase in BS 4987, except that it must substantially pass a 2.36 mm test sieve, and the quantity of minus 75 μm material must not exceed 22% by mass of the fine aggregate as determined by the decantation method.

The added filler, if required, is more closely controlled than in BS 4987. It is restricted to limestone or Portland cement; at least 85% of its mass must pass the 75 μm test sieve (75% is permissible, if approved), and its fineness is regulated by the requirement that its bulk density in toluene as determined by the method in BS 812: Part 2: 1975 shall be 0.5 to 0.9 g/ml inclusive, the finer fillers generally having the lower bulk densities.

(iii) Basecourse
Provision is made in BS 4987 for 40, 28, and 20 mm dense basecourse, 40 mm single course, and 20 mm open graded basecourse coated macadams. The DTp Specification also includes the HDM and DBM50 mixtures referred to above. The compacted thicknesses of these materials are in the range 45 to 140 mm, depending on nominal size. The midpoint aggregate gradings, except for those specifically for gravel mixtures, are plotted in Fig. 5.1, where it is possible to distinguish the different profiles of the dense and open graded materials. The requirements for aggregate type and quality for basecourse are the same as those specified for roadbase.

The bitumen contents for the crushed rock 40 mm and 28 mm coated macadam dense basecourse are significantly higher than those for roadbase with identical

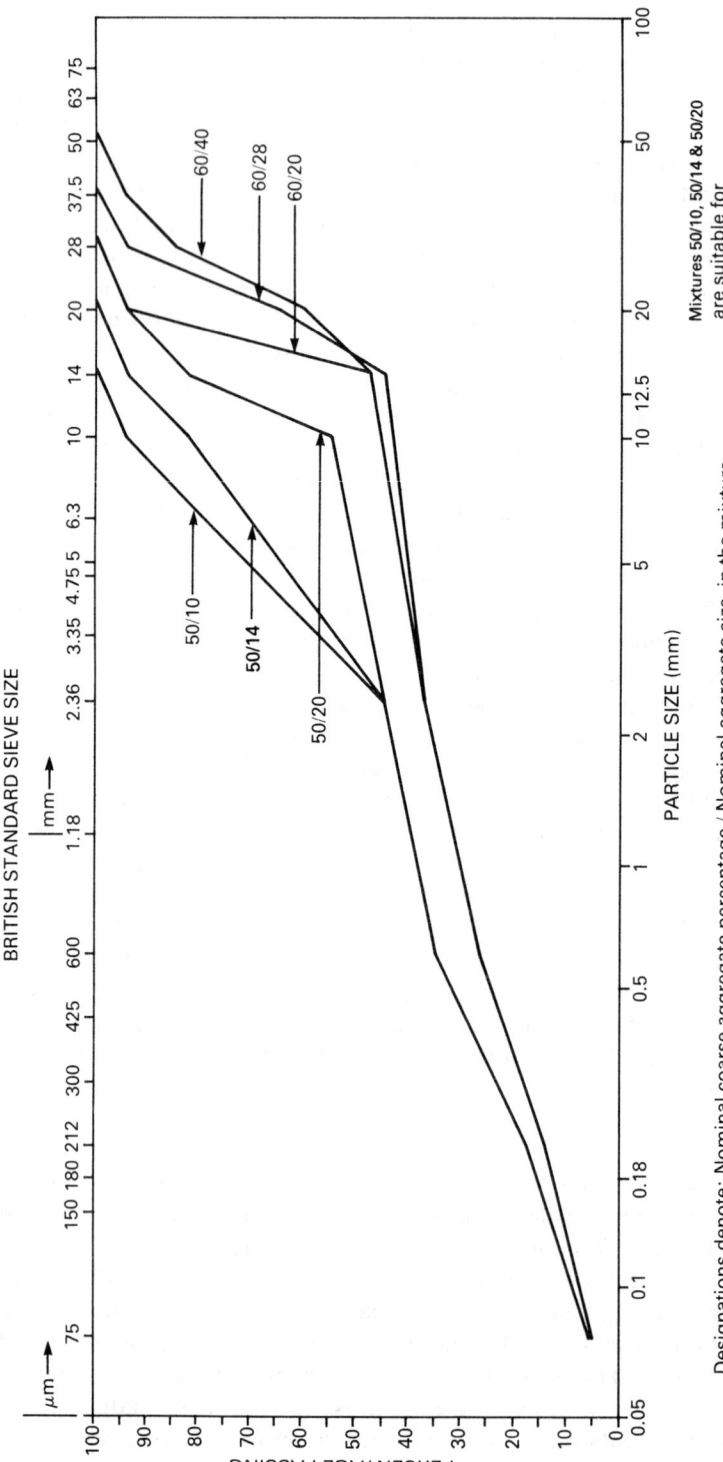

Fig. 5.3 — Midpoint gradings of BS 594: Part 1: 1985 Roadbase, basecourse, and regulating course mixtures.

gradings. Research at TRRL (Leech & Selves 1980) demonstrated marked improvement in the deformation resistance of 40 mm dense basecourse by reducing the binder content, and, taking other relevant factors into consideration, recommended a reduction of 1.0% to equal the 3.5% of roadbase. The recommendation is reflected in the DTp Departmental Standard HD 14/87 which specifies the use of dense basecourse only for traffic up to 3000 cv/d for new trunk roads and motorways; above this loading it is dispensed with and replaced by an upper roadbase of dense bitumen macadam (or HDM or DBM50).

There is no difference between the mix compositions of rolled asphalt basecourse and roadbase in Group 1 of BS 594: Part 1. The selection of the mix is guided by its suitability for the nominal thickness of the layer. Basecourse is generally laid 60 mm thick, so the appropriate mix is 60/20, which has 60% of 20 mm nominal size coarse aggregate and is suitable for laying to nominal thickness of 45 to 80 mm. The midpoints of the aggregate gradings for this material are shown in Fig. 5.3. Generous tolerances are permitted for the individual gradings of the coarse and fine aggregate, and these are reflected in the combined grading envelope given in the Standard, which allows ±17.5% passing the 14 mm and ±17% passing the 600 μm test sieves. Grading control is exercised predominantly at the 2.36 mm and 75 μm test sieve sizes, where the tolerances are ±7% and ±3% respectively.

The requirements for aggregate type and quality for rolled asphalt basecourse are similar to those for roadbase, and while it is normal to use 50 pen bitumen, 30 or 70 pen may be used in certain situations.

(iv) Wearing course

Coated macadam wearing course mixtures are specified in Groups 3 and 4 of BS 4987. The Group 3 wearing course mixtures are 14 mm and 10 mm open graded, 14 mm and 10 mm close graded, 6 mm medium graded, 6 mm dense, and 3 mm fine graded. Group 4 consists of two recently introduced pervious wearing course mixtures of 20 mm and 10 mm size. These have been designed with a view to reducing the incidence of wheel spray and skidding of fast traffic in wet weather by providing a permeable yet durable surface layer that absorbs the water in its porous structure and allows it to drain to the side of the carriageway (Daines 1986, Farrington & Roberts 1986). The midpoints of the gradings of these nine mixtures are shown in Fig. 5.4.

The specified minimum nominal layer thicknesses for these materials range from 15 to 45 mm which, apart from 2.3 for pervious wearing course and 5.0 for fine graded wearing course, give a minimum ratio of thickness to size of around 3. This is greater than the minimum ratio for basecourse and roadbase, and reflects not only the worth of having an adequate thickness to withstand traffic stresses and to promote durability, but also the need to ensure that these relatively thin layers of wearing course are well compacted. Increased thickness helps to conserve temperature and thereby extends the time available for compaction which should improve density, increase resistance to deformation, and minimize fretting.

Crushed rock (except schist) and slag may be used for coarse aggregate in all mixes except pervious wearing courses, for which limestone and blast furnace slag are not permitted. Limestone is excluded because of its low resistance to polishing. Blast furnace slags are excluded principally because their inadequate resistance to crushing during rolling reduces the voids and consequently the permeability of the

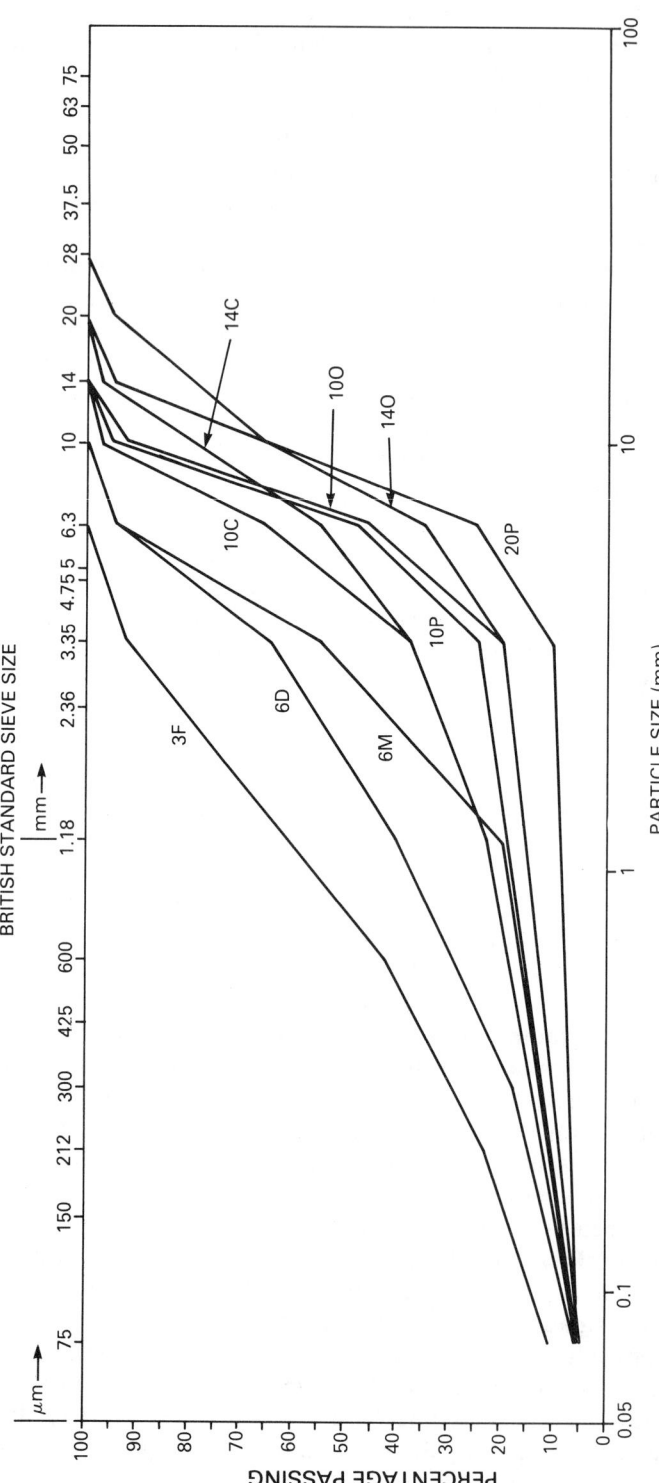

Fig. 5.4 — Midpoint gradings of BS 4987: Part 1: 1988 Coated macadam wearing course mixtures.

Sec. 5.1] **Properties and specifications of bituminous materials** 159

layer; variability of porosity and resistance to abrasion are other contributory factors. Gravel aggregates are permitted only in the close graded and medium graded mixes.

The requirements for aggregate type and quality specified in the Standard for wearing course are similar to those of coated macadam roadbase and basecourse, except that for pervious wearing course the flakiness index of the coarse aggregate must not exceed 25%, and 2% of hydrated lime by mass of aggregate has to be added to the mix. The aim of adding hydrated lime is to achieve the best combination of durability and resistance to deformation and stripping, using the specified 100 or 200 pen bitumen.

The Standard does not include requirements for aggregate properties such as strength, soundness, resistance to polishing or abrasion, etc. Nevertheless, reference is made in Appendix B5 to the need for minimum polished stone values and maximum abrasion values for wearing course aggregates in some situations. If required, they should be indicated by the purchaser, guidance being available in DTp Technical Memorandum H 16/76 (1976).

The properties of the 20 mm coarse aggregate specified by TRRL for the pervious wearing course trials on the A38 Burton by-pass, Staffordshire (Daines 1986) were a minimum polished stone value (PSV) of 60, a maximum aggregate crushing value (ACV) of 15, a maximum aggregate abrasion value (AAV) of 12, and a flakiness index not exceeding 20.

For friction course, which performs an approximately similar function to pervious wearing course on aircraft runways, the Airfields Branch of PSA (1979) specifies maximum values of ACV of 16 (standard test), flakiness index of 25, water absorption of 1.5%, and magnesium sulphate soundness loss of 18%, for the 10 mm coarse aggregate. In addition the aggregate has to comply with the requirements of a static immersion stripping test. The PSA Specification does not, however, restrict such requirements to friction course; the coarse aggregates used in Marshall asphalt, Marshall DTS (dense tar surfacing), rolled asphalt wearing course, and coated macadam basecourse all have limits prescribed for the results of these tests, and those specified in the 1979 edition are shown in Table 5.1. Fine aggregates are also limited to a water absorption of 2% and a magnesium sulphate soundness loss of 18%. The Specification is being revised, and variations to the magnesium sulphate soundness loss for coarse and fine aggregates have been issued for recent projects. In addition to the prescribed test limits there are also qualitative requirements for the aggregates which must not be overlooked.

Hot rolled asphalt wearing course was introduced at the beginning of this century as a blend of bitumen and natural, finely-graded sand that could readily be laid by hand and then compacted to a dense, impermeable layer. Subsequently a coarse aggregate was added to the mixture to improve resistance to deformation and wear.

There are two groups of rolled asphalt wearing course mixtures in BS 594 that differ according to the method of specifying the binder content. For Group 2 mixtures the binder content is determined for a given aggregate combination by the Marshall method of design described in BS 598: Part 3; for Group 3 mixtures the binder content is specified, in the traditional manner, as a component of the tabulated mix recipe. There are two types of mixture in each Group, characterized by the grading of the fine aggregate. The fine aggregate Type F has a fine grading typical

Table 5.1 — Test properties of coarse aggregates required by PSA Standard Specification Clauses for Airfield Pavement Works 1979. Part 4 Bituminous Surfacing

1	2	3	4	5	6	7	8	9	10
Surfacing material		Flakiness index % max.	Aggregate crushing value % max	Absorption % max	Magnesium sulphate soundness % max	Stability	Sulphur content % max	Bulk density kg/m^3 min	Stripping test
Marshall asphalt	R	30	30	2	18				6/150
	G	30	25	2	18				6/150
	S								
Marshall DTS	R	30	30	2	18				6/150
	G	30	25	2	18				6/150
	S								
Rolled asphalt wearing course	R	30	30	2	18				6/150
	G	30	25	2	18				6/150
	S	30	30	4	18				6/150
Coated macadam base course	R	30	30	2	—	A	2	1120	6/150
	G	30	25	2	—				6/150
	S	30	30	4	18	A	2	1120	6/150
Dense tar surfacing	R	30	30	2	18				6/150
	G	30	25	2	18				6/150
	S	30	30	4	18	A	2	1120	6/150
Friction course	R	25	16	1.5	18				3/150
	G								
	S								

Notes:
Column
2 R=Crushed rock; G=Gravel; S=Slag: all as described in Clause 1.4.2.1 of BS 594: 1973, Clause 1.4.2.2 of BS 4987: 1973 and Clause 4.2.3 of BS 5273: 1975, except that for friction course crushed rock is from the following groups only: basalt, gabbro, granite, hornfels, and porphyry. Gravel, when used in Marshall asphalt and Marshall DTS, shall have at least 75% by mass of particles with two or more crushed faces.
3 Flakiness index, in accordance with BS 812: Part 1: 1975.
4 Aggregate crushing value, in accordance with BS 812: Part 3: 1975.
5 Water absorption, in accordance with BS 812: Part 2: 1975.
6 Magnesium sulphate soundness, in accordance with Appendix 4B of PSA Specification Part 4, 1979. The overall percentage loss is based on the mass losses of the various nominal sizes tested from any one supply source.
7 Stability. A=stability to BS 1047: 1974, Appendix A.
8 Sulphur content in accordance with BS 1047: 1974.
9 Bulk density tested on 20 mm size sample complying with Table 1 of BS 63: Part 1: 1971.
10 Stripping test in accordance with Appendix 4A of PSA Specification Part 4, 1979. 6/150=not more than 6 out of 150 particles shall show any stripping.

of the fine sands used in the original, or traditional, rolled asphalt wearing course, while that of Type C has a coarser grading representative of that obtained from crushed rock or slag fines. It is not intended, however, that either type of mix should be exclusive to any of the specified fine aggregates: the designations F and C simply signify that the fine aggregate gradings are comparatively fine and coarse respectively. The midpoints of the aggregate grading of Type F mixes are shown in Fig. 5.5 and those of Type C in Fig. 5.6.

The fine aggregate grading of Type F mixtures is indicated by the mid-point grading of Mix 0/3 in Fig. 5.5 in combination with about 12% of added filler, assuming that the sand has 4.0% passing a 75 μm test sieve. It will be seen that the material substantially passes a 600 μm test sieve, so that when it is blended with a quantity of virtually single size 10, 14, or 20 mm coarse aggregate the combined grading has a conspicuous shortage of the intermediate sized particles. This gap-grading is a typical feature of traditional rolled asphalt mixtures. To prevent the use of too coarse a sand, the Standard stipulates additionally for each Type F mix containing coarse aggregate a maximum percentage for the 2.36 mm to 600 μm fraction that ranges from 18% for Mix 15/10 down to 9% for Mix 55/20.

The type and grading of the fine aggregate and filler (as well as the grade and quantity of bitumen) have a dominant influence on the service performance of rolled asphalt wearing course. Its strength and resistance to deformation can vary significantly with different sands having similar gradings, the variations being a function of particle shape, surface texture, and mineral composition. Grading differences, particularly on the 212 μm test sieve, can also cause variations in performance. Selection of the right sand for heavily trafficked roads is, therefore, a matter of prime importance.

Crushed rock and slag fines generally require a greater compactive effort than sand fines because of their rough surface texture and their angular and elongated shape, but they produce a more stable asphalt. Although their use was permitted, however, they were effectively excluded from the wearing course of BS 594: 1973 because, without possibly blending with sand, hardly any of them could comply with the specified fine aggregate stockpile grading. There were also doubts about the durability and resistance to wear of the wearing courses made with them, which may have arisen because of their inadequate compaction, inappropriate binder and filler contents, or, possibly the use of unsound aggregate. Nevertheless, as a result of evidence of their successful use in full-scale trials and routine surfacing, and the ongoing depletion and uneven distribution of sands suitable for high stability asphalt, a provisional recipe mix specification for a rolled asphalt wearing course having crushed rock fine aggregate was introduced as Clause 2801 in the DTp *Specification for Road and Bridge Works* 1976, Supplement No. 1, 1978. Subsequently, a provisional design mix specification (using the then current procedure in BS 594 of carrying out the Marshall test on the 'sand'/filler/binder mortar) was issued in Departmental Standard HD 2/79 as Clause 2809, the grading of the crushed rock fines being identical to that in Clause 2801. Table 5.2 shows this grading and, for comparison, the grading of the BS 594: 1973 fine aggregate. The former is notably coarser, and it is allowed a greater quantity passing 75 μm.

A fine aggregate grading compatible with crushed rock and slag fines was used to generate the range of total aggregate gradings (by wet sieving) of Type C mixes, the midpoints of which are plotted in Fig. 5.6. Mix 0/3 contains no coarse aggregate, so its

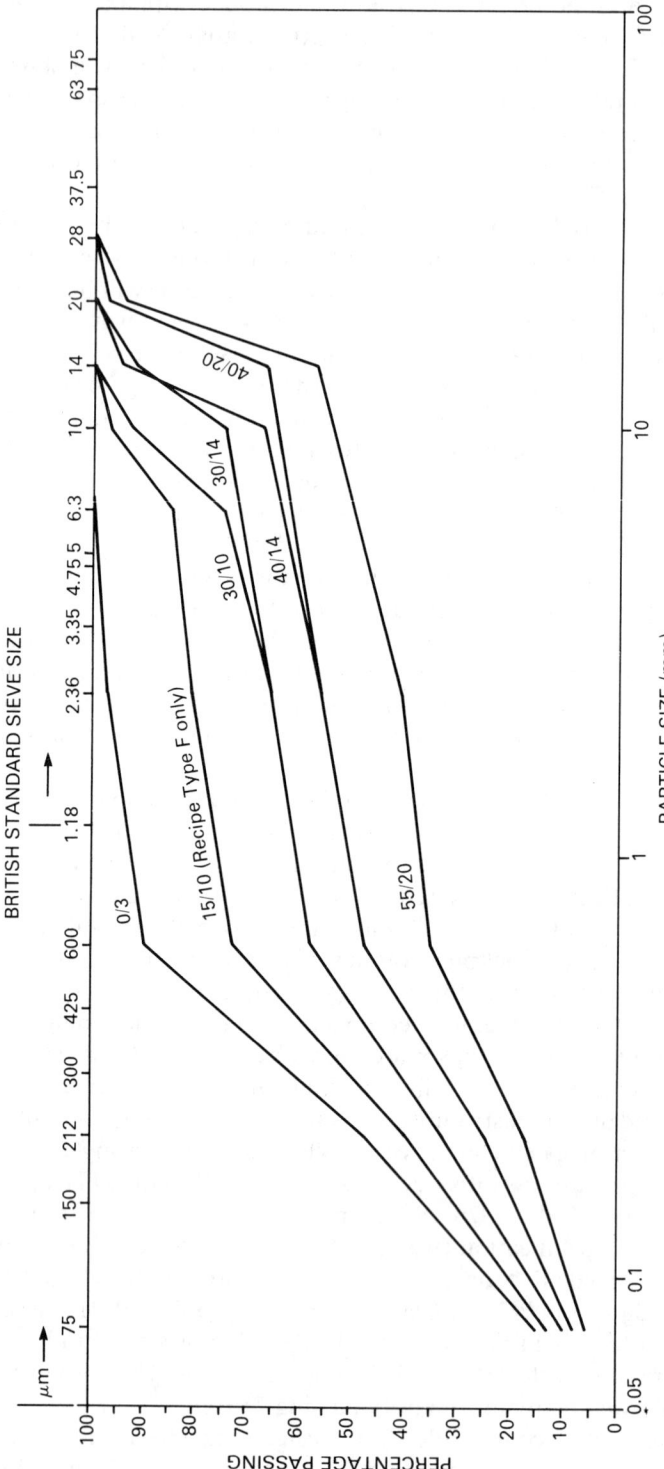

Fig. 5.5 — Midpoint gradings of BS 594: Part 1: 1985 Design type F and recipe type F wearing course mixtures.

Sec. 5.1]   **Properties and specifications of bituminous materials**   163

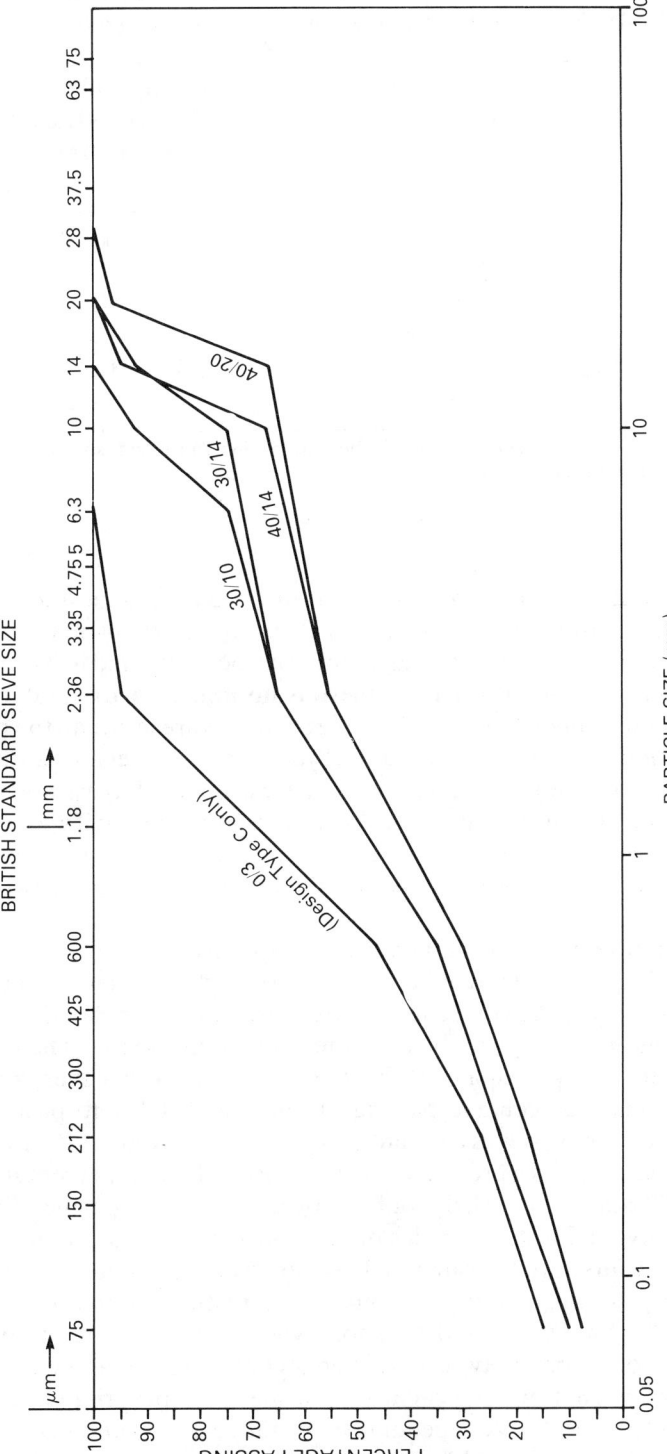

Fig. 5.6 — Midpoint gradings of BS 594: Part 1: 1985 Design type C and recipe type C wearing course mixtures.

Table 5.2 — Rolled asphalt wearing course. Gradings of fine aggregates by dry sieving in BS 594: 1973 and DTp *Specification for Road and Bridge Works* 1976

BS test sieve	BS 594: 1973 Fine aggregate % by mass passing	DTp Specification 1976 Clauses 2801 and 2809 Crushed rock fines % by mass passing
6.3 mm	—	100
5.0	100	—
2.36	95–100	80–100
600 μm	75–100	30–60
212	15–60	12–30
75	0–5	0–15

Note: The current editions of these specifications, published in 1985 and 1986 respectively, do not now specify separate gradings for fine aggregate.

midpoint grading would be the same as that of the fine aggregate if no filler were added. For the 30% and 40% coarse aggregate mixes, the midpoint quantity of minus 75 μm material required in Type C fine aggregate may be calculated to be 15 and 14% respectively by mass of fine aggregate if there is no minus 75 μm material in the coarse aggregate and no filler is added. These percentages are less than the maximum of 17% (decantation method) permitted in Type C fine aggregate, so if there is sufficient of this size available in the aggregate there is no need to use added filler to achieve the specified gradings of the mixes. This does not apply to Type F mixes, as the fine aggregate must not contain more than 8% of minus 75 μm material, and this is insufficient for it to comply with the specified aggregate grading without the addition of filler.

The aggregate gradings of Type C mixes are gap-graded but not to the same extent as those of Type F. Both are influenced by the single-size nature of the coarse aggregate, but for Type C the coarseness of the crushed rock or slag fines normally ensures that there is no shortage of the 2.36 mm to 600 μm fraction. The gradings of both types are controlled principally at the 2.36 and 75 μm test sieves, which have tolerances on the mass percentage passing of ±6% and ±2% respectively. The tolerances on the remaining sieves are much wider, ranging from ±11% to ±22.5%, but, as stated earlier, there is a proviso for Type F mixes that the percentage mass of the 2.36 mm to 600 μm fraction shall not be allowed to exceed specified limits.

The type, quality, and test properties of the coarse and fine aggregate and filler for BS 594 wearing course are the same as those for roadbase and basecourse except that, for fine aggregate, the quantity of minus 75 μm material found by decantation must not exceed 8% for Type F and 17% for Type C mixes. No requirements for PSV, AAV, or any other property such as strength or soundness are included in the Standard but it is expected that the client should specify appropriate test limits for coarse aggregate PSV and AAV, depending on traffic and site conditions, guidance being available in DTp Technical Memorandum H 16/76 (1976). Such limits would be applicable to the coarse aggregate of unchipped wearing course or the coated chippings of chipped wearing courses. In addition, the DTp Specification requires that coarse aggregate in design mix, chipped wearing course, which are laid on trunk roads and motorways having a minimum traffic loading of 100 cv/d (DTp HD 14/87) shall have a minimum PSV of 45. This minimum PSV is specified to ensure that there is no undue loss of skid resistance when the coarse aggregate becomes exposed, notably in areas which are not heavily chipped or where there has been wear or chipping embedment. The need for this intended safeguard is being studied by TRRL in a full-scale road trial of rolled asphalt wearing courses laid in 1983 on the A30 Staines by-pass, Surrey (Jacobs 1985). Five different coarse aggregates having PSVs ranging from 36 to 59 were used, and, for each aggregate, mixtures were made with a range of binder contents. After laying, 20 mm coated chippings with a PSV of 60 were spread at either the standard or a lighter rate and then rolled in. Visual assessments and regular measurements of sideway-force coefficient and texture depth are being made, and the final analysis should provide substantial evidence for evaluating the current 45 PSV limit.

The binder for rolled asphalt wearing course may be selected from a range of nine different types and grades all complying with either BS 3690: Part 1 or 3. They are bitumen of 35, 50, 70 and 100 penetration, lake asphalt-bitumen of 35, 50, or 70 penetration, and pitch-bitumen of 50 or 100 penetration. 40 penetration heavy duty (HD) bitumen is also permitted, with reservations.

In general terms, the harder grades are used for heavy traffic and warmer climates. In the UK, 50 pen is typically used, but 35 pen would be better at bus stops and probably 70 or even 100 pen in cold, wet climates or where the traffic is less dense.

Poor compaction and consequent permeability will increase the exposure of the bitumen to oxidation, and reduce the life of a wearing course. This may be instigated by the use of crushed rock or slag fine aggregates because mixtures made with them are more difficult to compact. Consideration should therefore be given to using a softer grade such as 100 pen, and to increasing the design bitumen content by 0.5% or more, particularly when laying these materials in cold weather.

The binder content of the Group 2 design mixtures (the 'target' binder content) is specified by the method of design ('Marshall method') described in BS 598: Part 3: 1985, while that of the Group 3 recipe mixtures is specified by reference to the recipe mix tables in the Standard — Table 5 for Type F mixtures and Table 6 for Type C mixtures.

The design method should be used for the wearing courses of heavily trafficked roads and is specified by the Departmental Standard HD 14/87 for all new rolled asphalt wearing courses on trunk roads and motorways with a minimum traffic loading of 100 cv/d, the coarse aggregate being restricted to crushed rock or slag.

The following is a brief outline of the procedure for determining the design binder content and the Marshall test data of an asphalt mixture using the method specified in BS 598: Part 3: 1985.

Representative samples of the coarse aggregate, fine aggregate, and filler are graded and combined, as prescribed, to make at least 27 batches, each weighing about 1100 grams. The proportions of the combination are defined by the mid-point mass percentages of the total aggregate grading of the required BS 594 mixture that are retained on or pass the 2.36 mm and 75 μm test sieves.

For the wearing course mix 30/14, for example, the proportions are:

Test sieves	% by mass
+2.36 mm	34
−2.36 mm	56
−75 μm	10
	100

Each batch of aggregate is heated, mixed with the required hot bitumen, and compacted in a mould by a free fall hammer, strictly in accordance with the specified equipment, temperature, and procedure. The specimen of asphalt is cylindrical, having the approximate dimensions of 100 mm diameter by 65 mm high. At least nine binder contents are used, spaced at intervals of 0.5% by mass and distributed above and below the expected optimum for the mix. Three specimens are made for each binder content.

The density of the mix and the compacted aggregate are determined for each specimen and averaged for each binder content.

Each specimen is then heated to 60°C and tested diametrically in compression at a constant rate of 50 mm/min, using a radiused testing head. The maximum load and deformation ('flow') immediately before failure are recorded. The maximum load is corrected, if necessary, for volume variations of the specimen, and is then termed 'stability'. The values of flow and stability are averaged for each binder content.

Graphs of mean stability, mix density, aggregate density, and flow are plotted against binder content.

The design binder content is calculated as the mean of the binder contents for maximum stability, maximum mix density, and maximum aggregate density, with the addition of a constant. This constant is 0.7 for the 30 and 40% coarse aggregate mixes and zero for the sand asphalt and 55% mixes. It is applied for the purpose of improving workability and durability for mixes of the type specified in BS 594.

The design binder content is compared with the minimum target binder content specified for the mix in Table 3 or Table 4 of BS 594: Part 1, and, subject to compliance, it is then usually adopted as the target binder content for the tested mixture. It is permitted, however, to add extra binder to the design value if further improvement in workability or durability is needed, before finalizing the target binder content.

The stability and flow for the target binder content are read off the design graphs and compared with the criteria given in Table 11 of BS 594: Part 1, reproduced here as Table 5.3, or with alternative criteria that may be specified in the contract. If satisfactory compliance is not obtained, then the mix has to be modified either by varying the grading or changing the constituents — usually the fine aggregate — and the design repeated.

The Marshall method has been used in the UK for the design of continuously graded 'Marshall asphalt' wearing course and basecourse mixtures for heavy duty

Table 5.3 — Criteria for the stability of laboratory design asphalt. (Reproduced from Table 11 of BS 594: Part 1: 1985)

Traffic (Commercial vehicles per lane per day)	Marshall stability of complete mix kN
Less than 1500	2 to 8[†]
1500 to 6000	4 to 8
Over 6000	6 to 10

[†]It may be necessary to restrict the upper limit where difficulties in the compaction of materials might occur.

Note 1: For stabilities up to 8.0 kN the maximum flow value should be 5 mm. For stabilities in excess of 8.0 kN a maximum flow of 7 mm is permissible.

Note 2: The stability values referred to in this table should be obtained on laboratory mixes.

aircraft pavements for more than thirty years. The procedure specified in the PSA Airfields Branch Specification Part 4 differs in detail from the more recent BS 598 method outlined above, and the test criteria are different, but the principle of determining the optimum binder content (without, however, the addition of any constants) for a combination of aggregates falling within a specified grading envelope remains the same.

Asphalt wearing course mixtures are laid to nominal thicknesses of between 25 and 50 mm, and, as shown in Table 5.4, the nominal size of the coarse aggregate is

Table 5.4 — Wearing course mixtures and nominal thickness of layer specified in BS 594: Part 1: 1985

Group 2 Design		Group 3 Recipe		Mixture designation[†]	Nominal thickness of layer mm	Ratio of nominal layer thickness to nominal aggregate size
Type F	Type C	Type F	Type C			
F	C	F	—	0/3	25	8.3
—	—	F	—	15/10	30	3.0
F	C	F	C	30/10	35	3.5
F	C	F	C	30/14	40	2.9
F	C	F	C	40/14	50	3.6
F	C	F	C	40/20	50	2.5
F	—	F	—	55/20	50	2.5

[†]The mixture designations signify the nominal coarse aggregate content of the mixture/nominal size of the coarse aggregate.

varied accordingly. The standard thickness of wearing course on major roads is 40 mm, consequently the most commonly used mix is 30/14. The thickness to size ratios of the coarse aggregate mixes, except those of 40/20 and 55/20, range from 2.9 to 3.6 which, in conjunction with the relatively high contents of fine aggregate, filler, and bitumen, contribute to their high level of impermeability and durability without causing problems with surface level tolerances, which may occur during compaction when the ratio is too large.

(v) Coated chippings

Coated chippings are applied to the surface of the hot asphalt by machine before rolling to provide a rough surface on rolled asphalt wearing courses having coarse aggregate contents of 40% or less. They are either 20 mm or 14 mm nominal size with a maximum flakiness index of 25. BS 594: Part 1 specifies no mechanical test properties for them, but the purchaser should state his requirements for PSV and AAV appropriate to the site conditions and traffic loading, guided by DTp Technical Memorandum 16/76 (1976). The grading must comply with Table 10 in the Standard, reproduced here as Table 5.5.

Table 5.5 — Gradings of chippings. (Reproduced from Table 10 of BS 594: Part 1: 1985)

BS test sieve	Percentage by mass passing BS test sieve	
	Nominal size of aggregate (mm)	
	20	14
28 mm	100	
20 mm	90 to 100	100
14 mm	0 to 25	90 to 100
10 mm	0 to 4	0 to 25
6.3 mm	—	0 to 4
75 μm	0 to 2	0 to 2

Note: Where the minimum texture depth recommended in sub-clause 6.3 of BS 594: Part 2: 1985 applies, not less than 75% by mass of the chippings shall be of the specified size.

Where a minimum texture depth is required on the asphalt surfacing complying with that recommended in BS 594: Part 2: 1985, then not less than 75% of the coated chippings must be of the specified size which is 20–14 mm for the nominal 20 mm size, and 14–10 mm for the 14 mm nominal size. The type, quality, and test properties of the aggregate are the same as those for wearing course coarse aggregate except for the special requirements given above.

The grading of coated chippings is more closely single-size than that required for surface dressing chippings in BS 63: Part 2: 1987 (see Table 5.6), but, whereas the quantity of minus 75 μm material must not exceed 2% by 'wet' sieving, that for surface dressing is only 1%, determined by the washing and sieving method in BS 812: Section 103.1. As the minus 75 μm material in some basic igneous rock aggregates may contain deleterious swelling clay which will cause loss of adhesion between the binder film and the chipping, it is advisable to ensure that the quantity and quality of this fraction in coated chippings are kept under closer control than might be inferred from the Standard. Chipping retention, fracturing during rolling, and rate of spread are affected by particle shape; for most types of aggregate a flakiness index somewhat less than the specified maximum of 25% will give improved performance. Other properties which have a bearing on coated chipping placement and performance are elongation index and resistance to crushing, and, although no limits are specified in the Standard, the elongation index should be less than 25 to 30% and the 10% fines value in excess of 160 kN.

The binder type may be any of those permitted for the wearing course asphalt, although refinery bitumen is almost invariably used. The grade, however, must be 50 pen irrespective of the grade used in the wearing course. This harder grade assists the free flow of coated chippings during spreading, particularly in hot weather. The amount of binder specified is 1.5±0.3%. This target may be increased to improve retention where surface texture is specified, provided, of course, that the coated chippings still flow freely and do not form clumps. Chipping retention may be impaired if the binder is overheated and unduly hardened during mixing. This is assessed by the Hot Sand Test described in BS 598: Part 3: 1985, by which a weighed sample of dry coated chippings is immersed for at least 10 minutes in single-sized (300–150 μm) hot sand at a temperature of 100–125°C. After cooling and abrading with silica sand (1.18 mm–600 μm) in the prescribed manner, the chippings are washed and dried and the mass of adhering sand and the mass of chippings having less than half sand cover are determined. For acceptance, the proportion of adhering sand must not be less than 40 g/kg for 20 mm and 50 g/kg for 14 mm chippings, and the mass of chippings with less than half sand cover must not be more than 7.5%

5.2 ORIGIN AND CLASSIFICATION OF AGGREGATES

5.2.1 Introduction

The mode of formation, composition, and history of a rock have a bearing on its performance as a road making aggregate. This is not to say that the performance of every type of rock is necessarily unique. It is the case, however, that some types of rock are acceptable as an aggregate in every layer of bituminous pavement construction; some of them can be used in the lower layers but not in the wearing course or running surface, and others are not suitable for use at any level. Schist, for example, is avoided primarily because it foliates when fractured, whereas gneiss, which is a coarser-grained form of banded metamorphic rock, is widely used. Limestone, typically, is not suitable in the running surface of heavily trafficked high speed roads because of its inadequate resistance to polishing and abrasion, but Carboniferous Limestone and the older limestones have a very satisfactory record of performance in the lower courses.

Table 5.6 — Gradings of surface dressing aggregates. (Reproduced from part of Table 2 of BS 63: Part 2: 1987)

BS test sieve	Percentage by mass passing BS test sieve			
	Nominal size of aggregate (mm)			
	20	14	10	6
28 mm	100	—	—	—
20 mm	85–100	100	—	—
14 mm	0–35	85–100	100	—
10 mm	0–7	0–35	85–100	100
6.3 mm	—	0–7	0–35	85–100
5 mm	—	—	0–10	—
3.35 mm	—	—	—	0–35
2.36 mm	0–2	0–2	0–2	0–10
600 μm	—	—	—	0–2
75 μm	0–1	0–1	0–1	0–1

Note: For all the gradings, not less than 65% by mass of the aggregate shall be of the specified size.

Flint gravels and some coarse-grained, acid, igneous rocks have poor adhesion to bitumen, and most mixes made with them will therefore benefit from the addition of Portland cement, hydrated lime, or a proprietary adhesion agent. This is not the case, however, for basic igneous rocks, such as basalt and dolerite, as they bond well to bitumen.

In consequence of the above, the classification of a rock, or rocks, from which an aggregate is derived provides useful data when assessing the serviceability of an aggregate in a given layer of bituminous pavement. This major section (5.2) begins with a brief account of the origins of the three principal geological rock groups, igneous, sedimentary, and metamorphic, and the methods of classifying the rocks in each group. This is followed by a description of the British Standard Group Classification of Aggregates. This was last published in BS 812: Part 1: 1975 but has been omitted from the latest edition of the Standard. It is still implied, nevertheless, by the descriptions of aggregates used in the latest revisions of BS 594 and BS 4987, as well as in other specifications for bituminous materials. For this reason alone there is justification in including a reference to it in this book, at least until it is superseded by an authoritative British or European alternative. The basis for such an alternative could be provided by CADAM, a system of classification and description of aggregate material proposed by the Geological Society of London, the essential features of which have been reproduced, with appreciation, from the comprehensive publication *Aggregates* (1985) edited by Collis & Fox.

5.2.2 Geological classification of rocks

Natural aggregates are obtained by the crushing and screening of quarried rock or by the excavation and screening of gravels and sands, which may be crushed or uncrushed.

Rocks are divided geologically into three large groups on the basis of their mode of formation:

Igneous rocks. Rocks that have solidified from hot viscous melts or magmas originating from beneath the earth's surface.
Sedimentary rocks. Rocks that have been formed at atmospheric temperatures by the consolidation or cementing together of deposited fragmentary material that has been eroded from pre-existing rocks, or by the accretion of organic material, or by chemical deposition.
Metamorphic rocks. Rocks formed by the mineralogical, chemical, and structural alteration of pre-existing rocks caused predominantly by the action of internal heat and/or stress arising from geological processes within the earth (endogenetic processes).

Igneous rocks are classified in terms of grain size or texture and mineral composition, which is directly related to the chemical composition. Grain size is a function of cooling history. As the magma is forced up into the earth's crust it may stop at a distance of, say, several kilometres below the surface where it solidifies at a slow rate which enables large crystals to form and produce a coarse-grained rock. Such rock formations are termed intrusive, and, because they have solidified at considerable depth, plutonic. Subsequently they may become exposed at the surface by erosion of the overlying strata, and, if they have a greater resistance to weathering than the surrounding country rock, will form features of higher ground, like, for example, the granite masses or bosses that lie between Dartmoor and Land's End in southwest Britain.

As the magma is forced closer to the earth's surface and is eventually extruded through weaknesses such as cracks and fissures or by volcanic activity, the rate of cooling and solidification is increased. Consequently the crystals are small and often visible only with a microscope; the rock is said to be microcrystalline. If the cooling is very rapid then the rock may be cryptocrystalline or even glassy, and also vesicular as a result of the formation of cavities by the release and expansion of gases under reduced pressure. Such rocks are exemplified by the basaltic lavas of Northern Ireland, Auvergne in central France, Iceland, the Deccan Plateau in India, etc.

Between the two extremes of the major intrusive and extrusive igneous rocks are the minor intrusive or hypabyssal rocks typified by those found in dykes or sills. These wall- or sheet-like structures are formed when the magma intrudes into vertical lines of weakness or more or less horizontal bedding planes respectively, and may vary from a few centimetres to hundreds of metres or more in width or thickness. Because of contact with the country rock, the edges of dykes and sills are fine to very fine grained, but the texture may change to medium or even coarse grained towards the centre owing to the slower rate of cooling there. There are many examples of dykes and sills in the British Isles, but the Cleveland Dyke (average width about

20 metres) and the Whin Sill (average thickness about 30 metres) are probably the largest and best known. Both are composed of rock in the dolerite group. Similar formations occur in geological domains all round the world.

The chemistry and mineralogy of most igneous rocks are complex and variable, but analyses indicate that the range of composition can be expressed roughly in the following terms (Read 1947):

		%w/w
Silica	SiO_2	35–75
Alumina	Al_2O_3	0–25
Iron oxides	FeO & Fe_2O_3	0–20
Magnesia	MgO	0–45
Calcium oxide	CaO	0–20
Sodium oxide	Na_2O	0–16
Potassium oxide	K_2O	0–12

The principal component of the magma is silica, with which the remaining six oxides combine to form the largest and most important group of rock-forming minerals, the silicates. The silicates comprise, in order of increasing complexity of atomic structure and decreasing mineral specific gravity: olivine, pyroxene (e.g. augite), amphiboles (e.g. hornblende), the biotite and muscovite micas, orthoclase and plagioclase feldspars, and quartz. This order also approximates to the sequence of crystallization that occurs as the magma cools. The minerals thus formed reflect the chemical composition of the magma as the quantity of total silica present, which may vary between 35% and 75% by weight. Silica content, therefore, provides a practical indicator of the composition of igneous rocks, and those with high silica content are termed acid (silica being an acid-forming oxide) and those with a lower silica content and a correspondingly greater proportion of basic oxides, basic. The percentage weights of total silica that are customarily used to classify igneous rocks by this means are:

	Total silica, SiO_2 (% w/w)
Acid	>66
Intermediate	66–52
Basic	52–45
Ultrabasic	<45

Acid rocks are typified by their containing free quartz — usually 10% or more — as well as silicate minerals. There is no free quartz in basic and ultrabasic rocks, but, as the name implies, there may be a small percentage in some intermediate rocks.

Table 5.7 provides a classification of the most frequently occurring igneous rocks based on their mineral composition expressed in terms of their silica content, and their texture expressed in terms of their mode of occurrence. The boundaries, particularly of grain size, are not rigidly adhered to in nature as the rate of cooling is unlikely to be uniform throughout the mass. Composition may also vary, sometimes

Table 5.7 — Broad classification of igneous rocks by mode of occurrence and silica content (excluding alkaline rocks)

Mode of occurrence	Acid >66% silica	Intermediate 66–52% silica	Basic 52–45% silica	Ultrabasic <45% silica
Extrusive (lava flows)	Rhyolite	Andesite	Basalt	Olivine basalt
Hypabyssal (dykes, sills, and bosses)	Quartz-porphyry	Porphyrite	Dolerite	Olivine dolerite
Plutonic (batholiths)	Granite	Diorite	Gabbro	Peridotite

significantly. Hence it is important, when considering the use of a rock as a roadstone aggregate from a given quarry, that its classification is up-to-date, as such variations can have a bearing on its engineering properties.

Sedimentary rocks may be divided into two groups according to their mode of formation — clastic rocks, and those formed *in situ*.

Clastic rocks comprise the consolidated particles eroded from pre-existing rocks. They are classified in decreasing order of grain size as conglomerate, breccia, gritstone or sandstone, siltstone, and mudstone or shale. With few exceptions, only gritstones and sandstones are employed as roadstone aggregate, but even their engineering properties can be variable, and careful selection is necessary if they are to be used with confidence in the surfacing of a well-trafficked bituminous pavement. This is evident from the detailed investigation of British arenaceous rocks for skid-resistant surfacings undertaken by Hawkes & Hosking (1972). Their petrographical studies indicated that specific values of mechanical and physical properties of arenaceous rocks were less dependent on composition and grain size than the degree of consolidation, where consolidation was taken to include the geological processes of compaction, diagenetic crystallization (i.e. crystallization occurring at the Earth's surface under atmospheric temperature and pressure, not metamorphic), cementation and tectonic compression caused by crustal forces. Sedimentary rocks of the pre-Cambrian era, which were the oldest rocks studied, were found to possess the necessary 'cohesion' (good resistance to abrasion and impact) for a wearing course aggregate, but the younger sedimentary rocks from the Cambrian to Carboniferous periods (Palaeozoic era) did not, unless they had been subject to additional cementation or tectonic compression effects.

The survey also showed that there are gritstone aggregates which possess high resistance to polishing, abrasion, and impact, ranking them amongst the best in the country for use as coated chippings and coarse aggregate wearing courses for high speed roads. There are also many others which, while not having a high resistance to polishing, are sufficiently indurated to make very satisfactory aggregates in the lower layers of a pavement.

Of the sedimentary rocks formed *in situ*, only limestone and flint are used extensively for roadstone. The origins of the limestones are chemical or organic or a combination of the two. They are composed mainly of calcium carbonate ($CaCO_3$) in the form of calcite, organic remains, and fossils, but they may also contain some magnesium carbonate ($MgCO_3$) as magnesian limestone. Dolomite is composed entirely of the double carbonate $MgCO_3.CaCO_3$, while dolomitic limestone contains both of the minerals dolomite and calcite. Subsequent to deposition, recrystallization or dolomitization may occur. In the latter process the original calcite is replaced by dolomite, and, because this is accompanied by a contraction in volume of about 12%, the altered rock may be more porous. Limestone often contains impurities such as clay or mud, and grains of quartz.

Limestone of the Carboniferous period, such as that quarried in the Mendips and Derbyshire, is extensively worked for roadstone, as is the Magnesian Limestone of the Permian period in Nottinghamshire, Yorkshire, and County Durham. Although they are eminently suitable for basecourse and roadbase, however, their inadequate resistance to polishing and abrasion generally excludes their use for coated chippings and the coarse aggregate for unchipped wearing courses on high speed roads. Nevertheless, their use in combination with a highly polish-resistant aggregate is not unknown in the wearing course of less important roads. The younger limestones such as the Jurassic oolites and the chalk of the Cretaceous period are too soft for use as roadstone.

Flints are irregularly shaped nodules of cryptocrystalline silica formed in the Upper and Middle Chalk from solutions that may have been derived from the siliceous spicules and skeletons of such primitive marine animals as sponges and radiolaria. They occur in horizontal layers and vertical joints in the chalk, and, on erosion, are transported, abraded, and fractured to be redeposited as beds of sand and gravel in river terraces and beaches, which are widely distributed throughout the southeast of England.

Flint particles are hard and brittle. Their shape and surface texture are influenced by their mode of occurrence and subsequent history so that, while many rounded and smoothed gravel particles exist, sub-angular forms and particles with cortex on them are also common. They break with sharp arrises and a conchoidal fracture that may distinguish them from the more or less flat fracture in chert, a similar type of sedimentary rock formed, for instance, in parts of the Carboniferous Limestone. The properties of flint are not ideal for the coarse aggregate of stable, durable, bituminous mixes for heavily trafficked roads and airfields, although their resistance to abrasion is in the highest category. To improve the stability of the mix, flint gravels should be crushed and their poor adhesion to bituminous binders enhanced by incorporating hydrated lime, Portland cement, or a suitable proprietary adhesion agent.

Metamorphism is the sum of the thermodynamic processes originating beneath the earth's surface that transform pre-existing rocks into distinctive new types — metamorphic rocks. The rocks thus formed may be broadly grouped into contact metamorphic rocks, whose alteration has been caused by the action of intense heat emanating from large igneous intrusions while cooling, and regional metamorphic rocks, whose alteration has been caused by the combined action of powerful stresses and heat while being buried deeply, or otherwise compressed, in the earth's crust.

With the principal exceptions of quartzite and marble, the contact metamorphic rocks are commonly termed hornfels. In hornfels, the action of heat transforms the softer minerals of the country rock into harder minerals such as hornblende and feldspar, and the ensuing recrystallization usually results in an interlocking fine-grained texture. As a consequence, hornfels rocks are usually tough and hard, but their distribution is not widespread and they are rarely used for roadstone in the United Kingdom.

The process of recrystallization of sandstone yields contact-metamorphic quartzite which is tougher than the quartz sandstone from which it is formed, but such quartzites are found only very locally. The gravels of the Bunter Pebble Beds of the Triassic period in Staffordshire and Cheshire in England are an example, and they have a long history of use as roadstone.

The principal rocks in the regionally metamorphosed group are schist and gneiss. Both exhibit a banded texture that is related to the tendency for the minerals to orientate themselves with their long axes perpendicular to the direction of applied stress. Schists have finely spaced bands and are largely composed of platey, micaceous minerals and prismatic minerals such as hornblende, all of which combine to cause the rock to laminate and form thin flaky particles, thereby rendering them unsuitable for use as roadstone.

Although gneiss has a banded texture it is much coarser than that of schist, and it has a lower content of platey minerals. Its composition frequently approximates to that of granite; consequently it is much less prone to lamination and usually provides a source of good quality roadstone. In the UK schist and gneiss occur mostly in the metamorphosed Scottish Highlands and in Anglesey, but they are also to be found in Cornwall, Devon, and Worcestershire.

Granulites are metamorphic rocks that have a granular texture, but with no preferred orientation of the minerals, probably because platey minerals, other than some micas, are poorly represented, and the applied stress was insufficient to cause melting and recrystallization. Many granulites have compositions similar to granite, but when they comprise only granular quartz (quartz granulites) they are then described as quartzites.

5.2.3 Group classification of aggregates

Rarely if ever does the highway engineer need to be familiar with the extensive nomenclature used by petrographers to describe and classify the great variety of technically distinguishable rock types that may have the potential for use as road making aggregates. Nevertheless, a practical system of classification of aggregates for engineering purposes is of benefit to the processes of specification, assessment, and selection of aggregates and to deciding on the suitability of a particular source for a specific purpose such as roadstone. To meet this demand BS 63: 1913 *Specification for sizes of broken stone and chippings* introduced a list of twelve trade groups comprising eleven groups of naturally occurring aggregates and one artificial group (essentially blast furnace slag). In 1943 the number of groups was reduced to eleven and the revised classification was published as BS 812: 1943 *Sampling and testing of mineral aggregates sands and fillers* as follows: Artificial, Basalt, Flint, Gabbro, Granite, Gritstone, Hornfels, Limestone, Porphyry, Quartzite, and Schist.

The list of groups was supplemented by lists of the petrographic names of the most commonly occurring rock types to be included in each group, allocated in accordance with the not entirely compatible criteria of petrography and roadstone performance, the latter prevailing in cases of conflict. It was retained, with minor modifications to the supplementary lists, in each revision of the Standard up to and including the 1975 edition. The full classification is given in Table 5.8.

Table 5.8 — Group classification of aggregates. (Reproduced from BS 812: Part 1: 1975, Clause 6.4)

1. *Artificial group* crushed brick slags calcined bauxite synthetic aggregates	6. *Gritstone group* (including fragmental volcanic rocks) arkose greywacke grit sandstone tuff

2. *Basalt group*
andesite
basalt
basic porphyrite
diabase
dolerites of all kinds
    including theralite
    and teschenite
epidiorite
lamprophyre
quartz-dolerite
spilite
	7. *Hornfels group*
	contact-altered
	    rocks of all kinds
	    except marble
	
	8. *Limestone group*
	dolomite
	limestone
	marble

3. *Flint group* chert flint	9. *Porphyry group* aplite dacite felsite
4. *Gabbro group* basic diorite basic gneiss gabbro hornblende-rock norite peridotite picrite serpentinite	granophyre keratophyre microgranite pophyry quartz-porphyrite rhyolite trachyte 10. *Quartzite group* ganister quartzitic sandstones recrystallized quartzite
5. *Granite group* gneiss granite granodiorite granulite pegmatite quartz-diorite syenite	11. *Schist group* phyllite schist slate all severely sheared rocks

The name of each natural rock group typifies the major rock within the group. There are four igneous groups, which are loosely based on their silica content and texture:

Sec. 5.2] Origin and classification of aggregates

      Granite   acid, coarse
      Porphyry   acid, fine
      Gabbro   basic, coarse
      Basalt   basic, fine.

Four sedimentary groups are represented by the following rock types:

      Flint   nodules of cryptocrystalline silica
      Gritstone   arenaceous rocks
      Limestone   carbonate rocks including marble
      Quartzite   including recrystallized quartzite.

Two metamorphic rocks account for those metamorphic rocks not assigned to the groups above:

      Hornfels   contact-metamorphic rocks
      Schist   dynamic and regional metamorphic rocks

Because of the intention to associate rocks of similar performance in a group, some of the rock types are petrographically misplaced. Gneiss and granulite, for example, are included in the Granite group and marble in the Limestone group. Moreover, although the Hornfels group is for 'contact-altered rocks of all kinds except marble', recrystallized quartz is included in the Quartzite group which is consequently catering for both sedimentary and metamorphic forms. It is also apparent that the classification does not identify separate intermediate or ultrabasic rock groups, and that the grain size of the igneous rocks can only be inferred, and then rather broadly as coarse or fine.

Despite the original intentions of the Group Classification, the 1975 edition of the Standard included the statement 'The petrological classification does not take account of suitability for any particular purpose, which should be determined in accordance with the appropriate British Standard'. This qualification draws attention to the fact that the many factors inherent in the derivation of aggregates, ranging from the composition, mode of formation, and subsequent history of the rock to the quarrying and processing of the aggregates themselves, will generate a wide range of properties within a group and, to a lesser extent, in every rock type. Consequently, the properties of a road making aggregate and, in particular, one intended for the wearing course of a major road or aircraft pavement, should be determined not only by authoritative petrographic examination but also, very importantly, by relevant physical and mechanical test procedures specified in a British Standard, or other appropriate document specified by the purchaser, and its quality and fitness for purposes judged by the test results obtained on representative samples.

Since its introduction, the Group Classification has done much to clarify the confusion that allowed most hard rocks to be referred to, habitually, as 'granite'. Its 'broad-brush' approach to the categorization of British aggregates, when supported by physical and mechanical test results, has served as an indicator of their engineering properties, has facilitated the writing and enactment of specifications for pavement materials, and, overall, has contributed to the present-day appreciation by engineers, producers, and contractors alike of the influence of aggregate petrology on performance. Its imperfections, however, have been the object of criticism for many years, and as a consequence it was deleted from BS 812: Part 1 in 1984, along with a comprehensive glossary of rocks and minerals and definitions of textural terms and particle geometry. At the same time, the section (in Part 1) dealing with the sampling and submission of samples was superseded by Part 102 methods for sampling, which includes, in Appendix A, Table 2, a short glossary comprising 29 petrological terms and definitions for rock types commonly used for aggregate, selected from BS 6100: Section 5.2: 1984. The table, which is reproduced here as Table 5.9, is adequate for recording a nominal description of the aggregate on the sampling certificate. It is to be hoped, however, that a more comprehensive list of terms and definitions will be issued in BS 812 to provide a standard reference for the petrologist's report and for the appreciation of its findings by the engineer and producer.

Although the Group Classification is no longer specified in BS 812, the two major Standards for bituminous materials, BS 594: Part 1: 1985 and BS 4987: Part 1: 1988 have maintained the Group terminology for specifying the aggregates used in rolled asphalt and coated macadam. It is also still used in the Property Services Agency Airfields Branch Specification Part 4 (1979) for describing the aggregates used in bituminous surfacings for aircraft pavements. The Department of Transport Specification (1986) requires that the bituminous materials comply with the appropriate British Standards such as BS 594 and BS 4987, which implies the use of Group terminology, but as the materials for a scheme have to be prescribed in Appendix 7/1 of the Scheme Specification, the actual terminology used will be influenced by the preferences of the Specifying Authority.

5.2.4 Classification and description of aggregate materials using the CADAM system

The Engineering Working Party of the Geological Society has proposed a system for the Classification and Description of Aggregate Materials (Collis & Fox 1985). Known by the acronym CADAM, the system has the objective of providing a simple, rational, and descriptive classification of aggregate material that is based on sound geological principles and is capable of being generally understood. The Working Party aimed at complying with several basic requirements which they considered were necessary for a classification system if it were to be of value for both commercial and contractual purposes; the system should:

'(i) Be simple in concept and yet at the same time use only terms which have a precise meaning capable of application in contract specifications.
(ii) Use properties relevant to aggregate materials as the basis for grouping them into classes.
(iii) Have no bias except that which may reflect on the importance of the material as an aggregate.
(iv) Be capable of being further expanded, as necessary, by the use of supplementary information relating to the particular material or its intended use.'

Sec. 5.2] **Origin and classification of aggregates** 179

Table 5.9 — Rock types commonly used for aggregates. (Reproduced from BS 812: Part 102: 1989, Appendix A, Table 2)

Petrological term	*Description*
andesite[†]	a fine grained, usually volcanic, variety of diorite.
arkose	a type of sandstone or gritstone containing over 25% feldspar.
basalt	a fine grained basic rock, similar in composition to gabbro, usually volcanic.
breccia[‡]	rock consisting of angular, unworn rock fragments, bonded by natural cement.
chalk	a very fine grained Cretaceous limestone, usually white.
chert	cryptocrystalline[§] silica.
conglomerate[‡]	rock consisting of rounded pebbles bonded by natural cement.
diorite	an intermediate plutonic rock, consisting mainly of plagioclase, with hornblende, augite or biotite.
dolerite	a basic rock, with grain size intermediate between that of gabbro and basalt.
dolomite	a rock or mineral composed of calcium magnesium carbonate.
flint	cryptocrystalline[§] silica originating as nodules or layers in chalk.
gabbro	a coarse grained, basic, plutonic rock, consisting essentially of calcic plagioclase and pyroxene, sometimes with olivine.
gneiss	a banded rock, produced by intense metamorphic conditions.
granite	an acidic, plutonic rock, consisting essentially of alkali feldspars and quartz.
granulite	a metamorphic rock with granular texture and no preferred orientation of the minerals.
greywacke	an impure type of sandstone or gritstone, composed of poorly sorted fragments of quartz, other minerals and rock; the coarser grains are usually strongly cemented in a fine matrix.
gritstone	a sandstone, with coarse and usually angular grains.
hornfels	a thermally metamorphosed rock containing substantial amounts of rock-forming silicate minerals.
limestone	a sedimentary rock, consisting predominantly of calcium carbonate.
marble	a metamorphosed limestone.
microgranite[†]	an acidic rock with grain size intermediate between that of granite and rhyolite.
quartzite	a metamorphic or sedimentary rock, composed almost entirely of quartz grains.
rhyolite[†]	a fine grained or glassy acidic rock, usually volcanic.
sandstone	a sedimentary rock, composed of sand grains naturally cemented together.
schist	a metamorphic rock in which the minerals are arranged in nearly parallel bands or layers. Platy or elongate minerals such as mica or hornblende cause fissility in the rock which distinguishes it from a gneiss.
slate	a rock derived from argillaceous sediments or volcanic ash by metamorphism, characterized by cleavage planes independent of the original stratification.
syenite	an intermediate plutonic rock, consisting mainly of alkali feldspar with plagioclase, hornblende, biotite or augite.
trachyte[†]	a fine grained, usually volcanic, variety of syenite.
tuff	consolidated volcanic ash.

[†]The terms microgranite, rhyolite, andesite, or trachyte, as appropriate, are preferred for rocks alternatively described as porphyry or felsite.
[‡]Some terms refer to structure or texture only, for example breccia or conglomerate, and these terms cannot be used alone to provide a full description.
[§]Composed of crystals so fine that they can be resolved only with the aid of a high power microscope.

The assessment of an aggregate material should be made on a representative sample ideally, but not essentially, by a qualified and experienced geologist. It requires the compilation of three sets of data:

  (i) The Form: for example, crushed rock, gravel, sand, etc.
 (ii) The Class: Carbonate, Quartz, or Silicate.
(iii) The Geological Age; Colour; Grain Size and Foliation or Fissility.

The data may be assembled on a standard form, an example of which is illustrated in Fig. 5.7. The rules and guidelines for supplying and entering the data are tabulated in Fig. 5.8.

AGGREGATE FORM	Crushed Rock	Gravel	Natural Crushed Mixed	Sand	Natural Crushed Mixed	Land-won Marine dredged
CLASS (or Miscellaneous)	Carbonate class	Quartz class	Silicate class: Igneous	Sedimentary	Metamorphic	Miscellaneous material (Correct name to be given below)
Petrological name (if known)						
GEOLOGICAL AGE/ COLOUR/ GRAIN SIZE/ FISSILITY						
Comment (if any)						

Compiled by: ______________ Date: __________

CADAM — CLASSIFICATION AND DESCRIPTION OF AGGREGATE MATERIAL

LOCATION AND SAMPLE DETAILS	Quarry/Pit address ______ Operator: ______ Sample: Type ______ Size ______ Preparation ______ Supplied by ______	Grid Ref. ______ Date of sampling ______	Date Rec'd ______ Sampling Cert. No. ______

Fig. 5.7 — Typical form used for assembling data for CADAM (from Collis & Fox 1985, Table 6.6).

The form also makes provision for entering the particulars of the sample, which are obtained from the certificate of sampling. These are supplementary to the CADAM data, but they are nevertheless an essential requirement in that they relate the sample to the source material. The certificate should include the name, address, grid reference, and operator of the quarry; the name and address of the supplier; the sample type and its location; the quarry processing experienced by the sample and whether or not it is typical of that used or to be used for the commercial product; and the extent and purpose of the investigation.

An important feature of the CADAM system is the grouping of the material in three Classes that are dependent only on the predominant mineral content — Carbonate, Quartz, and Silicate. The classification of a sample is, therefore, made initially without considering the geological origin of the material.

The Carbonate Class consists of aggregate material in which calcium carbonate or calcium magnesium carbonate is the predominant mineral, and includes both sedimentary limestones and metamorphic marble. The Quartz Class contains all aggregate material in which quartz or free silica predominates, such as flint,

Sec. 5.2] Origin and classification of aggregates

		Crushed rock				Gravel	Sand
				Silicate class (d)			
		Carbonate class (b)	Quartz class (c)	Igneous	Sedimentary	Metamorphic	Natural, crushed, mixed	Natural, crushed, mixed
1	Form of Aggregate (a)							
2	Class			Miscellaneous materials (e)			Class of major constituents as for crushed rock aggregates, with petrol. name (if known).	
				Petrological name (if known)			Description of minor constituents, e.g. mica, shell fragments, etc. (Mainly applicable to natural sands) (f)	
	Geological Age (g)	Essential for sedimentary rock materials		—	Essential	—	Geological age of deposit if other than Quaternary (h)	
	Colour (i)	Optional	Optional	Essential	Optional	Essential	Description of major constituent(s) as for crushed rock aggregates	
3(1)	Component (j) Grain size	Optional	Optional	Essential	Optional	Essential		
	Foliation (k)	If applicable	If applicable	—	If applicable	Essential		
4	Other (m) Comment	Any further information likely to be of help in assessing the material, e.g. for a Gravel–Glacial deposit.						

Notes:
(a) Indicates the nature of the aggregate material being considered. In case of Gravel and/or Sand an indication of whether the material is land-won or marine-dredged should be included.
(b) Includes ALL materials composed predominantly of calcium and/or calcium magnesium carbonate, irrespective of origin.
(c) Comprises ALL materials in which quartz or free silica is the dominant mineral, and includes materials such as flint, vein quartz, quartzites, many sandstones and gritstones, some greywackes and the like.
(d) Comprises ALL materials in which the rock forming silicates are the dominant minerals. These are subdivided according to origin into igneous, metamorphic and sedimentary in order to differentiate between those containing mainly high temperature silicates and those containing a high proportion of clay mineral silicates.
(e) Materials not covered by (a), (b) and (d) above are NOT allocated a class but the correct scientific name for the material MUST be recorded.
(f) In the case of gravel and particularly in the case of sand, the presence of certain materials, even in small quantities, can be important. Note should be made of the occurrence of such materials.
(g) The geological age scale to be used in the UK is as follows:— Quaternary (to include Pleistocene and Holocene), Tertiary, Cretaceous, Jurassic, Triassic, Permian, Carboniferous, Devonian, Silurian, Cambrian and Precambrian. Where alternative or different age terms are applicable in countries outside the UK then these should be used as is appropriate.
(h) The majority of sand and gravel deposits worked in the UK are of Quaternary age and unconsolidated. The older the deposit the more likely it is to be consolidated.
(i) Colour descriptions should be to a common standard. Reference is made to the Munsell Colour Chart system. (Geological Society of America, 1963), which is useful for both field and laboratory assessments. Should a more accurate colour description be required then this should be carried out under laboratory conditions and it is suggested that the Lovibond–Schofield Tintometer or similar method be used (Hosking and Ritson, 1968). Normally, however, such a detailed description would not be included in the CADAM.
(j) A standard component grain size notation should be used. Grains greater than about 2 mm nominal diameter are termed coarse: those which are less than 2 mm, but which can still be distinguished with the unaided eye, are termed medium; and those which cannot be distinguished with the unaided eye are termed fine. The component grain size is applied to the matrix.
(k) Foliation and fissility are particularly important with respect to certain metamorphic and certain sedimentary rock materials. Evidence of its occurrence should be recorded as appropriate.
(l) Entries to be made with respect to geological age/colour/component grain size/foliation for the different classes are indicated in the above table as essential, optional or not applicable.
(m) Additional comments should only be added when that information is likely to be of help in assessing the material and is not to be included in any subsequent more detailed description of the material.

Fig. 5.8 — Rules and guidelines for supplying and entering data for CADAM (from Collis & Fox 1985, Table 6.5).

quartzite, granulite, and most sandstones and gritstones. The Silicate Class comprises all aggregate material in which the predominant minerals are the rock-forming silicates such as olivines, pyroxenes, amphiboles, micas, and feldspars. With a few exceptions, the Silicate Class consequently contains those aggregate materials in which carbonate and quartz are subordinate minerals, and, because it can be of some importance to know the geological origin of such aggregates, the Class is sub-classified into igneous, sedimentary, and metamorphic.

Having classified the material, the addition of its petrological name is encouraged but is not considered essential by the Working Party. There are, however, a few aggregate materials such as barytes which cannot be classified as above. These are entered under the Miscellaneous heads and must be given their geological name or term, otherwise their identity will be lost.

The third set of data calls for the geological age of materials derived from sedimentary rocks because it often attests to their physical condition and strength, which are important characteristics of an aggregate. For example, in the UK Carboniferous limestone is widely used as a roadstone, whereas the more recent Jurassic limestone is generally unsuitable. For the remaining materials, geological age is an optional entry.

Colour and component grain size are essential entries for Silicate Class igneous and metamorphic rocks because they both provide useful petrographic data. It is recommended that their descriptions should conform to a standard system. For colour the Munsell rock-colour chart of the Geological Society of America (1963) is recommended. For the component grain size of the matrix the description is related to the terms applied to engineering soils in BS 1377, such that coarse grained is applied to grain size predominantly greater than 2 mm, medium grained from 2 mm to 0.2 mm, and fine grained to less than 0.2 mm which is approximately the limit of resolution by the naked eye. (In BS 1377, 2 mm separates fine gravel from coarse sand and 0.2 mm separates medium and fine sand).

Fissility is a function of foliation, the banded structure characteristic of schist and gneiss that results from the re-orientation and the formation of new minerals into parallel layers during intense regional metamorphism. It may also occur in some sedimentary rocks. As fissility contributes to the formation of platey or flaky aggregates its presence or not is an essential entry for the Metamorphic Silicates; it should also be recorded if found in other materials.

From the foregoing it may be seen that CADAM provides a system for classifying and describing in correct petrographic terminology the basic petrological/geological characteristics of an aggregate material that may be determined principally by visual inspection, in a form that should be generally understood by those concerned with its production and use. The classification is simple and rational, and reference standards are prescribed for describing colour and component grain size. This requires at least a basic understanding of petrology, and the inspection should therefore be undertaken by a competent person, preferably a qualified geologist. Despite this, the allocation of the correct petrological name to the material is only encouraged and is not an essential requirement of the system. It is unlikely, however, that an engineer concerned with selecting aggregates for roadstone or concrete will be satisfied with knowing only the Class of the material; the rules ought, therefore, to place greater emphasis on the provision of the petrological name.

Direct reference to engineering properties is avoided, and, in this respect, CADAM eschews any comment on either shape or surface texture on the stated grounds that these properties are influenced as much by processing as by the geological properties of the parent material, even though this is hardly the case with natural gravels and sands. While these properties should be quantified or described in subsequent physical tests or petrological examination they are, nevertheless, quite capable of being given a visual assessment under CADAM. They are of considerable significance to the engineer in his initial evaluation of an aggregate for roadstone, concrete, railway ballast, etc., and it is unreasonable to exclude comment on them on principle in a classification system designed specifically for aggregates.

As the Group Classification system has been withdrawn from BS 812 there is a need for a replacement. The basis for this could be CADAM, but some modifications to the existing system and the specimen standard form are required in order to improve its effectiveness for highway and materials engineers.

5.3 TESTS AND STANDARDS FOR AGGREGATES IN BITUMINOUS MATERIALS

5.3.1 Introduction

For an assessment of the quality of roadmaking aggregates both the petrological characteristics and the engineering properties have to be considered. Broadly they are interdependent and to some extent complementary. Petrographic analysis of representative samples of rock or aggregates and a prediction therefrom of performance by an experienced geologist are often essential steps in the evaluation of a potential source of roadstone; they are also necessary when significant changes are observed in the character of the rock at the quarry face or in the properties of the aggregates during processing or in stockpiles. It is the engineering properties, however, as well as the long-term soundness, of an aggregate that are of particular interest to the highway engineer because, in most cases, they provide a positive indication of the behaviour to be expected of the aggregate during the design, manufacture, and placement of the bituminous mix and during the service life of the pavement. Furthermore, they can be evaluated either directly or indirectly by means of physical and mechanical laboratory tests.

Qualitatively, a good road making aggregate for a wearing course, friction course, or surface dressing should be durable, dense, angular, non-flaky, and preferably medium textured, and resistant to crushing, impact, abrasion/attrition, polishing, and stripping. It should be screened and graded to comply with the relevant specification. With the exception of resistance to abrasion and polishing, which are properties specific to the interaction of the vehicle tyre and the wearing course surface, these properties are also desirable for basecourse and roadbase aggregates. Because they are not so severely exposed to traffic stresses and weathering as those in the wearing course, however, their quality need not be so high, and this may be indicated in the product specification by setting less stringent limits for laboratory test values.

With few exceptions, the physical and mechanical test methods for evaluating the engineering properties of aggregates in the United Kingdom are, or are intended to be, specified in BS 812 published by the British Standards Institution. This Standard is undergoing a comprehensive revision at the present time (1990), and it is expected that, when completed, the new edition will comprise more than 20 Parts, each dealing with a particular procedure or test method.

A number of additional tests are programmed for inclusion, and, in accordance with the policy adopted by the BS Technical Committee responsible for the revision of this Standard, both the existing and the new test methods will include reliable estimates of repeatability and reproducibility calculated from the results of trials specially commissioned by the Committee for the new edition.

This section is concerned with the testing of aggregates for bituminous mixtures used in flexible pavement construction, mainly in the United Kingdom. The comprehensive range of rock types, climate, and traffic conditions in the British Isles, and experience of using the tests on aggregates in many parts of the World, more than suggest, however, that the tests may be used internationally for aggregates for bituminous mixtures. The methods are described in sufficient depth to provide an adequate appreciation of the procedure to be followed, but with not enough detail to perform them with precision, for which purpose reference should be made to the published standard method. Most of the methods are described in BS 812, but other sources such as the American Society for Testing and Materials (ASTM), Transport and Road Research Laboratory (TRRL), Airfields Branch of the Property Services Agency (PSA), and published books and papers are referred to for useful supplementary or complementary tests for properties, such as the adhesion between aggregates and bitumen, that are not published in BS 812.

A number of test procedures that are directly concerned with the soundness, and hence the durability, of aggregates are given. They include the ethylene glycol and methylene blue tests, which are aimed specifically at detecting potential unsoundness in basic igneous rocks, as well as the controversial but more universally applied magnesium and sodium sulphate soundness tests. Many of the tests for soundness have been developed abroad in, for example, America, Australia, and South Africa where, until recently, there has been greater cause for concern about unsound roadstone than in the United Kingdom, with the possible exceptions of Northern Ireland because of the extensive use of basalt aggregates, and the British Airports Authority and Airfields Branch of PSA because of the danger that loose particles of wearing course aggregate arising from unsoundness may be ingested by jet engines. To meet the need for a British Standard for roadstone, a magnesium sulphate soundness test method is specified in BS 812: Part 121: 1989.

In this section the tests have been grouped broadly in terms of their engineering properties — physical, mechanical, and soundness, etc. — and, where applicable, test results prescribed by various specifications to indicate acceptable material are given. Included in each section are some non-standard test methods, references to investigations, and research published in the UK and abroad, and other information and comments which, *inter alia*, contribute to a better appreciation of the qualities required of a roadmaking aggregate and to the interpretation of test results. The sources of this information as well as those of the Standard tests are recorded, with appreciation, in the text and the references.

A list of the tests described in this section is given in Table 5.10.

Table 5.10 — List of aggregate tests

Section No.	Topic
5.3.2	Sampling of aggregates
5.3.3	Particle size and shape
5.3.3.1	Particle size distribution or grading
5.3.3.2	Particle shape
5.3.3.3	Flakiness index
5.3.3.4	Average least dimension
5.3.3.5	Elongation index
5.3.3.6	Angularity number
5.3.3.7	Surface texture of coarse aggregate
5.3.3.8	Shape and surface texture of fine aggregate
5.3.4	Physical properties
5.3.4.1	Relative density and water absorption
5.3.4.2	Bulk density of aggregate
5.3.4.3	Bulk density of filler in toluene
5.3.4.4	Voids of dry compacted filler
5.3.5	Mechanical properties
5.3.5.1	Aggregate crushing value
5.3.5.2	Ten per cent fines value
5.3.5.3	Aggregate impact value
5.3.5.4	Los Angeles abrasion value
5.3.5.5	Aggregate abrasion value
5.3.5.6	Polished stone value
5.3.6	Soundness
5.3.6.1	Soundness by the use of sodium or magnesium sulphate
5.3.6.2	Soundness by freezing and thawing
5.3.6.3	Soaked ten per cent fines and modified aggregate impact values
5.3.7	Soundness of basic igneous rock aggregates
5.3.7.1	Secondary mineral content by point counting
5.3.7.2	Soundness by ethylene glycol
5.3.7.3	Soundness by methylene blue adsorption
5.3.8	Adhesion of bitumen to aggregates
5.3.8.1	Static water immersion tests
5.3.8.2	Immersion wheel-tracking test
5.3.8.3	Immersion mechanical tests

5.3.2 Sampling of aggregates

The sampling of aggregates is dealt with comprehensively in Chapter 2, so the following notes are included only for the purpose of emphasizing some of the important requirements of BS 812: Part 102 and the influence of sampling in the validation of test results.

Sampling must be methodical and take into consideration:

(i) The variability of the material, which is a function of the source rock and the quarry processing.
(ii) The form of the mass to be sampled, which may be a stream on a moving belt, a stockpile, the discharge from a hopper, or a lorry load, etc.

(iii) The reason for sampling. Samples may be requested by a buyer for a comprehensive assessment of quality for a long-term supply contract; by the quarry manager for routine checking of production quality; for determining the variability of one or more test properties within a batch; for the identification of deleterious or unsound particles.

Any of these reasons may necessitate some modification to the sampling procedures specified in BS 812: Part 102. Any deviation from the standard procedure, however, must be clearly thought through and documented, so as to ensure that the worth of the subsequent test results is not debased by careless sampling methods.

The aim of the methods for sampling aggregates prescribed in BS 812: Part 102 is, except where explicitly stated to the contrary, 'to obtain a test portion that is representative of the average quality of the batch'. The Standard requires that the 'quantity of material to be represented by the bulk sample shall be clearly defined' and that the increments which are aggregated to form the bulk sample shall be taken 'from different parts of the batch in such a way as to represent the average quality'. The minimum number of sampling increments for different nominal sizes of aggregate is given in the Standard and is reproduced in Table 5.11. The Standard also gives a diagram to illustrate the form and dimensions of suitable sampling scoops.

Table 5.11 — Minimum number of sampling increments. (Reproduced from table 1 in BS 812: Part 102: 1989)

Nominal size of aggregate	Minimum number of sampling increments		Approximate minimum mass for normal density aggregate kg
	Large[†] scoop	Small[‡] scoop	
28 mm and larger	20	—	50
5 mm to 28 mm	10	—	25
5 mm and smaller	10 half scoops	10	10

[†]Large scoop to hold a minimum volume of 2 L (about 3 kg) of aggregate.
[‡]Small scoop to hold a minimum volume of 1 L (about 1.5 kg) of aggregate.

In drawing up sampling plans, BS 812: Part 102 advises that the general principles given in BS 5309 *Methods for sampling chemical products* Parts 1 and 4 should be followed. These require consideration of such interrelated factors as the heterogeneity of the product, sampling frequency, batch size, number of increments, and the acceptable sampling error.

Random sampling should minimize bias, and the effect of random errors will be reduced by increasing the number of increments and the frequency of sampling and testing. High rates of sampling and testing can markedly increase costs, however, and they should be employed only after careful consideration of all relevant factors, with the aim of achieving the optimum combination for the particular aggregate. This aspect also requires the attention of the specifier, who should understand the risks entailed in the sampling and testing of a naturally occurring heterogeneous material and make due allowance for them when specifying limits of compliance. He is not specifying an intrinsic fundamental property of the aggregate, but the result of a test made on an average sample drawn from a batch.

Further information on sampling is given in BS 812: Part 101: *Guide to sampling and testing aggregates*, and in Chapter 2 of this book.

5.3.3 Particle size and shape
5.3.3.1 *Particle size distribution or grading*
The particle size distribution or grading of an aggregate by sieving is found by shaking the sample through a series of square-hole test sieves complying with BS 410: 1986, in accordance with the methods specified in BS 812: Section 103.1.

The test sieves are manufactured in three diameters, 200, 300, and 450 mm. Coarse mesh sizes of 4 mm and larger are manufactured from perforated plate, and the fine mesh sizes are of woven wire; woven wire is also used for the coarser sizes up to 16 mm. The sizes of a given series are often interrelated in terms of one or more geometrical progressions, although the practice is not at present followed consistently in the UK. The sieves used in the UK for the grading of aggregate for bituminous mixtures and surface dressing are usually selected from:

450 or 300 mm diameter perforated plate:

    63.0, 50.0, 37.5, 28.0, 20.0, 14.0, 10.0, 6.3 and 5.0 mm

300 or 200 mm diameter wire cloth:

    3.35, 2.36, 1.7, and 1.18 mm, and 600, 425, 300, 212, 150, and 75 μm.

In Europe, at present, there is no universally accepted single series of sieve aperture sizes in use for aggregate, but it is an aim of the European Committee for Standardization (CEN) to reach agreement on a single unified series by 1992. The International Standard ISO 6274 *Concrete — sieve analysis of aggregates* already specifies a preferred Series A that comprises a selection of preferred aperture sizes taken from Series R 40/3 (or R 20) of ISO 565 *Test sieves — woven metal wire cloth, perforated plate and electroformed sheet — nominal size of apertures*. Series A is a geometrical progression, based on 1 mm, with a common ratio of 2:

    63.0, 31.5, 16.0, 8.00, 4.00, 2.00, 1.00, 0.500, 0.250, 0.125, 0.063 mm.

It has the benefit of simplicity, and it could provide an acceptable basis for compromise in CEN for the grading of aggregates used in both concrete and bituminous materials. However, while the sizes for specifying and analyzing the grading of the fine aggregate fraction would be suitable for dense bituminous mixtures, not all of them are necessary for open graded mixtures, and, for this purpose, some could well be omitted. On the other hand, the sizes are too widely spaced in the coarse aggregate fraction to accommodate and control the range of gradings used in bituminous mixtures and single sized aggregates for surface dressing, and it would therefore be advisable to introduce intermediate sizes. These could be provided by further selection of preferred sizes from the ISO R 40/3 (or R 20) series using the ratio of $\sqrt{2}$, e.g. 5.6, 11.2, 22.4 and 45 mm.

Whichever single unified series is eventually adopted, most European countries will, as a consequence, have to make some changes to screen sizes used in the aggregate processing and sieve sizes for laboratory analysis. If, for example, the sieves referred to above, including the intermediate sizes, were adopted, then changes would be necessary in the UK, where the apertures used are mostly non-preferred sizes in the ISO R 40/3 and R 20 series.

It used to be the normal practice under BS 812 to dry sieve aggregates. Moreover, aggregate gradings for all the British Standard bituminous mixtures used also to be specified on a dry sieving basis. Hence, when a BS 598 mix analysis procedure using a wet sieving technique for grading the extracted aggregates was used, a factor to convert the quantity passing 75 μm to a dry sieving basis was applied before checking compliance with the specification.

The emphasis was changed in 1985 when BS 812: Section 103.1 declared that grading analysis by washing and sieving was to be the preferred method, and that dry sieving was an alternative method which may be used for coarse and fine aggregate free from particles which cause agglomeration. This change was endorsed in subsequent revisions of BS 63, BS 594, and BS 4987 by specifying all their aggregate gradings on the basis of analysis by wet sieving. The one exception is BS 63: Part 1 which continues to specify dry sieving if the portion passing 75 μm is not to be determined.

The washing and sieving method requires the test portion to be dried and weighed. It is then immersed in water, agitated to bring the fines into suspension, and the suspension decanted over a 75 μm test sieve guarded by a 1.18 mm test sieve. After drying to constant mass, all the aggregate retained on 75 μm is dry sieved through the required series in order of decreasing aperture size (including 75 μm) and the material on each sieve weighed.

The minimum mass of test portion for each nominal size of aggregate is specified, from 100 g for passing 3 mm size to 50 kg for 63 mm. When these masses are compared with the minimum masses of test portion specified in BS 598: Parts 101 and 102 for the analysis of bituminous mixtures, there is reasonable agreement at the 14 mm aggregate size. For the smaller sizes BS 812 allows smaller portions than BS 598, but for the larger sizes they are significantly heavier, such that for a 40 mm material 15 kg are required compared with only 2.5 kg in BS 598. The Foreword to BS 598: Part 2 acknowledges these differences but indicates that smaller samples are acceptable because of the much reduced risk of segregation when sampling coated materials. This would seem to be a reasonable assumption, although limited precision data in the two Standards do not entirely support it. If a further comparison is made by referring to the sizes of test portions for sieve analysis required by ASTM C136, they are four times the mass of those specified in BS 812 for 10, 14, and 20 mm nominal size aggregate, but about half of those specified by BS 812 for 50 and 63 mm sizes. Such comparisons point to a certain arbitrariness in the fixing of minimum masses for test portions for sieve analyses, and the situation might benefit from study and harmonization.

It is important not to overload the sieves, and a table of maximum masses permitted to be retained on each of the sieves at the completion of sieving is given in BS 812: Section 103.1, 1985, Table 3. As a rough guide, the quantity of coarse aggregate retained on a sieve should not exceed a single layer of particles. This is not a practical restraint for fine aggregate although its retained mass should be reduced as the particles become finer, and this principle is adopted in the Standard. It is also observed in ISO 6724, *Concrete — sieve analysis of aggregate* which specifies that the retained fraction on sieves with apertures less than 4 mm must not exceed $(A\sqrt{d})/300$ g, where A is the sieving area in mm^2 and d the aperture size in mm. ASTM C136, however, does not differentiate between the various particle sizes, specifying instead that the retained fraction on any fine sieve must not exceed 4 g/in^2 (620 mg/cm^2). Clearly, these specifications are not in agreement, and, taking a 200 mm diameter, 600 μm sieve as an example, the maximum permissible residues are:

BS 812	100 g
ISO 6724	81 g
ASTM C136	195 g.

The ISO formula is obviously more useful in that, within reason, it can be applied to any sieving area and fine aggregate sieve aperture size. In its present form it gives rather low residues, but these could be increased by substituting an appropriately smaller denominator of, say, 250. For field applications the Standard already permits a value of 200.

BS 812 requires that the grading be reported in terms of the cumulative mass passing each test sieve expressed as a percentage of the total mass of the sample. The cumulative percentage passing method of reporting a grading has long been the practice for coated macadam and rolled asphalt, but it was not until the 1987 revision of BS 63 that the method was adopted for single-sized road aggregates. Until then they were specified and reported as the percentage by mass of 'specified size', 'oversize' and 'undersize', each 'size' being defined by particular test sieves. This method was useful and precise for defining surface dressing chippings but awkward for general application. The revised method is preferable, although it is still necessary to control the 'specified size' fraction by specifying a minimum value (between 60 and 70%).

The fine aggregate gradings for mastic asphalt used for paving and buildings are still specified in their respective Standards on a 'passing–retained' basis, but the gradings of the coarse aggregate incorporated in road and paving mastic asphalt (BS 1447: 1988) have to comply with BS 63: Part 1: 1987. Cumulative percentages are not specified for the overall grading of coarse and fine aggregate, but mix composition would be more readily controlled if they were.

An impression of the types of aggregate grading employed in UK bituminous surfacings can be obtained by referring to Fig. 5.9. The graphs, which are plotted semi-logarithmically, show calculated midpoint gradings for 14 mm nominal size

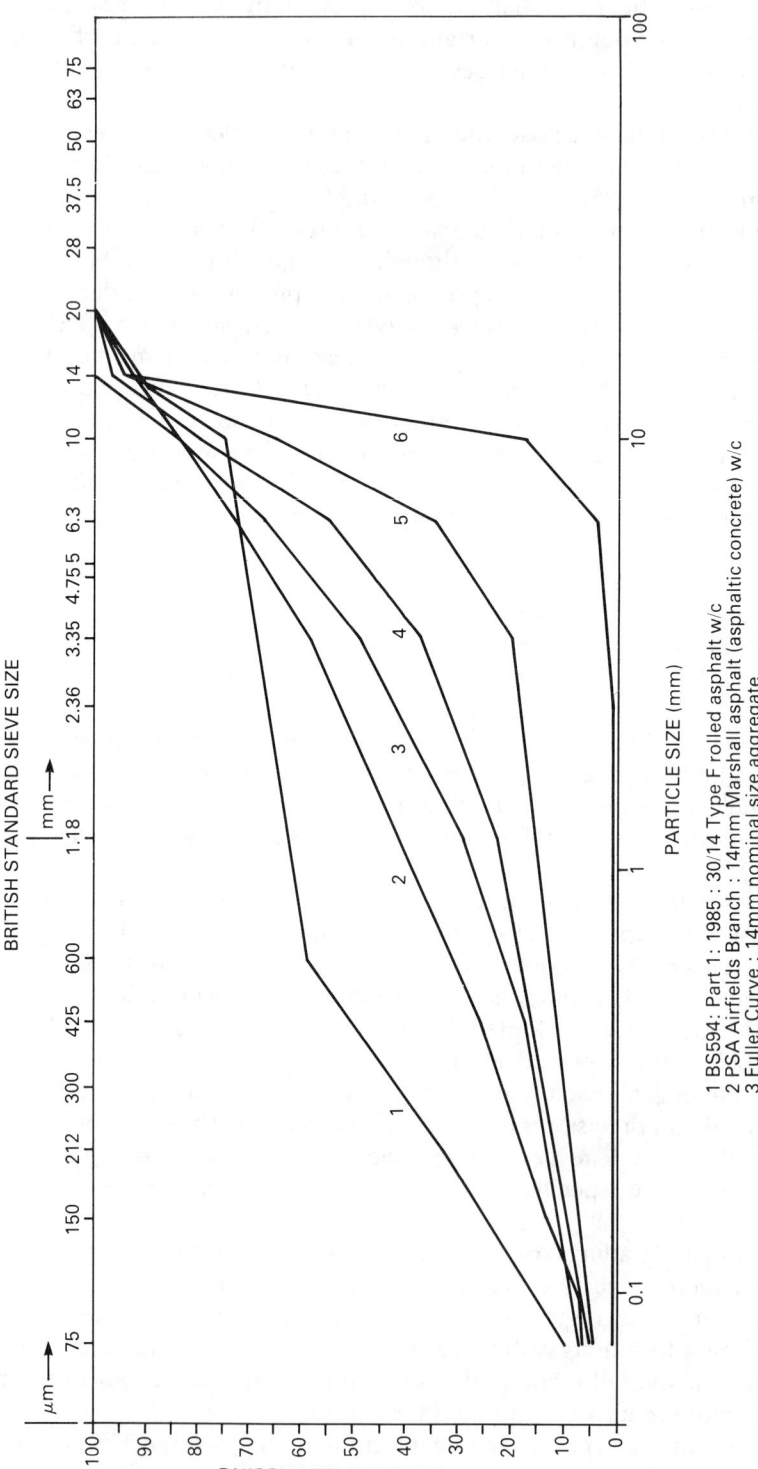

Fig. 5.9 — Types of 14 mm aggregate midpoint gradings used in UK bituminous surfacings.

rolled asphalt, close and open graded coated macadam, Marshall asphalt for heavy duty aircraft pavements, and single-sized surface dressing aggregate. In addition, the Fuller curve for 14 mm nominal size aggregate is shown. The grading of the close graded coated macadam tends to follow the Fuller curve more nearly than the other materials, the Marshall asphalt being finer, except for filler content, and the open graded macadam considerably coarser. The gap grading of the rolled asphalt can be judged by the small quantity of material between the 10 mm and 600 μm sieves, and the overall shape of the curve confirms that the main source of the strength and stability of rolled asphalt is not the mechanical interlock associated with dense, continuously graded mixes of the Fuller type, but the composition and properties of the bitumen-filler-fine aggregate mortar.

The separation between coarse and fine aggregate is made on a 2.36 mm sieve for rolled asphalt and the 3.5 mm sieve for coated macadam. The use of different sieves is rooted in the historical origins of the materials, the former being the traditional sieve of the sand and gravel industry and the latter of the quarrying industry. It would not be unreasonable to standardize on one sieve, and this will probably be achieved with European harmonization, but in view of the variety of sieves used for this purpose in Europe it remains to be seen what the ultimate choice will be.

There are no separate fine aggregate gradings specified for rolled asphalt or coated macadam (as there are, for example, for concreting sands in BS 882) as control is effectively exercised by the analysis of the total mix. To assist this control, however, limits are specified for the mass of material retained on the 2.36 mm sieve and that passing the 75 μm sieve, both being determined by the decantation or sedimentation process. Some authorities would like to take this a stage further by having limits prescribed for minus 20 μm (medium silt size and finer) and even 2 μm (clay size).

The procedure for determining material finer than 20 μm is described in BS 812: Section 103.2, *Sedimentation Test*. It relies on the principle that different sizes of particle in a suspension settle at different rates, and, if the particles are assumed to be spherical, Stokes' Law can be applied to grade them. For fine aggregate the silt and clay size particles are thoroughly separated from the dried and weighed test portion by agitation in a prescribed solution of a reagent in distilled water. The suspension is decanted over a 75 μm test sieve into a sedimentation tube, made up to 1000 ml with water and uniformly dispersed in the tube by inversions. After allowing the suspension to settle under steady conditions for 4 minutes, a sample of known volume is withdrawn at a depth of 100 mm by a variable position sedimentation pipette. The sample is dried, the residue weighed, and the percentage mass of minus 20 μm material, M, calculated from:

$$M = 1000 \left(\frac{M_2 - M_r}{V} \right) \frac{100}{M_1}$$

where M_1 is the mass (g) of the original sample, M_2 is the mass (g) of the dried residue, V is the volume (ml) of the pipette, and M_r is the mass (g) of reagent/pipette volume of solution.

If the quantity of clay size particles is required, then the settling time has to be extended for another six to seven hours when the sampling, drying and weighing procedure is repeated. With skill, time, and a stable sedimentation temperature, the test provides results of an acceptable repeatability.

The BS 812 version of the test is deemed by BS 4987 to indicate the quantity of material finer than 75 μm, although the test procedure actually determines the minus 20 μm material. The specification also contends that there is no acceptable test method for determining the clay content of aggregates. The argument hinges on the premise that, although the particles finer than coarse silt can be detected by the sedimentation test, and that they may in some circumstances inhibit good adhesion between the binder and the aggregate, they are not, *per se*, deleterious to a properly made bituminous mix. The harm is considered to be caused, in the main, by particles which soak up water and swell, and although these exist in the clay sizes (and only rarely in the silt sizes) clay minerals vary widely in their ability to do so, their order of activity, according to Perrin (1976), being:

smectites>illite>kaolinite.

Until the extent of the activity of the particles has been established, it can be argued, not unreasonably, that the material should not be condemned simply on the grounds of particle size.

5.3.3.2 *Particle shape*
The shape of an aggregate particle is a function of the petrology of the rock and the quarrying and aggregate production process. Although the particle shape of crushed aggregates is affected to some extent by the nature of the rock and the type of crusher used, it is largely determined by the reduction ratio in the final stage of crushing, and this should not exceed 4:1. Choke feeding also results in a more cubical shape as the particles are then broken against each other rather than between the crushing surfaces. Shape is therefore predominantly a function of process control and the selection of the right type of crushing plant for the particular rock; the available choice has been considerably improved by the relatively recent development in New Zealand of the horizontal centrifugal crusher.

BS 812: Part 102 classifies aggregate particles into six shapes, giving a brief description and photograph of each one. The shapes described are: rounded, irregular, angular, flaky, elongated, and flaky and elongated. This classification provides a standard reference and encourages the sampler or tester to make a visual appraisal of the material.

Extremes of shape of an aggregate have a bearing on the performance of the bituminous mix during both construction and service. Hard, smooth, rounded gravel gives a very workable mix but one that has low stability; crushing will improve the stability but it must be done efficiently. Very flaky and elongated particles have a larger specific surface and therefore a higher bitumen demand; they are more difficult to compact properly, and they fracture under the roller. Angular, cuboidal aggregates with a medium texture give optimum performance.

5.3.3.3 Flakiness index

An aggregate particle is said to be flaky when it has one dimension significantly less than its other two dimensions. This shape factor is numerically standardized in the flakiness index test of BS 812: Section 105.1: 1989 by defining a flaky particle as one whose smallest dimension is less than 0.6 times the arithmetic mean of the aperture sizes of the two sieves that limit the size fraction of the particle.

For example, consider a 14–10 mm particle.

$$\text{The mean sieve size} = \frac{14+10}{2} = 12 \text{ mm}$$

The particle is defined as flaky if the thickness is less than

$$0.6 \times 12 = 7.2 \text{ mm}$$

The flakiness index of an aggregate sample is determined by the following procedure:

(i) Discard aggregate retained on 63 mm and passing 6.3 mm test sieves and ensure that sufficient remains for test — for example, at least 5 kg for 28 mm nominal or 1 kg for 14 mm nominal, etc.
(ii) Sieve into size fractions and discard fractions of less than 5% by mass of the test portion.
(iii) Test each remaining fraction separately either by shaking through the appropriate slotted sieve and gauging the residue by hand, or gauging all by hand using the appropriate thickness gauge.

$$\text{Flakiness index } (I_F) = \frac{\text{total mass of particles passing gauges}}{\text{total mass of particles tested}} \times 100\%$$

Alternatively, if I_F is determined for each size fraction individually, then the overall I_F for the aggregate tested is found either by summing the masses of particles passing and tested for the size fractions and determining the percentage ratio as before, or by calculating the weighted average I_F of the individual size fractions.

The flakiness index is reported to the nearest whole number together with the sieve analysis of the aggregate sample tested.

The precision values for the test determined by eight laboratories on duplicate samples of 14 mm single size basalt and sandstone aggregate are given as:

Repeatability, $r_1 = 4.9$
Reproducibility, $R_1 = 9.1$

The range of flakiness indices specified for the aggregates in UK bituminous mixes and surface dressing is given in Table 5.12. While the limit of 25% for surface dressing and friction course should provide a satisfactory cuboidal aggregate, the value of 45% for rolled asphalt and coated macadam is far from onerous.

The definition of a flaky or a flat particle is not the same in every country. Different sieve sizes and dimension ratios are used, and in some cases the principal dimensions of each particle are checked with a proportional caliper. For these tests, limits for the width:thickness ratio ranging from 1.58 to 3 or even 5 may be set for flaky or flat particles, and the ratio of length:thickness, which is sometimes termed the shape coefficient, may have limits from 2.5 to 5. Clearly, there is scope for harmonization in Europe and the acceptance of a simple, rapid, and reasonably precise test (using flake-sorting sieves) with comprehensible and relevant limits that will facilitate quality control and the exchange of information.

Table 5.12 — Maximum flakiness index of coarse aggregates for bituminous mixes and surface dressing

Aggregate specification	Maximum flakiness index (I_F max) %
BS 63: Part 2. Aggregates for surface dressing	25
BS 594: Part 1. Coated chippings	25
BS 4987: Part 1. Pervious wearing course	25
PSA Airfields Branch. Part 4. Friction course	25
PSA Airfields Branch. Part 4. Marshall asphalt and all other mixes except friction course	30
BS 63: Part 1. Aggregates for general purposes 28, 20, 14 and 10 mm	35
BS 63: Part 1. Aggregates for general purposes 50 and 40 mm	40
BS 594: Part 1. Crushed rock and crushed gravel	45
BS 4987: Part 1. Crushed rock and crushed gravel	45
BS 594: Part 1. Uncrushed gravel	50
BS 4987: Part 1. Uncrushed gravel	50

5.3.3.4 Average least dimension (ALD)

This shape factor is used in the design of surface dressings, especially in South Africa (National Institute for Transport and Road Research, 1986), and Australia (Dickinson 1984), and, on occasions, in the UK (Heslop *et al.* 1982). It was first proposed by Hanson (1934). The ALD is variously described as the smallest perpendicular distance between two parallel plates through which the particle will just pass (NITRR); or the average height of the particles of an aggregate when they are spread as a single layer on a horizontal surface, the assumption being that after spreading and trafficking the chippings tend to settle with their least dimension perpendicular to the surface.

The ALD may be determined directly on a sample of about 200 stones by measuring the least dimension of each stone with calipers and calculating the average for the sample. Alternatively, it may be derived by first determining the grading, median size and flakiness index of the sample, and then calculating or reading from a nomogram using the established relationships between these factors.

For surface dressings, the stone and binder application rates are both related to the ALD by such considerations as the chippings lying shoulder to shoulder in a single layer, an assumed percentage of voids contained within the ALD volume, and a given proportion of these voids being filled with binder.

Generally, higher ALDs mean better packing, higher binder demand, and possibly greater texture depth, depending upon abrasion value and embedment (Heslop *et al.* 1982).

5.3.3.5 *Elongation index*
An aggregate particle is elongated when it has one dimension significantly greater than its other two dimensions. This shape factor is numerically standardized in BS 812: Part 1 by defining an elongated particle as one whose greatest dimension is more than 1.8 times the arithmetic mean of the aperture sizes of the two test sieves that limit the size fraction of the particle.

For example, consider a 14–10 mm particle:

$$\text{Mean sieve size} = \frac{14+10}{2} = 12 \text{ mm}$$

The particle is defined as elongated if its length is more than

$$1.8 \times 12 = 21.6 \text{ mm}$$

The elongation index of an aggregate sample is determined by the following procedure:

(i) Discard aggregate retained on 50 mm and passing 6.3 mm test sieves and ensure that sufficient remains for test, for example, at least 5 kg for 28 mm or 1 kg for 14 mm nominal size aggregate.
(ii) Sieve into size fractions and discard fractions of less than 5% by mass of the test portion.
(iii) Test each remaining fraction separately by gauging by hand each particle against its appropriate length gauge.

$$\text{Elongation index } (I_E) = \frac{\text{total mass of particles refused by gauges}}{\text{total mass of particles tested}} \times 100\%$$

The elongation index is reported to the nearest whole number together with the sieve analysis of the aggregate sample tested.

The elongation index is rarely prescribed in Standards for bituminous mixtures and surface dressings, possibly because an elongated particle of aggregate has a tendency, on average, to be flaky, and the flakiness index using slotted sieves can be determined more quickly. For example, the 1987 revision of BS 63: Part 2 caters specifically for surface dressing aggregates for the first time, and yet, although flakiness index is specified, no limits are set for elongation index, nor are there any set for coated chippings in the 1985 edition of BS 594: Part 1. There is therefore a case to delete the elongation index test from BS 812.

However, it is also argued that because there is not a clear dependent relationship between elongation and flakiness for many aggregates and because elongated particles make poor chippings for surface dressing and coated chippings for rolled asphalt, it is appropriate to determine elongation index directly in such cases. The test would therefore serve a local rather than a national need, and limits would be prescribed in accordance with local requirements. Consequently, the test is being retained in the current revision of the Standard.

5.3.3.6 *Angularity number*

The angularity or lack of roundness of an aggregate affects the workability and stability of bituminous mixtures, rounded particles providing better workability and angular particles better stability.

The angularity number is a measure of the angularity of an aggregate relative to that of a smooth well-rounded gravel.

The test is based on the observation that the volume of air voids in a compacted angular aggregate is greater than that in a rounded aggregate. The air voids in a compacted, smooth, well-rounded single size gravel average 33% of the volume, and the angularity number is therefore defined as the amount by which the measured percentage of air voids exceeds 33 (BS 812: Part 1: 1975).

The test is intended essentially for comparing the properties of different aggregates for mix design. As it can be applied only to aggregates that do not break down under the vigorous compaction procedure, it has only limited application, and its results are not comparable with those of the bulk density test.

The test portion, which must be taken from the predominant single size fraction of the sample, is selected from the following list:

-20 mm $+14$ mm
-14 mm $+10$ mm
-10 mm $+6.3$ mm
-6.3 mm $+ 5$ mm.

The oven-dried aggregate is compacted in three layers in a calibrated 0.003 m^3 volume (150 mm×150 mm) steel cylinder, using 100 blows of the tamping rod for each layer. The aggregate in the cylinder is weighed and, after repeating the same procedure twice, the mean mass of the three determinations is calculated. Then, where:

M is the mean mass (g) of the aggregate in the cylinder,
C is the mass (g) of the water to fill the cylinder,
G_A is the relative density of the aggregate on an oven-dried basis

$$\text{Volume of voids, } V = 100\left(1 - \frac{M}{CG_A}\right) \%$$

As the angularity number is the amount by which V is greater than 33%, that is $V - 33$:

$$\text{Angularity number} = 67 - \frac{100M}{CG_A}$$

The angularity number is reported to the nearest whole number and is stated in the Standard to range from 0 to about 12.

5.3.3.7 Surface texture of coarse aggregate

Representative, visual characteristics of coarse aggregate particle surface texture are given under six headings in BS 812: Part 102. They are intended to convey the impression gained from a visual inspection of hand specimens, and no attempt is made to call on petrographic terms. They are helpful to the practising engineer and technician, and the headings (with added examples in parentheses) are as follows:

Glassy:	Conchoidal fracture (e.g. black flint and vitreous slag).
Smooth:	Water-worn (e.g. gravels), or smooth owing to fracture of laminated or very fine-grained rock (e.g. slate, marble, some rhyolites).
Granular:	Fracture showing more or less uniform grains (e.g. sandstone, oolite).
Rough:	Rough fracture of fine-or medium-grained rock containing no easily visible crystalline constituents (e.g. dolerite, porphyry, Carboniferous limestone).
Crystalline:	Containing easily visible crystalline constituents (e.g. granite, gabbro, gneiss).
Honeycombed:	Containing visible pores and cavities (e.g. some blast furnace slag, foamed slag, pumice).

The surface texture of a coarse aggregate has a significant effect on its functions in a bituminous mixture. It influences binder–aggregate adhesion so that, for example, the adhesion between bitumen and the glassy, cryptocrystalline surface of fractured flint is poor, whereas that between bitumen and the rough surface texture of dolerite is good. There are other contributory factors, but most of the research indicates that the dominant property is that of surface texture. In association with particle shape, it affects the workability and the stability and flow of bituminous mixes and the performance of surface dressings. Furthermore, in conjunction with the macro texture of the wearing course, it has a direct and vital bearing on the wet-weather skid resistance of a bituminous surfacing.

In these circumstances it would seem beneficial to have a simple, rapid, and practical test to measure surface texture directly, but to date there are no standard methods in everyday use. The scanning electron microscope is used for this purpose in research on aggregates (Williams & Lees 1969), and micro-texture depths of between 10 and 100 μm are considered necessary for adequate wet-weather skid resistance in order to penetrate the thin film of water (less than 25 μm) remaining on the aggregate after the bulk water has been removed, and to establish real areas of contact between the tyre tread and road surface (Williams 1976). However, because of the difficulties in devising an appropriate test to measure texture in isolation and, not least, in applying the results to specific problems, those tests which are in regular use only indirectly assess its influence on the relevant and essential properties and performance of a bituminous mixture. Examples of such tests are coating and stripping, immersion shear, Marshall, creep, polished–stone value, and sideway-force coefficient.

5.3.3.8 Shape and surface texture of fine aggregate

There are no British Standard tests for classifying the shape or surface texture of fine aggregates. Microscopic examination used by geologists and research workers is too time-consuming for routine testing as well as requiring special skills, but is nevertheless very useful for proving quality.

One approach that has potential but has not been widely used relies on the principle that smooth, rounded sand particles flow more freely than rough, angular particles (Rex & Peck 1956). For the test, a weighed, washed, and dried sieve fraction of the sand sample is placed in a glass jar fitted with a metal conical screw cap having an accurately dimensioned orifice and a cork stopper. The jar is inverted and clamped and, after removing the stopper, the time required for the sand to flow through the orifice is measured.

The rate of flow is assumed to be a function of the mass of the test portion and its grading, particle shape, and surface texture. If the initial mass of every test portion is the same, then the difference in time required for a unit volume of various sands to flow through the orifice will reflect the differences in their shape and texture. This flow time has been termed the 'dry viscosity'.

If t is the time (s) of flow of the test portion, m is the mass (g) of the test portion, and d is the relative density of sand on an oven-dried basis, then:

$$\text{Dry viscosity} = 100 \frac{td}{m} \text{ s/100 ml of solid volume}.$$

In the procedure proposed by Rex & Peck (1956) of the US Bureau of Public Roads, the orifice diameter is $\frac{3}{8}$ in (9.5 mm), the sand fraction ASTM No 20–30 (850 to 600 μm), and the mass of test sample for each determination is 500 g. Three determinations are made, and their flow times, which will probably lie between 20 and 40 seconds, should not differ by more than 2 seconds. The dry viscosity is then the mean of the three determinations. This figure is divided by the dry viscosity of the same size fraction of a standard sand (Ottawa sand was used by Rex & Peck) to obtain the time index. Good correlation was found between the values of time index and the discernable surface characteristics of the sand particles, the indexes ranging from 1.12 to 1.53.

The resistance to deformation of asphalt mixtures is considerably influenced by the surface characteristics of the sand, but these are not necessarily uniform throughout the grading. There is therefore a case for determining the time index on more than one size fraction, ensuring that it is narrow in order to minimize the effect of grading variations in the test sample. The test has been employed in the UK, using an orifice size of ¼ in (6.35 mm) and a size fraction of 600 to 425 μm.

5.3.4 Physical properties

5.3.4.1 *Relative density and water absorption*
The relative density of an aggregate is the ratio of its mass to the mass of an equal volume of water.

In BS 812: Part 2 there are three methods for determining the relative density, all of which enable the water absorption to be found *en route* if required.

Apparatus	*Aggregate size*
Wire basket (not for friable aggregate)	+10 mm
Gas jar	−40 mm + 5 mm
Pycnometer or gas jar	− 5 mm + 75 μm

As aggregates are porous and often have a variable composition, different sizes and gradings of the same aggregate may yield different test results. Hence the same gradings must be used when comparing the relative densities and absorptions of different aggregates. Each test on a coarse aggregate should be made on as narrow a size range as practicable to minimize these variables and, typically, size fractions should not exceed 40–20 mm, 20–10 mm, 10–5 mm, 5 mm–75 μm, and −75 μm. It is not normally necessary to subdivide fine aggregate.

The BS method requires two representative portions to be tested, and the test result is the mean of the two determinations.

Test portions are pretreated by being thoroughly washed to remove all traces of undersize material, and, for fine aggregate, the decantation procedure is used to reject all material passing 75 μm.

It is standard BS practice not to pre-dry the test portion before testing for water absorption. Instead the test portion is immersed in water for 24 hours at 20±5°C immediately after pretreatment, after which it is dried to a saturated and surface-dried (SSD) condition and weighed. This procedure usually yields higher water absorptions than if the test portion were pre-dried typically at 105°C for 24 hours, which is the preferred procedure in some standards such as ASTM C127 and C128.

Failure to take due care in determining the correct SSD condition causes errors in the absorption value. Cloths must be dry, soft, and absorbent for each coarse aggregate test, and if there is difficulty in catching the 'just free-running' condition of a fine aggregate then assistance may be obtained by using either a 75 mm cone frustum mould for a miniature slump test, or an inverted glass funnel, both of which are referred to in the Standard.

After weighing, the SSD aggregate is oven-dried at $105 \pm 5°C$ for 24 ± 0.5 h, cooled, and weighed.

Taking the pycnometer or gas jar method as an example, then, where:

A is the mass (g) of the saturated and surface-dried (SSD) test portion in air,
B is the mass (g) of the pycnometer (or gas jar),
C is the mass (g) of the pycnometer (or gas jar) filled with water, and
D is the mass (g) of the oven-dried test portion in air.

$$\text{Relative density on an oven-dried basis} = \frac{D}{A-(B-C)}$$

$$\text{Relative density on a saturated and surface-dried (SSD) basis} = \frac{A}{A-(B-C)}$$

$$\text{Apparent relative density} = \frac{D}{D-(B-C)}$$

$$\text{Water absorption (\% of dry mass)} = 100\frac{A-D}{D},$$

Estimates of the precision are given in BS 812 for the SSD relative density by the wire basket method, and they are: repeatability 0.02 and reproducibility 0.04, which increase to 0.04 and 0.08 respectively for porous aggregates with a low density, less than 2.60. The repeatability and reproducibility for water absorption are quoted as 5% and 10% respectively of the recorded value, but these (particularly the latter) are considered by the authors to be optimistic figures.

The apparent relative density gives the highest value of the three expressions because the volume of aggregate used in the calculation does not include the volume of water-accessible or permeable pores. Furthermore, the test results are more reproducible than those of the oven-dried relative density because the procedure does not require the measurement of water absorption.

The oven-dried relative density is less than the apparent relative density because its volume does include the permeable pores, and the difference between the two expressions is an indication of the water absorption. Both values are used in bituminous surfacing calculations entailing mass–volume relationships and rates of spread, and in laboratory mix design procedures. The Asphalt Institute method of mix design uses the oven-dried relative density (ASTM C127 and C128 *Bulk specific gravity*) and also the 'effective specific gravity' which is calculated from a laboratory determination of the maximum specific gravity of the loose paving mixture (ASTM D2041) for the purpose of applying a correction for the absorption of bitumen by the aggregate.

The Property Services Agency Airfields Branch method of Marshall asphalt design employs the apparent relative density, the oven-dried relative density, and the water absorption. If the absorption is less than 1%, all mix design calculations are made on the basis of the apparent relative density, thereby assuming that bitumen absorption equates to water absorption. If the water absorption exceeds 1%, then the mean of the apparent and the oven-dried relative densities is used, thus assuming that bitumen absorption is half that of water absorption.

No British Standard for roadmaking aggregates, except BS 1047: 1983, specifies a maximum value for water absorption. When a specification does require absorption to be controlled, then a maximum value of 2% for both coarse and fine aggregate is usually imposed. This limit is applied to aggregates used in all bituminous mixtures on PSA airfields, except that for blast furnace slag it is 4% and for friction course 1.5%.

High and variable water absorption is undesirable for bituminous mix aggregates because of its effect on bitumen demand, aggregate drying (by both drum dryers and drum mix plants), and the problems arising therefrom. It may be indicative of low frost resistance and poor durability, but, by itself, it does not provide conclusive evidence of such weaknesses, and confirmation is required from an experienced geologist to relate it to the aggregate mineralogy.

The relative density of a filler (BS 812: Part 2) is determined by using a 50 or 100 ml density bottle and a liquid which does not react with the filler — preferably purified xylene. The standard procedure includes thorough release of all air trapped in the immersed sample, and this requires the use of a vacuum desiccator.

5.3.4.2 Bulk density of aggregate

The bulk density of an aggregate is its mass per unit volume. In the BS 812: Part 2 method of test for compacted bulk density, a cylindrical container of known volume is filled with either oven-dried or saturated and surface-dried (SSD) aggregate in three tamped layers, and the bulk density is calculated in kg/m^3 by dividing the mass of the aggregate by the volume of the container. Different size containers (0.003 and 0.03 m^3) and numbers of tamping strokes are specified for different nominal size aggregates. The method is similar for the uncompacted bulk density, except that the layers are not tamped. If the oven-dried relative density of the aggregate is known, then the percentage volume of air voids may also be calculated.

When dry sand is moistened, films of water form round the particles and cause an increase in volume, or bulking. The increase is progressive and may extend to 20 or 30% of the loose dry volume for a water content of 4 to 6%, depending on the type of sand and its water absorption. With further increases in water the bulking decreases until the sand is completely immersed or inundated, when its volume approximates to that of its dry condition. The bulk density test method may be used to determine percentage bulking by measuring the uncompacted bulk densities of oven-dried sand and of damp sand at a given moisture content, the volume increases being expressed as a percentage of the dry volume. Then, where:

d_1 is the uncompacted dry density of oven-dried sand,
d_2 is the uncompacted dry density of damp sand, and
m is the moisture content of damp sand as a percentage of oven-dried mass.

$$\text{Bulking} = \frac{d_1(100+m)}{d_2} - 100\% \ .$$

Because aggregate grading, container size and shape, state of compaction, and moisture content all have a bearing on the value of bulk density, the BS test is primarily intended for comparing the properties of different aggregates. If a practical mass/volume conversion factor or percentage bulking is required, then a test procedure appropriate to the application should be employed instead.

The compacted bulk density test is specified in BS 594 and BS 4987 as the basis for adjusting the binder contents of all recipe type mixes, from roadbase to wearing course, using blastfurnace slag coarse aggregate. In BS 594 the slag bulk densities range from 1440 to 1120 kg/m^3, and the binder content, which at 1440 kg/m^3 is 0.0 to 0.2% higher than that specified for crushed rock (depending on the mix), is increased by increments of 0.1 or 0.2% for every decrease in the slag bulk density of 80 kg/m^3. This results in an overall increase in the binder content of 0.4 to 0.8% for a total decrease in bulk density of 320 kg/m^3.

The effect of the change in slag coarse aggregate bulk density for BS 4987 coated macadam recipe mixes is more pronounced. This Standard specifies that for a decrease of 320 kg/m^3 between the same maximum range of bulk densities as in BS 594, the bitumen content should be increased by between 1.3 and 2.0%, depending on the type of mix.

5.3.4.3 *Bulk density of filler in toluene*
Filler in an asphalt mix is essentially the material passing 75 μm, and its grading and mineral composition have an influence on the performance of dense bituminous mixes.

The viscosity of a filler/binder system is inversely proportional to the effective pore size of the compacted filler which, in turn, is proportional to the fineness of the filler. The relative fineness of a filler can be expressed in terms of its bulk density in toluene. In this BS 812 test a 10 g portion of oven-dried filler is shaken in toluene in a 50 ml glass-stoppered measuring cylinder and allowed to settle for 6 hours. If its volume at the end of this period is V ml, then taking the mean of three determinations.

$$\text{Bulk density of filler in toluene} = \frac{10}{V} \text{ g/ml} \ .$$

The limits are specified in BS 594 as 0.5–0.9 g/ml, in PSA Airfields Branch Specification Part 4 for Marshall asphalt as 0.5–0.95 g/ml, and in BS 5273 for dense tar surfacing as 0.5–0.8 g/ml.

A low bulk density indicates a fine filler because its settled volume has a high voids content. The limit of 0.5 g/ml is intended to reject fillers that are too fine (clay size particles) and, therefore, difficult to incorporate in the mix. Hydrated lime has a low bulk density, although usually not less than 0.5 g/ml. As a result it has a greater stiffening effect on the filler/binder system than the crushed limestone customarily used. In the main, this is the reason for its use by the Property Services Agency and the British Airports Authority in friction course as it restrains stripping, enhances stability, and improves durability by permitting the use of a softer (200 pen) bitumen.

The upper limit of 0.95 g/ml is set because above this value the filler is too coarse to have much effect on the filler/binder viscosity. A reduced upper limit of 0.8 g/ml is set for dense tar surfacing as the flow characteristics of the tar-filler mixture, which is the initial basis of the design of the material, are highly dependent on the fineness of the filler, and when the bulk density exceeds 0.9 g/ml there is a rapid reduction in resistance to deformation (Lee & Rigden 1945). Consequently, setting the maximum at 0.8 g/ml is a safeguard against this occurring, and it is important to appreciate that it applies to both the added filler and the minus 75 μm material in the aggregate, particularly if present in a significant amount such as in quarry fines.

The PSA Airfields Branch Specification Part 4 emphasizes the role of filler by specifying that at least 60% of the material passing 75 μm in both Marshall asphalt and rolled asphalt mixes must be in the form of added limestone or Portland cement filler. BS 594 also specifies added filler to be limestone or Portland cement, but no minimum quantity is required other than that to achieve the aggregate grading for the specified mix. BS 4987 takes a more relaxed view for the dense as well as the open graded mixes by allowing that only 75% of added filler need pass 75 μm (instead of 85% for rolled asphalt) and by not specifying any limit on the bulk density in toluene. No minimum quantity of added filler is specified, but, if it is used, it must consist of crushed rock, crushed slag, hydrated lime, Portland cement, or other approved material.

5.3.4.4 Voids of dry compacted filler

In the design of dense tar surfacing, the fractional voids content of compacted filler is measured to give the required proportion of tar to be associated with the filler, and the critical minimum filler/sand ratio (RRL 1962). The 'Rigden' apparatus is used for this purpose and is illustrated in BS 812: Part 2. Briefly, it comprises a 25 mm diameter cylinder which can be raised and allowed to fall through 100 mm on to a base plate. About 10 g of filler is placed in the cylinder, which is then closed with a sliding cylindrical plunger that rests on the filler. The cylinder, filler, and plunger are raised through 100 mm, and released, 100 times. The volume and mass of the compacted filler are measured and the fractional voids content determined. This last is the volume of air expressed as a fraction of the bulk volume of the filler, which is calculated with the aid of the oven-dried relative density of the filler. At least three determinations are made, and the mean is calculated and reported to the nearest 0.01 as the voids of dry compacted filler. The fractional voids content should be in the range 0.3 to 0.4 for suitability as a filler in dense tar surfacing.

5.3.5 Mechanical properties

5.3.5.1 Aggregate crushing value

The aggregate crushing value (ACV) gives a relative measure of the resistance of an aggregate to crushing when subjected to a compressive force that is gradually increased to a specified maximum value in a prescribed time.

In the standard test in BS 812: Part 3, a measured volume and mass of sieved and surface-dried 14–10 mm aggregate is placed and tamped in three layers to a total uniform thickness of 100 mm in a 150 mm diameter hardened steel cylinder. It is then loaded through a close fitting ram with a force that is increased linearly from 0 to 400 kN in 10 minutes. The crushed aggregate is removed from the cylinder and the mass of material passing a 2.36 mm test sieve is determined. Then where:

A is the mass (g) of the surface-dried 14–10 mm aggregate test portion, and
B is the mass (g) of the crushed aggregate passing the 2.36 mm test sieve,

$$\text{ACV} = \frac{B}{A} \times 100 \ .$$

The test result is the mean of two determinations and is reported to the nearest whole number. The precision estimates of the test given in BS 812 are repeatability 0.8 and reproducibility 1.5.

Other single size aggregate fractions in the range of 28 mm to 2.36 mm may be tested in the same apparatus, using appropriate sieves ranging from 5.0 mm to 600 μm for separating the crushed aggregate. Such non-standard tests give higher (that is, weaker) results for fractions larger than 14–10 mm and lower (that is, stronger) results for smaller size fractions. These variations, whose amount depends on the type of aggregate, are believed to be because the minute fissures and planes of weakness created by the quarry crushers are less prevalent in the smaller sizes.

Fractions smaller than 14–10 mm may be tested in a 75 mm diameter cylinder, using a layer thickness of 50 mm and a maximum force of 100 kN. The result is then similar to the standard value, so the increased confinement of the smaller cylinder and the thinner bed of aggregate have a compensating effect, as the maximum applied stress of 22.64 N/mm^2 is the same for both sizes of cylinder.

Very strong aggregates have an ACV of about 10, and while it is possible for a value of this order to be achieved by the strongest example in most of the Groups, average values are considerably higher. Values for the commonly used roadstones range from about 15 for basalt and porphyry through 25 for limestone to 28 for blast-furnace slag (RRL 1959).

In general, an aggregate with an ACV of 25 or less will have the strength required for all roadmaking purposes. Some aggregates, notably blastfurnace slag, having ACVs of over 30, perform satisfactorily in a surfacing, but, as a rule, natural aggregates with values of this order are suspect or unacceptable on account of other weaknesses such as unsoundness or poor abrasion resistance.

British Standards are hesitant to specify minimum ACVs or other strength values for road aggregates used in bituminous materials, although there are exceptions. No values are specified in the current editions of Part 1 of the Standards BS 63, BS 594, and BS 4987, but, at one time, a maximum value of 25 was required for gravel coarse aggregate used in both rolled asphalt and coated macadam.

BS 63: Part 2 specifies that surface dressing aggregates shall have a 10% fines value (see next section) of not less than 160 kN, or 85 kN for blastfurnace slag. These values equate to ACVs of 25 and 31 respectively when the formula:

$$\text{ACV} = 38 - 0.08 \text{ (10\% fines value)}$$

of the South African Standard SABS 1083–1976 is applied.

The road and paving mastic asphalt Standards BS 1446 and BS 1447 do not specify any ACV values for the coarse aggregate, but a limit of 28 is required for the coarse aggregate in flooring mastic asphalt specified in Standards BS 6577 and BS 6925.

The PSA Airfields Branch Specification Part 4, 1979, specifies a maximum ACV of 30 for crushed rock and slag and 25 for gravel aggregates used in all bituminous mixtures except friction course (crushed rock), for which the value is reduced to 16. The open grading and low fines content of the comparatively thin layer (20 mm) of friction course increases any tendency to crushing of the coarse aggregate during compaction. Crushing would reduce permeability and, possibly, durability, as well as facilitating the loss of surface stones under traffic. Consequently a stronger aggregate, as indicated by a lower ACV, is specified.

The shape of an aggregate affects its ACV. Research carried out by Dhir *et al*. (1971) on fine-grained, basic intermediate rocks (basalt, andesite, and dacite) established that the ACV (and the aggregate impact value — AIV) has a component which is characteristic of the petrographic type and a component which is a function of the flakiness index. For these aggregates the mean relationship could be expressed as:

$$\text{ACV} = 13.5055 + 0.0677 \, I_F \ .$$

One of the objectives of the research by Dhir *et al*. was to study the effect of shape on the aggregate crushing value residue (ACVR), which is the percentage mass of the original size fraction retained on the 10 mm test sieve after conducting the ACV test. The mean relationship established for the same group of aggregates was

$$\text{ACVR} = 34.2024 - 0.2035 \, I_F$$

and it was concluded that ACVR was a more sensitive indicator of aggregate quality, particularly flakiness, than ACV.

As flakiness has a significant effect on the ACV (and AVCR) of an aggregate, it should be one of the properties examined with a view to improvement when marginal test results are obtained. It can be of particular relevance when aiming to achieve low values, such as that of 16 for PSA friction course aggregate, even though the associated maximum flakiness index of 25% already implies good control in processing the 10 mm nominal size aggregate.

If ACVR is a reflection of the extent of aggregate fracture during rolling, then minimizing flakiness would improve the quality of the compacted material by limiting this undesirable feature.

The discrimination of the ACV test between aggregate strengths becomes less precise above a value of about 25. Above about ACV 29, which is entering the realm of the weaker aggregates which, although of doubtful quality for surfacing courses could still be suitable for bases, the strengths are exaggerated. To avoid these problems the 10% fines test should be used.

5.3.5.2 Ten per cent fines value

The 10% fines value is described in BS 812:Part 3. It is the compressive force required to generate 10% by mass of crushed aggregate passing the 2.36 mm test sieve, when a 100 mm thick horizontal bed of 14–10 mm sieved and surface-dried aggregate, tamped in three layers, is loaded in the standard ACV apparatus at a rate that achieves the required force in 10 minutes. During the development of the method from the ACV test (Shergold & Hosking 1959) it was found that the curve of crushed fines against force was linear between 7.5 and 12.5% of fines, and that the average value of the intercept on the y-axis was -4%. Furthermore, the mean depth of penetration of the loaded ram into the cylinder at this stage was 15 mm for rounded aggregates, 20 mm for normal crushed aggregates, and 24 mm for honeycombed aggregates. Guided by these data, or by the alternative relationship that the required force in kN equals approximately 4000/(aggregate impact value (AIV)), two determinations are made at the same force to give a percentage fines within the range of 7.5 to 12.5%. Then where:

x is the maximum force (kN), and
y is the mean percentage fines of two determinations at x kN,

$$10\% \text{ fines value} = \frac{14x}{y+4} \text{ kN}$$

The result is reported to the nearest 10 kN for forces of 100 kN or more, or to 5 kN for loads of less than 100 kN.

The precision estimates of the test given in BS 812 are repeatability 7 kN and reproducibility 14 kN.

It is permissible to test non-standard size fractions of aggregate on the standard apparatus, using the standard volume of aggregate and, as for the ACV test, separating the fines on the appropriate test sieve specified in Table 1 of BS 812: Part 3. There are, however, insufficient published data to correlate the results with those obtained by using the standard aggregate size of 14–10 mm. Experimental work has also been conducted with a modified version of the standard apparatus that enables a smaller test sample to be used and less force to be applied (Shergold & Hosking 1963). The modified test employs a 3 in (76 mm) diameter cylinder, a 2 in (51 mm) deep tamped layer of 3/8 in (9.5 mm) size aggregate, and a BS 10 (1.70 mm) test sieve for separating the fines. The force generating 10% fines is then multiplied by four to give a modified 10% fines value, which is 'approximately equivalent to the 10% fines value obtained in the standard test'.

The study that established the standard test method confirmed that weaker materials are compacted to a dense mass before the ACV force of 400 kN is reached, and by filling the voids between the particles the rate of production of further fines is reduced. On the other hand the additional voids lying within the particles of vesicular or honeycombed aggregate such as blastfurnace slag ensure that there are more spaces for the crushed material to fill, and so the fall-off in the rate of fines production for this type of aggregate is delayed. As a consequence, soft, weak aggregates with small internal voids could be ranked incorrectly as being stronger than weak vesicular or honeycombed aggregates. By reducing the production of fines, the 10% fines test eliminated this anomaly. Moreover, test results on a range of aggregates with ACVs of less than 29 showed that the 10% fines test ranked them in a similar order of merit. These attributes, and the more logical arrangement of a higher result indicating a stronger material, support the view that the 10% fines test should become the preferred BS strength test, as well as one of the preferred tests in the forthcoming European Standards for aggregates.

If the relationship given in SABS Standard 1083–1967 of:

$$ACV = 38 - 0.08 \,(10\% \text{ fines value})$$

or

$$10\% \text{ fines value} = 12.5 \,(38 - ACV)$$

is applied, then an ACV range of 10 to 30, which covers the vast majority of aggregates suitable for use as roadstone, would correspond to a 10% fines range of 350 to 100 kN. Surface dressing aggregates to BS 63: Part 2 are specified to have a minimum 10% fines value of 160 kN, or 85 kN for blastfurnace slag. At present no other British Standard specifies the test for bituminous road making aggregates, but this would seem to be a reflection of the reservations held about the need for specifying nationwide aggregate strength limits in British Standards, and not an implied criticism of the test itself.

5.3.5.3 *Aggregate impact value*

The aggregate impact value (AIV) gives a relative measure of the resistance to crushing of an aggregate subjected to repetitions of a suddenly applied force.

In the standard test described in BS 812: Part 3, a measured volume and mass of sieved and surface-dried 14–10 mm aggregate is placed and tamped in a single horizontal layer having a thickness of the order of 27 mm in a 102 mm diameter × 50 mm deep hardened steel cylindrical cup. The cup is fixed firmly on the base of the impact test machine standing on a 450 mm thick, plane, and level concrete floor. The hammer, which has a 100 mm diameter cylindrical head and a total mass of 13.5–14.0 kg, is raised vertically in guide rods to a height of 380±5 mm above the surface of the aggregate and is released so that it falls freely and directly on to the aggregate. Fifteen blows are delivered. The crushed aggregate is removed from the cup and the mass of material passing a 2.36 mm test sieve is determined. Then, where

A is the mass (g) of the surface-dried 14–10 mm aggregate test portion, and
B is the mass (g) of the crushed aggregate passing a 2.36 mm test sieve,

$$AIV = \frac{B}{A} \times 100 \,.$$

The test result is the mean of two determinations and is reported to the nearest whole number.

The precision estimates of the test given in BS 812 are repeatability 1.0 and reproducibility 2.0.

Non-standard aggregate sizes that are smaller than 14–10 mm may be used, but, as for the ACV test, slightly lower results will be obtained.

For the majority of aggregates, although the AIV has a tendency to have slightly lower values, AIV and ACV are approximately comparable. This is not entirely to be expected, bearing in mind the different methods of loading, even though the AIV test was developed as an auxiliary to that of ACV, and similarity of test results was built into its design (Collis & Fox 1985). The AIV test does, however, distinguish brittle aggregates such as flint, chert, and quartzite which are less resistant to impact forces, in that they have AIVs about 5 points higher than their ACVs.

Compared with the ACV and 10% fines tests, the AIV test has the advantages of a smaller, simpler, and more portable machine, a quicker test procedure, and a smaller sample. The machine must, however, stand firmly on the specified base. The guide rods must be true and rigid and the cup must be exactly located so that the hammer falls freely and cleanly on to the aggregate. If these precautions are not taken, the results will be unreliable (Ramsay et al. 1973).

Spence (1979) confirms the necessity for these precautions, but he found that by using a concrete floor in excess of specified minimum thickness of 450 mm the samples were poorly compacted and the fines were concentrated in the bottom of the cup. He attributed this to observed hammer rebound and movement of the machine across the floor. When using thinner floors the samples were found to be highly compacted with a uniform distribution of fines throughout; the results obtained were lower and more repeatable than those for the thick floor, and they correlated more closely with the ACV. When a large softwood block was placed under the test machine, with the grain vertical, the tested samples were similar in appearance and gave similar results to those tested on the thinner floors, irrespective of the type or thickness of the floor used. The incorporation of such a block in the standard test equipment would improve the portability of the apparatus which is at present inhibited by the requirement to mount it on a concrete block or floor having a minimum thickness of 450 mm.

As with the ACV test, the cushioning effect of the generated fines can cause anomalous results with weaker and vesicular aggregates that have an AIV above about 25. To improve the discrimination at this level, a modified impact test was developed (Shergold & Hosking 1963) whereby the number of blows is reduced to generate between 5 and 20% fines, and this amount is then adjusted proportionately to obtain the 15-blow equivalent. If $x\%$ by mass of fines is generated by n blows:

$$\text{Modified AIV} = \frac{15x}{n}.$$

As for the aggregate crushing value (ACV), Dhir et al. (1971) indicated that the AIV is composed of a component that is characteristic of the petrographic type and a component that is a function of the flakiness index (I_F). For the deliberately narrow range of fine grained basic–intermediate rocks studied in the investigation, the relationship was found to be:

$$AIV = 9.7813 + 0.0571\, I_F\ .$$

Likewise the effect of shape on the AIV residue retained on the 10 mm test sieve (AIVR) was also studied, and, for the same rocks, this relationship was:

$$AIVR = 54.1757 - 0.3067\, I_F\ .$$

It was concluded that the AIVR (and the ACVR) should be included in an analysis of the properties of an aggregate, as it was a more sensitive indicator of aggregate quality than the standard test. Although this recommendation has not been adopted in BS 812, it is clear that more information about the properties of aggregates could be obtained by minor extensions to the current procedures of the ACV and AIV tests, and, possibly, the 10% fines test. Such information could be of use in the final analysis when selecting aggregates for bituminous surfacings, particularly chippings for rolled asphalt and surface dressing on highly stressed sites.

5.3.5.4 Los Angeles abrasion value

The Los Angeles abrasion value (LAAV) test is described in ASTM C131 for coarse aggregates smaller than 37.5 mm, and in ASTM C535 for coarse aggregates larger than 19 mm. The method specified in test C131 is the one normally used for roadstone.

The test is not included in BS 812, but it is a well established and internationally recognized test procedure which assesses the resistance of an aggregate to attrition by impact and abrasion forces.

The test machine comprises a closed, hollow steel cylinder with internal dimensions of 711 mm diameter by 508 mm length, fitted with a 150 mm wide covered opening and an 89 mm wide by 508 mm long removable steel shelf, which is fixed to the internal surface of the cylinder, parallel to the axis, and projecting radially inwards. The cylinder is mounted horizontally on stub axles and is rotated at a speed of 30 to 33 rpm. For test C131 it is charged with twelve to six 47 mm diameter steel balls having a total mass of 5000 to 2500 g respectively. The size of the charge depends on which of the four prescribed gradings, A to D, is tested. Twelve balls are required for grading A (37.5 to 9.5 mm), eleven for B (19.0 to 9.5 mm), eight for C (9.5 to 4.75 mm), and six for D (4.75 to 2.36 mm). The mass of the test portion is 5000 g, irrespective of the grading.

The aggregate is washed (if necessary), oven-dried, and sieved. The test portion, of required mass and appropriate grading, is prepared and then tumbled in the machine with the relevant charge for 500 revolutions. On removal from the machine, the mass of aggregate retained on a 1.70 mm test sieve is determined (by wet sieving, if necessary). Then, where:

M_1 is the mass (g) of the test portion, and
M_2 is the mass (g) of aggregate retained on the 1.70 mm test sieve after test,

$$\text{LAAV} = \frac{M_1 - M_2}{M_1} \times 100$$

The precision of the test is given as the difference between two results expressed as a percentage of their average value. For 19.0 mm nominal size aggregate (grading B), having a LAAV of between 10 and 45, repeatability is 5.7% and reproducibility is 12.7%.

Additional information about the uniformity of the sample can be obtained, in addition, by determining the loss after 100 revolutions. For material of uniform hardness, the ratio of loss at 100 revolutions to that at 500 revolutions should not greatly exceed 0.2.

The test method is simple, although noisy. The speed of rotation must be uniform, and the shelf, which is subjected to severe wear and impact, must be carefully maintained. The results of the test for aggregates for corresponding size have been stated to be numerically similar to the ACV for values up to about 30, that is, for all aggregates that would normally be used in bituminous surfacings in the UK (RRL 1962), but for weaker aggregates the LAAV is higher. The US Federal Aviation Administration Standard for Airports, 1974, specifies maximum LAAVs of 40 for bituminous surface course aggregates (Item P-401) and 50 for base course (Item P-201) ('base' in UK terminology).

Recent work on Northern Ireland basalts (Woodside & Peden 1983) has also shown good correlation between LAAV and both ACV and AIV. In 1979, it was proposed that the LAAV test should be standardized internationally (PIARC 1979), and it is likely to be included as one of the preferred tests in the forthcoming European Standards for aggregates.

5.3.5.5 *Aggregate abrasion value*

The aggregate abrasion value (AAV) is an important aggregate property in the Department of Transport *Specification for Highway Works*, 1986. It is used in the selection of wearing course aggregates for high speed roads, to ensure that they have adequate resistance to surface wear by traffic in order to maintain good macrotexture, which, at high speeds, is an essential factor in the dispersion of water at the tyre/wearing course interface, and the prevention of aquaplaning.

The British Standard AAV test uses a similar machine to the Dorry abrasion test, which was specified before the 1960 edition of BS 812 for the determination of the coefficient of hardness of cylindrical rock specimens. The machine comprises a horizontal 600 mm diameter steel or cast iron grinding lap equipped with diametrically opposed test specimen holders and sand feeder hoppers. Each test specimen is pressed onto the lap by a total load of 2 kg and fed across its full width with 600 to 425 μm Leighton Buzzard silica sand at a rate of 700 to 900 g/min, while the lap is rotated at 28 to 30 rpm. A test specimen consists of at least 24 representative, non-flaky, washed, and surface-dry particles of 14–10 mm aggregate, of measured SSD relative density, and having a mass of about 90 g. The particles are mounted in resin, shoulder to shoulder, in a flat 92 mm by 54 mm single layer, with their flattest surface exposed. In the 1960 version of test the volume of the aggregate had to be a measured 33 ml, but that volume is now assumed when the particles are selected and mounted as specified. Each test specimen is weighed before test.

The test specimens are abraded simultaneously on the machine for 500 revolutions of the lap, and, after abrasion, are removed from the machine, cleaned, and weighed again. The AAV of each specimen is calculated as the percentage loss in volume of the original aggregate (of assumed volume 33 ml). Hence, where:

A is the mass (g) of a test specimen before abrasion,
B is the mass (g) of a test specimen after abrasion, and
d is the relative density of the aggregate on a saturated surface-dried basis,

$$AAV = 3(A-B)/d .$$

The AAV is reported as the mean of the two determinations to two significant figures. BS 812 advises that the repeatability and reproducibility of the test are 1.5 and 3.0 respectively.

As for the aggregate crushing value and the aggregate impact value, a numerically lower result indicates a greater resistance in the test. Values range from 1 or 2 for very hard aggregates such as flints to above 20 for soft aggregates (Salt 1977), but values in excess of 14 have been considered to indicate inadequate abrasion resistance and to be unsuitable for road surfacing (Hawkes & Hosking 1972).

In their recommendation for defining levels of resistance to skidding, Salt & Szatkowski (1973) proposed that the coarse aggregates with AAV not exceeding 12 and 10 should be used in wearing courses of roads carrying traffic up to 1750 and in excess of 1750 commercial vehicles per lane per day respectively. Subsequently, in 1976, the Department of Transport published Technical Memorandum H 16/76 which specified, *inter alia*, maximum AAVs for surface dressing chippings, coated chippings applied to dense wearing courses, and coarse aggregates for coated macadam wearing courses. These values are shown in Table 5.13, which is taken from Appendix 2 of H 16/76 and relates the maximum AAVs to traffic in commercial vehicles per lane per day calculated to be using the lane at a time two-thirds of the way through the anticipated life of the surfacing.

The values are based on the recommendations of LR510, but a distinction is made in H 16/76 between chippings and coarse aggregates for coated macadam wearing courses, in that the latter are permitted to have higher values than chippings owing to their lower level of exposure to traffic wear. Moreover, there is no maximum requirement specified for wearing course coarse aggregates in rolled asphalt and dense tar surfacing. The values were confirmed in the Department of Transport *Notes for Guidance on the Specification of Highway Works*, 1986.

The patented proprietary surfacing material known as Delugrip (Williams 1975) employs a blend of two different size coarse aggregates with different rates of wear. The smaller size stone is specified to have an AAV of at least 1.3 times that of the larger size harder stone, with the objective of maintaining, under traffic, the largely interconnected macrotexture that is a design feature of the material. A mixture of

Table 5. 13 — Traffic loadings and maximum aggregate abrasion values for flexible surfacing. (Reproduced from Technical Memorandum H16/76, Department of Transport)

Traffic in commercial vehicles/lane/day	Under 250[†]	Up to 1000	Up to 1750	Up to 2500	Up to 3250	Over 3250
Maximum AAV for chippings	14	12	12	10	10	10
Maximum AAV for aggregate in coated macadam wearing courses	16	16	14	14	12	12

[†]For lightly trafficked roads carrying fewer than 250 commercial vehicles/lane/day, aggregates of higher AAV may be used where experience has shown that satisfactory performance is achieved by aggregate from a particular source.

Note: A replacement for H 16/76 is being prepared by the Department of Transport, but it is probable that the same values for AAV will be retained. It is also expected that the traffic loading will be that calculated to be using the lane at the end of the anticipated life of the surfacing, instead of the two-thirds life.

two aggregates yields a resistance to skidding and a depth of surface texture that is approximately equal to the mean of those given by the constituents on their own (Salt 1977). If this finding is applicable to Delugrip, it would be advisable, as a first estimate, for the AAV of the larger size, harder stone not to be greater than 10.4 if the maximum AAV of 12, specified by the Department of Transport for traffic of more than 2500 commercial vehicles per lane per day, is not to be exceeded. This value would be reduced if the AAV of the softer stone were more than 1.3 times that of the harder stone.

Resistance to abrasion is a function of the textural strength and mineralogy of an aggregate. Because of the crystal intergrowth, fresh igneous and metamorphic rock aggregates have good abrasion resistance, given a preponderance of hard minerals, particularly quartz, and a uniform texture of medium to small grain size. Consequently, hornfels, quartzite, and porphyry, in that order, have low AAVs of between about 2 and 7, while a large majority of aggregates in the basalt group, which contains less silica, lie within the range 3 to 11 with a mean of about 6 or 7.

Tests made on nine samples of sound and inert basalt aggregate taken from different quarries in the Antrim plateau in Northen Ireland (Woodside & Peden 1983) had an AAV range of 4.6 to 9.4 with an average of 6.3. Four severely weathered, vesicular samples, taken from a tenth quarry to represent materials of inferior standard were also tested, and these had an AAV range of 6.0 to 8.4, with an average of 7.4. All these values are less than the severest value of 10 specified by the Department of Transport, and the difference between the average values for sound and unsound aggregates is only about 18%. So, taking into consideration the precision values referred to above, the AAV test has its limitations for ranking the harder aggregates, and for indicating unsound material and durability.

Of the sedimentary rock aggregates, flint, being composed of cryptocrystalline silica, has a very low AAV of 1 to 2, whereas sandstone and gritstone suitable for

Sec. 5.3] **Tests and standards for aggregates in bituminous materials** 213

roadstone have a wide range of about 3 to 14 because of their dependence on the nature of the cementing materials and the degree of consolidation. Higher values of up to more than AAV 50 were obtained by Hawkes & Hosking (1972) in their study of arenaceous rock aggregates, but, apart from their having high polished–stone values (PSVs) which are frequently associated with poorer abrasion resistance (higher AAV), these aggregates were ranked as being unsuitable for roadstone.

As a group, limestone has the poorest resistance to abrasion because of its low mineral hardness and cleavage. Geological formation and age, however, also have a strong influence, so that limestone aggregates older than the Carboniferous period have the lower AAVs, particularly if they contain some silica impurities.

5.3.5.6 Polished–stone value
The polished–stone value (PSV) is a relative measure of the resistance of a single-size roadstone to the polishing action of a tyre under conditions intended to simulate those that occur with traffic acting on a bituminous road surfacing. The test was introduced in BS 812: 1960, and the numerical value of the result then approximated to the mean skid-resistance value (SRV) of a road surfacing using similar chippings. Since that time modifications have been made to the accelerated polishing machine and the test and reporting procedures, but the present-day results relate closely to those obtained in the original test.

Minimum PSVs are specified in the Department of Transport Specification for chippings and coarse aggregate in bituminous wearing courses and in British Airport Authority Specifications for coarse aggregate in bituminous surfacings for aircraft pavements.

The latest major revision of the PSV test method is described in BS 812: Part 14: 1989. Prepared specimens of the aggregate are polished in an accelerated polishing machine and their wet state of polish is then measured by a pendulum type friction tester. The result is expressed as the polished–stone value after applying any necessary corrections that may arise from the simultaneous testing of the PSV control stone.

The aggregate for test, which must have been taken from a representative sample of a normal production run, consists of clean, dust-free, non-elongated particles of average surface texture all passing a 10 mm test sieve and retained on a 14–10 mm flake-sorting sieve that has a slot width of 7.2 mm. For each test specimen 35–50 particles are mounted in polyester resin shoulder to shoulder, with their flattest surface exposed in a 90.6 mm×44.5 mm single layer that is flat transversely but curved longitudinally. Four specimens are prepared per sample, as well as four specimens of the PSV control stone. Two specimens from each sample are clamped in a specified order round the periphery of a vertical 'road wheel', together with two control stone specimens, so as to form a continuous strip of particles of 406 mm diameter. Polishing is effected by bringing to bear on the aggregate a 200 mm diameter×38 mm width solid rubber tyred wheel with a force of 725 ± 10 N, and, at specified rates, continuously feeding the tyre with corn emery and water for 3 hours, followed by emery flour and water for 3 hours, while the road wheel is rotated at 320 ± 5 rpm (24.5 ± 0.4 km/h peripheral speed). The specimens are removed from the machine, washed and brushed, and immersed in water at $20\pm2°C$ for a period of between 30 minutes and 2 hours.

The state of polish of a saturated specimen is measured by a calibrated pendulum friction tester, at a temperature of $20\pm2°C$, in terms of the frictional resistance generated by the trailing edge of a 31.75 mm wide×25.4 mm deep, inclined, spring-loaded and wetted rubber slider. This is mounted on the end of the pendulum and traverses the surface of the wetted specimen when the pendulum has been released from the horizontal position. The specimen is rigidly located at the lowest point of swing, with its longitudinal axis and its direction of polish lying centrally in the track, and in the same direction as the swing, of the pendulum. The slider is located 510 mm from the point of suspension and it traverses the specimen with a mean nominal force of 22.2 N over a length of 76 mm. The height of the forward upward swing of the pendulum, after it has traversed the specimen, is shown by a friction pointer against a scale calibrated as a measure of the coefficient of sliding friction ($\times 100$).

The test specimens and control stone specimens are tested in this manner five times, and the mean of the last three readings is recorded for each. The procedure of polishing and friction testing is repeated on the remaining two specimens of each sample and control stone, and the mean calculated for each group of four specimens and recorded to the nearest 0.1 unit. Then, provided that the specified conditions concerning the ranges of the results are satisfied, the polished–stone value (PSV) is obtained by reference to an equation by which a correction is applied to the test result depending on the performance of the PSV control stone in the test. It is reported to the nearest whole number, a higher value signifying a greater resistance to polishing. Estimates of the precision of the test given in the Standard are repeatability 3 and reproducibility 5.

The PSV of an aggregate is not, and must not be confused with, the skid resistance value (SRV), the sideway-force coefficient (sfc), or the breaking force coefficient (BFC), all of which are measurements of the skid resistance of a wetted road surface.

The SRV is determined over a contact length of 126 mm with the pendulum friction tester in accordance with the method described in Road Note 27 (RRL, 1969). However, because of the small contact area of the slider, which tends to simulate the action of a patterned tyre, the relationship between texture depth and skidding resistance differs from that obtained with the smooth tyre of the Sideway-force Coefficient Routine Investigation Machine (SCRIM), especially on fine-textured surfaces. On coarse-textured surfaces the results can be misleading because of operational difficulties (Department of Transport HA 36/87, 1987). Consequently the method has its limitations and the results always require skilled interpretation.

The sideway-force coefficient (sfc) is a general term used for the ratio of sideway-force to normal force obtained with sideway-force road testing equipment. In the UK, sfc is determined with the SCRIM which measures skidding resistance by recording the axial force acting on a loaded, smooth-tread, rubber-tyred wheel running freely on a wetted road surface at an angle of 20° to the direction of travel. SCRIM readings, which are continuous records of the frictional resistance usually measured over subsection lengths of 10 m, are corrected for calibration, speed, temperature, and season, as necessary, and adjusted by the Index of sfc to relate the values obtained to those measured with the equipment at TRRL during the period 1963 to 1972.

The corrected and adjusted readings, expressed as a decimal fraction to two

Sec. 5.3] Tests and standards for aggregates in bituminous materials

decimal places, are termed SCRIM coefficients (SC). The speed for standard testing is 50 km/h, but, for sections of road with a radius of curvature of less than 100 m, a speed of 20 km/h is used with a subsection length of 5 m (Department of Transport HD 15/87).

The braking force coefficient (BFC) is a high-speed measurement of skidding resistance which is normally reported by TRRL at 130 km/h. It is determined from the torque acting on the brake of the wheel of a small, car-towed trailer when it is locked for approximately 2 seconds on a wetted surface. It is more difficult, however, to obtain consistent results at high speeds, and there are problems in testing surfaces with high macrotexture if the contact area and normal load between the tyre and surface are not reasonably constant and if the minimum free water film thickness cannot be maintained at different speeds (Davis 1979).

Correlations between SRV, sfc, and BFC have been proposed over the years. The pendulum friction tester is useful for areas such as accident and incident sites and where SCRIM cannot operate accurately, but its dependence on surface texture, referred to above, has confounded attempts to find a unique relationship with sfc. As a guide, an approximate correlation is SRV=105 SC, but this is applicable only to medium textured surfaces (HA 36/87). Measurements of BFC are usually lower than those of sfc, and, typically, BFC is approximately equal to SC×0.80 (Young 1982).

The sfc varies inversely with temperature, owing mainly to changes in tyre resilience. It is also lower in summer than in winter, probably because of an increase in weathering and a coarsening of the detritus, both of which occur in the colder months. The effect of these variables is reduced by averaging three readings of SC taken in a single year during the summer period May to September, after the surfacing is at least one year old. The resulting value is termed the Mean Summer SCRIM Coefficient (MSSC). The sfc also varies with the PSV of the surface coarse aggregate and inversely with traffic volume.

By analysing the results from 139 different sections of road the following correlation between MSSC, PSV, and traffic volume was established for a straight road by Szatkowski & Hosking (1972):

$$SFC_{50} = 0.024 - (0.663 \times 10^{-4}) q_{cv} + (1 \times 10^{-2}) p$$

where SFC_{50} is the equivalent of MSSC, q_{cv} is the traffic volume in commercial vehicles per lane per day, and p is the PSV determined in accordance with BS 812. This relationship, which indicates that a change of one unit of PSV corresponds to a change in SFC_{50} (MSSC) of 0.01, is developed in Table 5.14 taken from LR510 (Salt & Szatkowski 1973). It provides a method of specifying the PSV of the coarse aggregate needed to achieve the required MSSC on a straight road, provided that there is a realistic estimate of the volume of commercial traffic.

Typically, a difficult type of site such as a major priority junction requires a higher skidding resistance than a lightly trafficked straight road. In addition, there may be a greater risk of skidding at some sites of the same type because of an unusual or awkward layout or other hazardous feature. Both the type of site and the risk rating are taken into consideration for the determination of the required level of skidding resistance in Table 5.15 which is reproduced from LR510, 1973. Each site category,

Table 5.14 — PSV of aggregate necessary to achieve the required skidding resistance in bituminous surfacings under different traffic conditions. (Reproduced from TRRL Report LR 510, Table 3)

Required mean summer SCRIM coefficient (MSSC)[†]	PSV of aggregate necessary					
	Traffic in commercial vehicles per lane per day					
	250 or under	1000	1750	2500	3250	4000
0.30	30	35	40	45	50	55
0.35	35	40	45	50	55	60
0.40	40	45	50	55	60	65
0.45	45	50	55	60	65	70
0.50	50	55	60	65	70	75
0.55	55	60	65	70	75	
0.60	60	65	70	75		
0.65	65	70	75			
0.70	70	75				
0.75	75					
AAV	not greater than 12			not greater than 10		

 SFC values in these traffic conditions are sometimes achievable with aggregates of extreme hardness and very high resistance to abrasion, such as certain grades of calcined bauxite.

[†]Expressed in LR510 Table 3 as mean summer SFC at 50 km/h.

comprising a group of site types, is allocated a range of risk ratings from which one is selected for the site under consideration by reference to skidding records and knowledge and judgement of relevent hazards. Each risk rating corresponds to a

Table 5.15 — Minimum† values of skidding resistance for different sites. (Reproduced from TRRL Report LR 510, Table 2)

SITE		DEFINITION	\multicolumn{10}{c}{Mean summer SCRIM coefficient (MSSC) Risk rating}

SITE		DEFINITION	1	2	3	4	5	6	7	8	9	10
A1 (v difficult)	(i)	Approaches to traffic signals on roads with a speed limit greater than 40 mile/h (64 km/h)						0.55	0.60	0.65	0.70	0.75
	(ii)	Approaches to traffic signals, pedestrian crossings and similar hazards on main urban roads‡										
A2 (difficult)	(i)	Approaches to major junctions on roads carrying more than 250 commercial vehicles per-lane per day										
	(ii)	Roundabouts and their approaches						0.55	0.60	0.65		
	(iii)	Bends with radius less than 150 m on roads with a speed limit greater than 40 mile/h (64 km/h)				0.45	0.50	0.55	0.60	0.65		
	(iv)	Gradients of 5% or steeper, longer than 100 m										
B (average)	(i)	Generally straight sections of and large radius curves on: Motorways										
	(ii)	Trunk and principal roads	0.30	0.35	0.40	0.45	0.50	0.55				
	(iii)	Other roads carrying more than 250 commercial vehicles per lane per day				0.45	0.50	0.55				
C (easy)	(i)	Generally straight sections of lightly trafficked roads	0.30	0.35	0.40	0.45						
	(ii)	Other roads where wet accidents are unlikely to be a problem	0.30	0.35	0.40	0.45						

† 'Minimum' in this context is defined as the mean summer SCRIM coefficient (MSSC) (which is the average of 3 readings taken during the months May–September) in a year of normal weather conditions. (Expressed in LR510 Table 2 as SFC (at 50 km/h).
‡ Main urban roads would generally be included in Marshall road categories 1, 2, and 3.

specific SFC_{50} (MSSC) and, by reference to Table 5.14, the required PSV may be determined for the known or calculated traffic volume.

The results of the research at TRRL into skidding resistance were substantially used in the preparation of Department of Transport Technical Memorandum H 16/76. This document was superseded by the *Specification and Notes for Guidance for Highway Works*, 1986, but the requirements for determining the numerical value of PSV and AAV given in Appendices 1 and 2 of H 16/76 were retained. Appendix 1 is reproduced here as Table 5.16, which incorporates the necessary features of the two TRRL tables and, by allocating risk ratings based on layouts conforming to the Department's geometric standards, specifies the minimum PSV for the site category and the volume of traffic calculated to be using the lane at the end of the anticipated life of the surfacing. For the difficult sites in categories A1 and A2 the PSV includes an additional 5 points to accommodate the increased polishing effect of braking and turning traffic.

Note. The Department of Transport intend to issue a new Standard and Advice Note to replace Technical Memorandum H 16/76. It is expected that it will prescribe more but smaller traffic increments, and a revised list of site categories that conforms to Standard HD 15/87. It is also expected to relate the PSVs to investigatory levels of MSSC and risk ratings given in Standard HD 15/87, and that there will be a trend to higher PSVs.

Many sources of crushed rock aggregate can achieve the minimum PSV of 45 required for category C sites, but few can consistently attain the values of 65 to 70 that are required for the sites in categories A1 and A2. Although such high PSV aggregates are obtainable in the UK from some igneous rock formations in the Basalt, Granite and Porphyry groups, the principal sources, or potential sources, are the older rocks (pre-Carboniferous period) in the Gritstone group. However, because high resistance to polishing is generally associated with a low, or relatively low, resistance to abrasion, particularly with regard to the sedimentary gritstones and low-grade metamorphic rocks, there are, at present, very few sources of aggregate in the UK that satisfy both the requirements of PSV greater than 65 to 70 and AAV less than 10, and are, at the same time, economically workable.

An artificial aggregate which has a very high resistance to polishing and abrasion is refractory grade, calcined bauxite, of 3.35 to 1.18 mm size. The main source is Guyana, but it is also imported from China. It has a PSV in excess of 70 and an AAV of less than 5, these values being attributable to the uniformity and purity of the product (approximately 90% aluminium oxide — Al_2O_3) and the optimum control of temperature and atmosphere during calcination. It is used as a form of surface dressing by distributing it uniformly over a sprayed-on, resin-based binder and, at the end of the curing period and before the surface is trafficked, removing the surplus by vacuum sweeper or other means. No roller is applied.

The treatment provides a durable, highly skid-resistant surface that is used extensively in Greater London and elsewhere for heavily trafficked approaches to roundabouts, traffic signals, pedestrian crossings, etc. Because it cannot meet the specified texture depth requirements, however, it is not used, other than in

Sec. 5.3] Tests and standards for aggregates in bituminous materials

Table 5.16 — Categories of sites and minimum polished-stone value for flexible roads. (Reproduced from Department of Transport Technical Memorandum H16/76, Appendix 1)

1 SITE	2 APPROXIMATE PERCENTAGE OF ALL ROADS IN ENGLAND		3 DEFINITION	4 MINIMUM POLISHED STONE VALUE	5 REMARKS
A1 (difficult)	Less than 0.1%	(i)	Approaches to traffic signals on roads with 85%ile speed of traffic greater than 40 mile/h (64 km/h).	Less than 250 cv/lane/day: 60 250 to 1000 cv/lane/day: 65 1000 to 1750 cv/lane/day: 70 More than 1750 cv/lane/day: 75	Risk rating 6. Values include +5 units for braking/turning.
		(ii)	Approaches to traffic signals, pedestrian crossings and similar hazards on main urban roads.		
A2 (difficult)	Less than 4%	(i)	Approaches to and across major priority junctions on roads carrying more than 250 commercial vehicles per lane per day.*	Less than 1750 cv/lane/day: 60 1750 to 2500 cv/lane/day: 65 2500 to 3250 cv/lane/day: 70 More than 3250 cv/lane/day: 75	Risk rating 4. Values include +5 units for braking/turning.
		(ii)	Roundabouts and their approaches.		
		(iii)	Bends with radius less than 150 m on roads with an 85%ile speed of traffic greater than 40 mile/h (64 km/h).		
		(iv)	Gradients of 5% or steeper, longer than 100m.		*In (i), the 250 cv/lane/day applies to each approach.
B (average)	Less than 15%	(i)	Motorways	Less than 1750 cv/lane/day: 55 1750 to 4000 cv/lane/day: 60 More than 4000 cv/lane/day: 65	Risk rating 2.
		(ii)	Trunk and principal roads		
		(iii)	Other roads carrying more than 250 commercial vehicles per lane per day.		
C (easy)	Less than 81%	(i)	Generally straight sections of lightly trafficked roads, i.e. less than 250 cv/day.	45	No risk rating applied. Many local aggregates have a PSV well above 45 and normally these should be used.
		(ii)	Other roads where wet skidding accidents are unlikely to be a problem.		

Notes
1. The volume of traffic given for the PSVs in column 4 should be that calculated for the lane at the end of the anticipated life of the surfacing.
2. The PSVs in column 4 apply only to roads constructed within the geometric design standards of 'Layout of roads in rural areas' and 'Roads in urban areas' (as amended) except if otherwise defined in column 3. For other layouts and abnormal risks, the Engineer must select a PSV appropriate to the layout and based, where possible, on experience of similar situations in the past.

exceptional circumstances, on high-speed roads. It is very expensive, the cost of the aggregate alone being at least £220 per tonne (Q1 1990).

From the results of SCRIM and BFC tests, carried out by TRRL, mainly on high-speed roads surfaced with BS 594 rolled asphalt, it has been determined that there would be no significant reduction in skidding resistance between 50 and 130 km/h provided that an average minimum texture depth of 2 mm, measured by the sand patch test procedure, was maintained. However, the surface texture of rolled asphalt is obtained by coated chippings which are applied to the surface and rolled in simultaneously with the compaction of the layer. This procedure, even with the specified well-shaped, single-size chippings, militates against the achievement of a texture depth of 2 mm without the high risk of a substantial loss of chippings under traffic. As a consequence, a minimum average texture depth of 1.5 mm for each 1 km section of carriageway lane is specified by the Department of Transport for which, it has been estimated, the reduction in skid-resistance between 50 and 130 km/h is only 10%.

By specifying and achieving a minimum aggregate PSV and texture depth, it is the intention to obtain the desired MSSC at the design stage. Although this procedure may require considerable effort in the initial selection and testing of aggregates and the control and supervision of the construction process, it is justified by the eventual performance of the surfacing. The alternative is to place undue reliance on subsequent testing of sfc which, because of the lapse of time and the influence of such variables as trafficking, weathering, temperature, seasonal variations, and PSV, cannot be considered an appropriate method for controlling a contract. Consequently, the PSV, used in conjunction with AAV and texture depth as necessary, is one of the most important test properties of an aggregate for bituminous surfacings on high-speed and heavily trafficked roads. Considerable care is therefore required when carrying out the tests to ensure that the results are as representative and as accurate as possible.

Since about 1981, the British Airports Authority (BAA) has specified a minimum PSV of the order of 60 for the coarse aggregate used in dense bituminous surfacings and friction course for some aircraft pavements. While the traffic volume in movements/day is considerably less than on the vast majority of trunk roads, the introduction of a minimum PSV has been considered beneficial in the maintenance of skid resistance. For a similar reason the fine aggregate is required to contain about 40% of crushed rock fines derived from rock that has a specified minimum PSV.

Probably because of the relatively low number of military aircraft movements, the Property Services Agency does not specify a minimum PSV for the bituminous surfacings of military aircraft pavements.

The minimum PSVs specified in Technical Memorandum H16/76 have to be applied to wearing course coarse aggregates used in new motorway and trunk road construction, and they are also recommended for new construction of other types of road. Outside the mandatory situations there is possible scope for the highway authority to choose a minimum PSV appropriate to local circumstances: furthermore, there exists the possibility of blending aggregates of different PSVs for other roads provided that the combined value is equal to or greater than the specified minimum. While the use of blended aggregates for coated chippings would incur too much risk of undetected uneven distribution, there is less likelihood of this occurring in a plant-mixed wearing course using weighed quantities of different size aggre-

Sec. 5.3] Tests and standards for aggregates in bituminous materials 221

gates. It is, for example, a feature of Delugrip, which employs aggregates of differing AAVs. It would therefore seem feasible to extend the principle to differing PSVs of the coarse aggregate in coated macadam and rolled asphalt wearing courses, if required.

If it is assumed that a mixture of two or more coarse aggregates of normal surfacing quality yields a resistance to skidding that is approximately equal to the weighted mean of that given by the constituents on their own (Salt 1977), then applying this to the PSV of a mixture of two aggregates, the quantity of the higher PSV aggregate required to achieve the specified PSV would be:

$$x \text{ (\% by volume)} = 100(P_S - P_L)/(P_H - P_L)$$

where P_S is the PSV required by specification,
P_L is the PSV of the aggregate with the lower resistance to polishing, and
P_H is the PSV of the aggregate with the higher resistance to polishing.

If $P_S = 60$, $P_L = 52$, and $P_H = 65$, then the required percentage of the higher PSV aggregate would be 62% by volume.

That the combined PSV may not be the exact mean of the two values is the view taken by the Texas Highway Department (1978) who have introduced into their formula in Test Method Tex-438-A/1978 an addition of 2 units to the specified PSV, as follows:

$$P_R \text{ (\% by volume)} = 100[PV_S + 2) - PV]/[PV_R - PV]$$

where PV_S is the polish value required by specification,
PV_R is the polish value of non-polishing aggregate (P_R), which is defined as an aggregate used to improve the polish value of the aggregate mix, and
PV is the polish value of polishing aggregate to be improved.

If $PV_S = 60$, $PV = 52$, and $PV_R = 65$, as for the previous example, then the required percentage of 'non-polishing aggregate' would be 77% by volume.

From the typical values quoted it is apparent that by adding an allowance or 'factor of safety' of only 2 units to the specified PSV, an increase of 15% is required in the quantity of the higher PSV aggregate. If this is put into the context of plant-mix grading variations and the PSV test repeatability and reproducibility of 3 and 5 respectively, it is apparent that some caution is required in fixing the aggregate proportions to achieve the specified PSV. It would also be a wise precaution to carry out sfc measurements on the finished surface at an appropriate time to confirm whether or not the required standard has been attained.

In addition to the theoretical calculation for the determination of the proportion of two aggregates with different PSVs required to obtain a specified PSV, the Texas Highway Department Test Method Tex-438-A/1978 also specifies a PSV test procedure using blends of the aggregates. The aggregate particles in their required proportions are mounted in the mould at random, but, in other respects, the method follows their standard procedure. When *Synthesis* 49 was published, this was based on the original BS 812 procedure, except that seven specimens for each aggregate were used, and, before polishing for nine hours, an initial friction value was determined for reference purposes. As a useful extension to the test, the rate of wear may also be assessed by a dial gauge attached to the frame of the accelerated polishing machine, the heights of selected and marked particles being measured before, during, and after polishing.

5.3.6 Soundness

Soundness is the term widely applied to express the resistance of aggregates to the effects of weathering or degradation due to the influence of the atmosphere and hydrosphere; particularly wetting and drying, freezing and thawing, and thermal cycling at elevated temperatures. Weathering takes place to a varying degree in the rock mass and continues after the rock has been crushed to form road aggregates and incorporated in the bituminous pavement.

Weathering is described as disintegration when the rock is physically degraded into small particles by, for example, the action of rain, wind, temperature, and the growth of crystals of ice and salts in pores, cracks, and fissures. Physical weathering does not change the minerals of a rock. The gravels and sands which, with time, are progressively formed, have qualities that are essentially similar to those of the parent rock.

Diurnal temperature variations of the exposed surfaces of rocks that may in the extreme approach 100°C in desert areas (Weinert 1980) give rise to temperature gradients, differential expansion and contraction, and consequential stresses that fracture the rock. Frost action freezes interstitial water, causing an expansion of 9% which may fracture both rocks and aggregates, depending on their strength and the nature and distribution of pores and fissures. The growth of sulphate, chloride, and other salt crystals in the pores of porous rocks and aggregates may also cause mechanical breakdown. This can be an important failure mode in desert areas, particularly where a high, saline water table provides constant replenishment of the evaporated water by capillary action (Fookes & Higginbottom 1980). Further internal stressing may be caused by the thermal expansion of the salts or by cycles of hydration and dehydration associated with wide ranges of temperature and moisture changes. The latter process depends on the quantity of water taken up/released during hydration/dehydration, and this may be appreciated by an inspection of the chemical notations of the hydrated salts of, for example, calcium sulphate ($CaSO_4.2H_2O$), magnesium sulphate ($MgSO_4.7H_2O$), and sodium sulphate ($Na_2SO_4.10H_2O$).

Chemical weathering is referred to as decomposition, and the process causes the breakdown of mineral particles to form new compounds, or secondary minerals. The most active agent causing decomposition is water, which intrinsically contains minute quantities of dissociated and chemically active hydrogen and hydroxyl ions as well as other ions and radicals that can penetrate the crystal lattice and, by hydrolysis and also hydration, cause disruption and volume change as the mineral alteration proceeds. Traces of impurities are also always present in natural water such as rain and groundwater. These substances are generally acidic, and, although carbonate rocks are those that are the most susceptible to the solvent action, many other rocks are affected in that such substances are an additional source of hydrogen ions. Oxidation, or the formation of oxides by the combination with oxygen in the atmosphere, is another contributor to decomposition, the reaction being considerably intensified by the presence of water.

Mineral alteration by chemical weathering principally affects crystalline igneous rocks; metamorphic rocks are less influenced, particularly with regard to their use by the highway engineer. The carbonate formations of sedimentary rocks are seriously affected by solvent action, but although the mineral may be altered, the products are usually transported away from the parent rock (erosion) to be redeposited elsewhere. Consequently the highway engineer, being concerned with the residual limestone rock for aggregates, is little troubled with the unsoundness that is associated with the integral presence of inferior secondary minerals.

Despite the fact that it is not a form of chemical weathering in the accepted sense, chemical decomposition can occur during the formation of igneous rocks. Such rocks start their formation at very high temperatures, and during cooling the various minerals crystallize sequentially. During this process, minerals which have been formed at the highest temperature may be susceptible to chemical attack from the remaining magma, so that even before the commencement of chemical weathering, alteration has been initiated. An example of this process is olivine, a typical primary mineral of basic and ultra-basic rocks, which, after being the first to crystallize may be altered to the softer mineral serpentine by subsequent magma reactions.

The approximate order of resistance to decomposition by the principal rock-forming minerals (which has an association with the inverse order of their sequence of crystallization from the magma) is quartz, mica, feldspar, amphibole (hornblende), pyroxene (augite), and olivine. The position of feldspar requires qualification in that the potassium feldspars such as orthoclase, which typically occur in a granite, have a resistance on a par with the micas, whereas the sodium and calcium feldspars, which are found characteristically in intermediate and basic rocks respectively, have a decreasing resistance as sodium gives way to calcium, such that the latter have a resistance more akin to that of pyroxene. Furthermore, it should be noted that quartz is practically totally resistant to decomposition. The order of resistance of these minerals gives an indication of the resistance of the rocks themselves, and it is evident that with respect to decomposition the basic rocks have a tendency to be inferior to acid rocks.

Climate has a major influence on the type and rate of weathering and hence the soundness of rocks and road aggregate, and the most important of the climatic factors are rainfall and temperature and their mutual association.

In South Africa an intensive study was undertaken in the 1970s to determine why the durability of weathered Karoo dolerite, and subsequently all other basic igneous rocks and their metamorphic derivatives employed in the various layers of bituminous roads, was markedly worse in the eastern, by comparison with the western, parts of the country (Weinert 1980). It was appreciated that the dominant influence was climate, but it was very soon realized that the original idea of poor durability being simply linked to high rainfall or any other single climatic factor provided an inadequate explanation of the overall situation. Consequently, all climatic factors were investigated, particular attention being paid to the warmest and coldest months of January and July respectively, and related to the boundary of satisfactory and unsatisfactory performance. Ultimately a climatic index of weathering was derived and has become known as the N-value which is given by the expression

$$N = 12\, E_J/P_a$$

where E_J is the computed evaporation from a shallow free-water surface during the warmest month January (July in the northern hemisphere) and P_a is the total annual rainfall. These two parameters take into consideration both the availability of water to promote decomposition, and temperature which has a direct bearing on evaporation and the rate of chemical reaction.

By determining the N-values at various locations throughout the country the contour of $N=5$ acquired outstanding significance. It coincided with the acknowledged boundary separating the areas of satisfactory and unsatisfactory performance of crystalline rocks. These areas were also found to correlate well with the two modes of weathering of both basic and acid rocks. In those areas which were satisfactory ($N>5$), physical weathering or disintegration predominated with the formation of residual gravelly soils containing oxidized minerals of the parent rocks and only a limited quantity of secondary minerals. In those areas which were unsatisfactory ($N<5$), chemical weathering or decomposition was prevalent, resulting in the formation of residual clay soils composed of montmorillonite from basic rocks and kaolinite from acid rocks. Where the N-value exceeded 10, no secondary minerals developed, the mode of weathering being entirely that of disintegration.

The calculation of the N-value may be adapted for locations in latitudes other than those of South Africa, but where the mean annual temperature does not rise above 0°C or where rocks do not noticeably decompose, the N-value loses its significance (Weinert 1980).

When a source rock or gravel is being investigated for potential use as a roadstone, its petrologic composition and its physical and mechanical properties, as determined by laboratory tests, should provide sufficient indication of its suitability and probable durability when the results are related to the conditions of service and the environmental factors at the place of use. However, laboratory tests for roadstone should be, and usually are, devised with the intention of providing simple, quick, and meaningful methods of establishing and confirming as a matter of routine the required properties of an aggregate. As a consequence, the more specialized and generally more time-consuming requirements for petrological examination, which would in most instances indicate the potential durability, provided that the interpretation and application of the results are made in conjunction with competent inspections of the quarry face, are reduced to a minimum or even omitted after the initial study. Furthermore, while most UK project specifications require some measure of regular physical and mechanical testing of roadstone, it is rare for them to specify petrological examination other than at the time of source approval.

This situation has prevailed for a number of reasons which include the fact that the vast majority of UK aggregate production has provided, and continues to provide, durable roadstone. Nevertheless, there is a need for simple, rapid, and reliable soundness test procedures to supplement and, for routine quality control, to replace petrological examination so that unsound aggregates can be identified without delay. Such tests would benefit both the producer and the engineer, the former by facilitating the selective working of a marginal source and eliminating remedial costs, and the latter by minimizing the risk of incorporating unsound aggregates in a road or runway that could have serious consequences with regard to durability and safety, particularly in a wearing course.

Soundness may be viewed as a measure of a road aggregate's resistance to disintegration and decomposition caused by climatic factors, while durability should include additionally an allowance for their resistance to the simultaneous action of mechanical forces arising from the effects of the designed traffic loading. Clearly there can be no single laboratory test capable of making a reliable assessment of either the soundness or the durability of the various road aggregates, but a number of tests do exist from which a judgement of soundness is intended to be made, and in the account which follows a brief description is given of some of them. Their application is not restricted to aggregates for bituminous roads; on the contrary, most of the tests may be employed for any of the normal uses of an aggregate in construction, in spite of the fact that they are quoted more frequently in highway and aircraft pavement specifications.

5.3.6.1 *Soundness by the use of sodium or magnesium sulphate (sulphate soundness test)*

The sulphate soundness test measures the resistance of specified sizes of aggregate when subjected to repeated cycles of immersion in a saturated solution of sodium or magnesium sulphate followed by oven-drying.

A form of the test is believed to have been used in France in the 1820s to detect the inability of building stones to resist frost action, and in 1931 a standardized test procedure was published in the United States as ASTM C88 to estimate the soundness of concrete aggregates with respect, in particular, to freeze–thaw cycles.

In the late 1950s a test similar to ASTM C88 was specified nationally for the first time in the UK for estimating soundness of aggregates in concrete pavements for military aircraft, and its application was subsequently extended to coarse and fine aggregates for bituminous pavements for similar purposes. It is still included in the Property Services Agency Airfields Branch Specifications for concrete and bituminous surfacings for military airfields, and is also required by British Airports Authority (BAA) specifications for civil aircraft pavements. In both cases, however, only magnesium sulphate solution is required. A version '... adapted to British practice' is given in Appendix B of BS 6349: *Code of Practice for Maritime Structures*, Part 1: 1984, *General Criteria*, and yet another, using only sodium sulphate and with major variations in test method, is given in Appendix B of BS 1438: 1971 (1983) *Specification for Media in Percolating Biological Filters*. In October 1989, BS 812: Part 121, *Method for determination of soundness*, was published. The test procedure, which is referred to later in this section, is also a variant of ASTM C88.

Internationally, the most widely applied procedure is that given in ASTM C88: *Standard test method for soundness of aggregates by use of sodium sulfate or magnesium sulfate*, and, apart from the preparation of the solutions, the method is identical for the two salts. Washed and dried aggregate fractions of specified size and mass (available in amounts of 5% or more) are placed in separate perforated containers and immersed in a prepared saturated solution of magnesium or sodium sulphate of specified density at a temperature of $21\pm1°C$ for a period of 16 to 18 hours. The fractions in their containers are then removed from the solution, drained for 15 ± 5 minutes, dried to constant mass in an oven of specified performance at $110\pm5°C$, and cooled to room temperature. This cycle of alternate immersion and drying is repeated until the required number of cycles, which is usually five, is obtained. After removing all traces of salt from the aggregate by washing, and testing the wash water by reacting it with barium chloride, each fraction is dried at $110\pm5°C$, cooled, and resieved. Each coarse aggregate fraction is sieved over a sieve size that is less than that on which it was originally retained — the required sizes are specified in ASTM C88 — and the fine aggregate fractions over the same sieves. The aggregate retained on each sieve is weighed and each loss in mass is expressed as a percentage of the initial mass of the fraction tested. By a system of weighted averages, the percentage sulphate soundness loss is calculated for the laboratory sample based on the grading as received or, preferably, on the average of the supply material of which the sample is representative. This system makes provision for calculating and incorporating assumed losses for the untested sizes that may be present in the sample. These are — sizes weighing less than 5% of the total, undersize (<4.75 mm) of coarse aggregate, oversize (>9.5 mm) of fine aggregate, and the minus 300 μm of fine aggregate which is assumed to have zero loss. If there is more than 10% of undersize or oversize it must be tested and reported separately. All aggregate particles initially retained on the 20 mm sieve must be examined qualitatively and the form of distress recorded, and it is recommended that this should also be carried out for the smaller coarse aggregate fractions.

The precision of the test is given as the difference between two results expressed as the percentage of their average. For coarse aggregate with weighted average sulphate soundness losses in the ranges 6 to 16% for sodium and 9 to 20% for magnesium, the values are:

Repeatability	68% for sodium and 31% for magnesium sulphate; and
Reproducibility	116% for sodium and 71% for magnesium sulphate.

The definitive method for determining soundness specified in BS 812: Part 121 requires the test to be made on two 425 ± 5 g, washed and dried, 10–14 mm size aggregate fractions using a saturated solution of magnesium sulphate of specified density at a temperature of $20\pm2°C$. When required, or if the definitive size is not available, other sizes may be tested by the procedure given in Appendix A of the Standard.

The method is broadly similar to that in ASTM C88. However, the periods of immersion and draining are $17\,h\pm30$ min and $2\,h\pm15$ min respectively, and drying is carried out at 105 to 110°C for at least 24 hours; each specimen is then cooled for $5\,h\pm15$ min. The duration of one cycle is, therefore, approximately 48 hours. At the conclusion of the fifth cycle each specimen is thoroughly washed, dried to constant mass at 105 to 110°C, and cooled in a desiccator. It is then hand sieved on a 10 mm test sieve and the mass of material retained is expressed as a percentage of the original mass to give the soundness value, S, of the specimen.

If S_a and S_b are the soundness values of the two specimens, then:

$$\text{Magnesium sulphate soundness value (MSSV)} = \frac{S_a + S_b}{2}$$

expressed to the nearest whole number.

Precision values at various levels are given for the test. For a value of MSSV=91.0, for example, repeatability $r_1=5.2$ and reproducibility $R_1=8.2$.

Essentially, the soundness test subjects the various aggregate particles to internal pressures caused by the salt solution entering the pores or fissures and expanding by crystallization, to alternate wetting and drying, and to periods of high temperature, all of which are intended to generate in a matter of days an artificial and accelerated physical type weathering. The form of disintegration promoted by the test is specialized and occurs only on a larger scale in desert regions or, according to Weinert (1980), where the N-value is greater than 10. For such regions the use of the test is reasonably appropriate in estimating durability because of the importance of salt weathering as a cause of deterioration (Fookes & Higginbottom 1980). The original reasons for the introduction of the test, however, were to estimate the resistance of concrete aggregates to frost attack as it was assumed that the growth of sulphate crystals in the pores of the aggregate reproduced the disruptive forces of ice formation. Since then a number of investigators have question this assumption, although the accent has been on relating the test result to the performance of concrete rather than asphalt aggregates. Hudec (1983), for example, reports that Larson *et al.* (1964) concluded that it was not a reliable procedure for predicting the durability of aggregate when confined in concrete, and that Verbeck & Landgren (1960) could not relate it to the performance of aggregate when frozen in concrete. The present-day tendency for aggregates, however, is to use the test to detect unsoundness in general, irrespective of whether or not the material is likely to be subjected to freezing temperatures; this is also questionable because various aggregates require different numbers of test cycles to obtain an interpretable result (Weinert 1980).

A major criticism of the test is its poor precision. Repeatability and reproducibility are better with magnesium sulphate than with sodium sulphate, but, even so, the variability of test results gives cause for concern particularly when nearly two weeks are required to repeat the test to obtain confirmation. The duration could be decreased by heating and drying in a microwave oven, provided that the test results were not distorted, and tests on Northern Ireland basalt have indicated that a period of twenty minutes is equivalent to twenty-four hours, which is the standard period used in UK testing (Woodside & Woodward 1989). Six days would nevertheless be required for five cycles.

The differences between the various test procedures and methods of reporting results, excluding BS 1438 which is unique, make unqualified comparison between reported values unsafe. In this respect the differences in sample preparation and the events in the cycling programme are probably insignificant, although this may not always be the case for every type and condition of aggregate. Of some significance, however, is the use by ASTM C88 of a sieve size smaller than the original for obtaining the residual mass — e.g. 8 mm instead of 10 mm. This helps to compensate for possible differences in sieving techniques used in obtaining the test specimen and the residue, and for undue degradation of the specimen caused by 'handling' during the cycles and the subsequent washing and drying. It also tends, nevertheless, to reduce the soundness loss when compared with the BS 812 test and those specified by the PSA and BAA, which use the original sieve size. The reporting of only a single size of 10–14 mm by BS 812 is simpler than that of the system of weighted averages required by ASTM C88, PSA, and BAA, but it is by no means clear what effect this has on the result. Moreover, it must be borne in mind that BS 812 reports MSSV as the mean result of tests on duplicate specimens in terms of the percentage retained, instead of a single test result in terms of percentage loss.

On the whole, and despite the reservations, the test is capable of distinguishing between aggregates of superior and inferior soundness, but it is unreliable when used for rating the service performance of many intermediate aggregates. The range of limits prescribed in various specifications is a reflection of this. A maximum limit of 18% magnesium sulphate soundness loss for coarse and fine asphalt aggregates used to be specified by the PSA Airfields Branch for bituminous surfacings, but within the last few years the limit for fine aggregate has been lowered to 15%. Further changes in the limits are under consideration, as well as modifications to the method of calculating the weighted average test result for fine aggregate. Selective, in addition to representative sampling from stockpiles for separate testing of apparently different types of aggregate or suspect particles, has been introduced, together with acceptance limits. The ASTM limits for coarse and fine aggregates for bituminous materials specified in ASTM D692 and D1073 are, respectively, 18 and 20% loss for magnesium sulphate and 12 and 15% loss for sodium sulphate. The US Federal Aviation Administration Advisory Circular 150/5370-10 specifies for the coarse aggregate of bituminous surface courses a maximum value of 12% for magnesium sulphate and 9% for sodium sulphate; no limits are prescribed for the fine aggregate. None of these quoted limits differentiates between the type of rock or the climatic conditions in service; all of them relate to five cycles of immersion and drying.

Appropriate limits have yet to be published in national standards or specifications for the BS 812 method of test. Before they are it is desirable that they be nationally related to service requirements, test precision, and, preferably, to aggregate type. In the meantime, locally prescribed limits should, logically, be guided by similar considerations.

5.3.6.2 Soundness by freezing and thawing

Although the sulphate soundness tests may attempt a form of disruption akin to the freezing action of water, there are tests that do this directly. One such test is that specified by the American Association of State Highway and Transportation Officials (AASHTO) in T.103-86 (1986), *Standard method of test for soundness of aggregates by freezing and thawing*. The method describes three procedures for testing specified fractions of coarse and fine aggregate: total immersion-freezing of previously soaked aggregate while totally immersed in water, and partial immersion-freezing of previously vacuum-saturated aggregate while partly immersed either in water or in a 0.5 mass % solution of ethyl alcohol in water. The samples are frozen in a cabinet set at a temperature not higher than −26°C and then thawed in water or alcohol–water solution, the duration of each cycle being sufficient to achieve both freezing and thawing. The number of cycles is not standardized but is at the discretion of the specifying authority: 50 cycles for total immersion in water and 25 and 16 cycles for part immersion in water and in alcohol–water solution respectively, have been required by some US authorities. After freeze/thaw cycling the aggregate is dried, sieved, weighed, and analysed in a similar manner to the sulphate soundness tests. Previous editions of the standard method stated that 'the results of the method are considered more reliable for determining the quality of aggregates than those obtained by other methods of soundness tests on discrete particles of aggregate'. This statement is omitted from the current edition which, nevertheless, still points out that the test 'furnishes information helpful in judging the soundness of aggregates subjected to weathering, particularly when adequate information is not available from service records of the behavior of the aggregate'.

Another freeze/thaw test, albeit for concrete aggregates, is given in the Federal Republic of Germany DIN 4226 (1971). It describes two procedures: total immersion freezing and thawing, and air freezing followed by thawing in water. In the former procedure the temperature of the immersed sample in the freezing cabinet is gradually reduced over a period of 7–10 hours and then held at a temperature of −15 to −20°C for at least 4 hours. Afterwards the test portion is thawed in water at 20°C for 5 hours. This cycle is carried out 10 times. Each fraction is dried and sieved over the next smaller sieve size, and the percentage loss in mass is determined. For the air freezing procedure aggregate soaked for 2 hours is drained and placed in a freezing cabinet at a temperature of −15 to −20°C for 6 hours and then thawed in water at 20°C for 1 hour. The cycle is carried out 20 times and the subsequent treatment of the test sample is the same as that for the total immersion test.

Having a duration of about a week or more, these freeze/thaw tests suffer from a disadvantage similar to that of the sulphate soundness tests, and they are, as a consequence, unsuitable for rapid site testing. Nevertheless, by using water instead of a saturated sulphate salt solution, they should more nearly demonstrate the frost resistance of an aggregate provided that the frequency of the accelerated rate of freeze/thaw cycling does not stimulate anomalous behaviour and that the number of cycles has relevance to anticipated service life and climate.

One of the properties of an aggregate that, customarily, has been thought to provide an indicator of freeze/thaw resistance is water absorption as determined by the standard 24-hour immersion test. There are, however, many exceptions that do not necessarily show a susceptibility to freeze/thaw disintegration when the absorption is relatively high, and it is recognized that aggregates having a high proportion of small pores suffer the most: it is microporosity rather than merely porosity that influences the performance adversely. By the use of a constant-temperature humidity cabinet, at 95% relative humidity, to determine the adsorption of water vapour on internal pore surfaces, Hudec (1983) obtained an approximation to the internal surface area of aggregates and hence an indication of the dominant pore size that decreases with increasing surface area. When the relationsip between these results and those of a series of other, mostly non-standard, tests made on aggregates representing a wide range of rock types from Ontario were analysed, a very strong correlation was found between freeze/thaw deterioration and adsorption, and the ratio of adsorption to absorption. He suggests that the adsorption test could be used in place of, or as a back-up for, freeze/thaw tests.

5.3.6.3 Soaked ten per cent fines and modified aggregate impact value tests

The standard and modified 10% fines test and the modified AIV test, described earlier, are suitable for testing weak (as well as strong) surface-dry aggregates. This is because the quantity of fines produced is limited, thereby preventing the 'cushioning' of the applied force and the consequent distortion of the test result that starts to occur at, or a few points above, the 25% level of fines.

When these tests are conducted on saturated and surface-dried (SSD) samples of aggregate previously soaked in water for 24 hours, and the results compared with those obtained on dried samples, it is found that many of the weaker aggregates, particularly those derived from sedimentary rocks, exhibit a significant drop in strength as measured by the respective test procedures. Furthermore, when a range of 21 UK, low-grade, argillaceous and gritty rock aggregates of known performance in lightly trafficked, unsurfaced roads and sub-bases was tested in the SSD condition, the results correlated better with the local engineers' experience of these materials than did those obtained on the dried aggregate (Shergold & Hosking 1963). By drawing a line between unacceptable performance and acceptable or borderline performance, tentative acceptance limits were suggested for the modified 10% fines and AIV test results on SSD samples of aggregate for these purposes of >5 tons (50 kN) and <40% respectively. Aggregates acceptable on these terms had decreases in strength of up to 50%.

Although the strength testing of SSD aggregate is not yet a standard procedure in BS 812, it is specified in the Department of Transport *Specification for Highway Works* 1986 for wet-mix macadam roadbase aggregate, which must have a 10% fines value in the SSD condition of not less than 50 kN. To achieve this condition the aggregate sample is required to be soaked in water at room temperature for 24 hours without previous oven-drying and its surface then dried, presumably in the normal manner by draining and blotting with a damp cloth. After crushing, the sample is oven-dried at 105±5°C before sieving out the fines. In South Africa, where the air-dry and soaked 10% fines tests (10% FACT) are widely used on roadmaking coarse aggregates, the sample for the soaked test is immersed in water for 24 hours, drained until the flow of free water ceases, and then tested by the same method as that used for the air-dry test (Weinert 1980). No problems seem to be experienced as a result of omitting surface drying, but a more consistent test sample should be obtained if it is included.

Indeed, there is possibly more potential for variable results inherent in the BS 812: Part 3; 1975 instructions for preparing the test sample for ACV, AIV, and 10% fines value, because, with respect to moisture content, it prescribes only that 'The aggregate shall be tested in a surface-dry condition. If dried by heating the period of drying shall not exceed 4 h, the temperature shall not exceed 110°C, and the aggregate shall be cooled to room temperature before testing'. The only essential requirement is, therefore, surface dryness, and, provided that this has been achieved, the moisture content of the aggregate may range from zero to virtual saturation without being quantified and without having to be referred to in the test report. In the light of the effect that water can have on the strength of some aggregates, the desired moisture condition should be specified and controlled more precisely. It would therefore be preferable always to carry out the test on an oven-dried sample, as required by the Los Angeles abrasion test, which also specifies washing and drying before and after test unless the aggregate is essentially free from adherent coatings and dust.

Essentially, the 24-hour soaked 10% fines test is not a test for aggregate soundness because the period of immersion is too short and the use of (distilled) water at atmospheric pressure and temperature does not attempt to generate accelerated disruptive tensile stresses from within the pores of the aggregate. Extending the period of soaking to 20 days for the modified AIV test on SSD aggregate has given results of similar ranking to those of the sodium sulphate soundness test on aggregates susceptible to decomposition (Hosking & Tubey 1969), but the method is neither sufficiently rigorous nor reliable for universal soundness testing, and there is little practical merit in prolonging test periods unnecessarily. Nevertheless, for many aggregates the soaked 10% fines test does indicate a reduction in strength in the presence of water which would otherwise remain undetected by using only dry aggregate, and Weinert (1980) advises that the test should always be conducted on both air-dry and soaked samples. As stated earlier, with the exception of BS 63: Part 2 which specifies for surface dressing aggregates a minimum surface-dry 10% fines value of 160kN (85 kN for blastfurnace slag), no strength or soundness test values are specified in British Standards for aggregates used in bituminous materials construction for highways. In South Africa, however (Weinert 1980), although sulphate soundness tests have been omitted from the test specifications of the South African Bureau of Standards and the Department of Transport, recommendations are given for minimum air-dry 10% fines values for various groups of natural aggregates used in bituminous materials, and the soaked test result must not be less than a specified percentage of the air-dry result. For example, aggregates from basic and acid crystalline rocks and from carbonate rocks should achieve minimum air-dry 10% fines values of 210 kN for rolled-in chips, 160 kN for bituminous surfacing mixtures, and 110 kN for bases, and the soaked test results must not be less than 75% of the dry test results. For surface treatments the minimum values should be slightly greater than 210 kN and 75%.

The dry and SSD 10% fines tests are also used on sealing (surface dressing) aggregates in Australia. Minimum values are specified for the SSD strength, and the dry/SSD strength decrease must not exceed a stated percentage of the dry strength. The values proposed by Dickinson (1984) for different classes of traffic expressed in terms of the average annual daily traffic in both directions on a two-lane road (AADT), are shown in Table 5.17. It is worth noting that for traffic Class A the derived minimum dry 10% fines value is 154 kN, which is only 6 kN less than the minimum value for UK surface dressing aggregates, and that even at this quality level up to 35% decrease in wet strength is acceptable. Nevertheless, Dickinson comments, when referring to resistance to degradation or decomposition on exposure, that the wet/dry strength is believed to be of little value for assessing altered basalts (in Australia), some of which degrade rapidly in service but give acceptable test values.

The combined degradation effect of water and self-abrasion on fresh crushings of

Table 5.17 — Strength required for sealing aggregates. (After Dickinson (1984))

Traffic class AADT	A >5000	B 300–6000	C <300
10% fines value minimum wet strength, kN	100	80	70
Decrease in wet strength as a per cent of dry strength (maximum), %	35	40	45

igneous and metamorphic source rock may be indicated by the modified Washington degradation test (Dickinson 1984). The bulk sample is sieved into four size fractions that are uniformly distributed across the range of 13.2 to 2.0 mm. The fractions are washed and oven-dried, and a 250 g portion is taken from each to form a 1 kg test portion. This is placed in a plastic container with water and shaken for 20 minutes by a mechanical sieve shaker. The suspension of minus 75 μm particles is separated out and, after the addition of a flocculating solution, is agitated and transferred to a measuring cylinder, filling it to the 380 ml mark. The suspension is thoroughly redispersed and allowed to settle for 20 minutes. If the height of the column of sediment at this time is H mm, then:

$$\text{Degradation factor} = \frac{380-H}{380+1.75H} \times 100$$

Minimum values of degradation factor are specified in Australia as 45 or 50 for wearing course coarse aggregates, and 30 or 35 for intermediate course and base, depending on rock type. At the same time the secondary mineral count for basaltic aggregates must not exceed 25% and 35% respectively.

5.3.7 Soundness of basic igneous rock aggregates

The soundness of basic igneous rock aggregates has been a matter of serious concern in many countries. Considerable research has been undertaken in, for example, Australia, France, Northern Ireland, South Africa, and the United States to elucidate the problem and to devise practical methods of detection that are capable of identifying unsound, or potentially unsound, materials before their being incorporated in the works. In Britain, although increasing attention is being focused on the detection of unsoundness, particularly with regard to wearing course aggregates for motorways and airports, the incidence of unsoundness to date, although important, is localized or relatively infrequent.

In Northern Ireland approximately 6.7 Mt, or 37% of quarry production, is basalt of Tertiary origin. The basalts comprise two main layers, upper and lower, with an intermediate layer of contemporaneously weathered material. The lava flows are vesicular in their upper and lower portions and have been affected by secondary mineralization. There is widespread oxidization and weathering of flow tops. As a consequence, there are quantities of unsound, or potentially unsound, rock in the mass, some of which may be incorporated in roadstone. The presence of unsound aggregate in bitumen bound surfacings causes breakdown of the bitumen–aggregate bond, degradation, and loss of surface particles. In wet conditions within sub-bases, the degradation of unsound aggregate creates a coating of clayey slurry that reduces the stability of the unbound layer (Stewart & McCullough 1985, Woodside & Woodward 1989).

In the United States, Higgs (1976) draws attention to the increasing problem of slaking basalts in the western United States, and refers to reports describing problems encountered with basalt aggregates in asphalt surfacings on the Nestucca road in northwestern Oregon, near Orofino in Idaho, and in southern California, as well as in revetment stone used on the Santa Maria breakwater in southern California and at Keene Creek dam in southern Oregon.

The principal cause of degradation of basic igneous rocks is attributed to the pressures developed by the swelling of secondary clay minerals, mainly those of the chlorite–smectite group such as montmorillonite. The formation of these minerals occurs initially during the later stages of consolidation of the magma or lava, but it is possible that as the rock starts to fail weathering causes further alteration. Higgs (1976) refers to the remark of Van Atta and Ludowise that the aggregate in the asphalt pavement and stockpiles used on the Nestucca road had an increase of alteration products of up to 100% when compared with the amount found in the parent rock. Solutions to this problem have been sought by techniques of petrographic analysis such as X-ray diffraction and point counting to identify and quantify the presence of secondary minerals, and by soundness tests that either accelerate the swelling action of the clay by using organic polar liquids such as ethylene glycol, or give an overall assessment of the clay characteristics of the aggregate by using methylene blue.

5.3.7.1 Secondary mineral content by point counting
The point count method examines a thin section of the rock by attaching a device to the stage of the microscope which enables the whole area of the section to be covered in preset equal point step intervals along traverse lines set at appropriate equal spacings. Three push buttons on the point counter are allocated for recording the number of primary and secondary minerals, and voids (vesicles, microcracks, etc.). As the stage is moved in steps along each line, the mineral or void at the intersection of the ocular cross hairs is identified when viewed generally through crossed nicols, and appropriately recorded. A minimum of about 600 points should be identified and the percentages of primary and secondary minerals and voids (if required) calculated. Two, or preferably three, slides from each sample should be analyzed.

Having prepared the slides, point counting provides a rapid method of determining the secondary mineral content of an igneous rock. It is not always necessary to identify the type of secondary mineral; if required this may be done, but the time needed for each examination is then considerably increased. The secondary mineral percentage indicates the degree of decomposition of an igneous rock, and this knowledge has been used to confirm or predict the performance of an aggregate in highway works. It has been suggested by Knight & Knight (1948) that the best quality igneous rocks have a decomposition figure of not more than 15%; they also pointed out that it is necesssary to take into account both the amount and distribution of the decomposition products.

Weinert (1964) found that dolerites with a satisfactory service record in road bases usually had less than 30% of identifiably altered mafic minerals. Later Weinert (1980) related durability of aggregates in bases and sub-bases to both the percentage of secondary minerals and the N-value (see 5.3.6) and showed that percentages of 30 or more would be durable only for N-values greater than 4 to 5 for bases and 2 to 3 for sub-bases. At N-values less than 1, the percentages of secondary minerals decrease to approximately 15% for bases and 25% for sub-bases. Hosking & Tubey (1969) conducted a limited number of point counts on weathered dolerite gravels from Basutoland, and, although there were wide differences between different petrological sections from the same sample, they considered the method to be adequate for giving a first indication of the degree of weathering. Higgs (1976) concluded, however, that the total amount of weathering product has no effect on rock durability; rather it is the amount of discrete montmorillonite, of which as little as 11% was sufficient to cause the degradation of the basalts that he investigated.

5.3.7.2 Soundness by ethylene glycol

Ethylene glycol is one of a number of polar liquids that easily penetrate and expand the basal spacing of the lattice of swelling clays of the smectite or montmorillonite group, which are found as products of alteration in basic igneous rocks. If the accessibility, distribution, and existing state of expansion of the clay within a sample of rock or aggregate are amenable then it is to be expected that the sample will be fractured and probably fragmented by the tensile forces induced by the expansion, if the aggregate contains more than a critical proportion of the clay. Consequently, immersion in ethylene glycol provides an accelerated form of weathering or soundness test.

One of the earliest test procedures is that in CRD-C 148-69 *Method of testing stone for expansive breakdown on soaking in ethylene glycol* issued by the Concrete Research Division, US Army Corps of Engineers Waterways Experiment Station, Vicksburg (1969). By this method, a 5 kg sample of 76–19 mm particles is dried, weighed, and totally immersed in ethylene glycol at room temperature in a container having a tight-fitting cover. The sample is inspected every 3 days and after 15 days it is removed from the reagent, washed over the 19 mm test sieve, and the residue dried to constant mass. The mass percentage retained, and qualitative inspections, are then used to assess the performance in service of the material.

Of the 10 basalt samples investigated by Higgs (1976), one disintegrated after half an hour, and at the end of the day it had slaked to mud. However, some of the samples with a known poor performance had not cracked by the end of the prescribed 15 days, and so the period of immersion was extended to 30 days. By this time only the two samples of known good performance remained unaffected. Higgs concluded that 30 days rather than 15 days should be the prescribed test period.

To accelerate the reaction of the ethylene glycol, Fielding & Maccarrone (1982) of the Country Roads Board, Victoria, Australia (from 1 July 1983 the Road Construction Authority) investigated the effects of increased temperature and pressure and arrived at two procedures which were issued by the Board in November

Sec. 5.3] Tests and standards for aggregates in bituminous materials

1981 as Draft Test Methods CRB 376.01 *Accelerated Soundness Index — Reflux Method* and CRB 376.02 *Accelerated Soundness Index — Pressure Vessel Method*.

Both methods test a carefully prepared sample of 50 washed and surface-dry particles passing the 13.2 mm and retained on the 11.2 mm test sieves as well as on the 9.66 mm wide slotted sieve, as experience has shown that the test is size-dependent. For the reflux method, the sample is immersed in ethylene glycol in a 250 ml flask fitted to a Liebig or reflux condenser, and then subjected to 5 daily cycles of 8 hours boiling at 197°C and 16 hours soaking to achieve the required total of 40 hours' boiling. For the pressure vessel method, the sample is immersed in a solution of 80% v/v ethylene glycol in water, and then maintained at a temperature of 232°C and a gauge pressure of 1300 kPa for 3 hours. Thereafter, for both methods, the cooled sample is filtered and dried; all friable particles are broken with the fingers, and the masses of the material retained on and passing the 6.7 mm test sieve are weighed. Then, where:

M_r = mass of material retained in 6.7 mm test sieve, and
M_p = mass of material passing 6.7 mm test sieve

the accelerated soundness index by the Reflux Method (ASI_r) or by the Pressure Vessel Method (ASI_p) is:

$$ASI_r \text{ or } ASI_p = 100 \, M_r/(M_r + M_p) \, .$$

Based on a study of field data for non-durable aggregates, and the amount and mode of failure in the reflux test, the following limits have been proposed:

Source rock classification	Accelerated soundness index — reflux (ASI_r)
Sound	Greater than 94
Marginal	94–90
Unsound	Less than 90

Only a limited proportion of marginal material should be allowed in the quarry product, commensurate with expected service life, usage, and availability of better quality material.

By increasing the temperature of immersion of 197°C for 8 hours during each daily cycle of the reflux test, the degradation of an altered basalt sample can be increased by a factor of about 20, which suggests that the duration of the reflux method is equivalent to about 35 days' immersion at room temperature. Three hours' duration in the pressure vessel approximates to 40 hours' boiling, but an interesting feature of the pressure vessel method is that it is necessary to use a solution of ethylene glycol and water (80% v/v) because, with the reagent alone, many of the samples tested during the development of the test showed no degradation unless they were subsequently allowed to soak in water overnight. The authors conclude that re-entry of water into the lattice of the smectite clay is required for degradation to occur, and this is achieved by using the ethylene glycol solution.

The reflux method has been used in Victoria since 1978. It has been shown to be related to the secondary mineral content, the total surface area, and the total moisture loss to 900°C (assumed to be related to the relative amount of hydrated mineral present), and to provide a sensitive means of characterizing the durability of basaltic rock materials. Nevertheless, for specification purposes other properties must still be taken into consideration when selecting materials for a particular use (Fielding & Maccarrone 1982). The repeatability and reproducibility values of the test are not stated, so no comparison can be made in this respect with the poor precision of the sulphate soundness tests normally conducted on random samples.

5.3.7.3 Soundness by methylene blue adsorption

The continuing search for a rapid and effective means of detecting unsoundness in aggregates used in bituminous construction for highways and airports has drawn attention to the methylene blue adsorption test. Methylene blue is a dark green, organic, crystalline substance forming a blue aqueous solution that is used, for example, as a stain in microscopic and biological work. It is adsorbed by clay minerals, and the extent to which this is achieved provides a measure of the total, that is the external and internal, surface area of the mineral. Active swelling clays of the smectite group, which are the principal product of decomposition of basic igneous rocks causing unsoundness, have a very high specific surface (surface area per unit volume). The purpose of the methylene blue test is to obtain an assessment of the quantity of these clays in a sample of rock or aggregate by measuring the amount of methylene blue adsorbed.

The test as applied to fine aggregate was described and discussed by Tran Ngoc Lan (1980), and it is now a standard test in France where it supplements, or is an alternative to, the sand equivalent test. It is specified by the Department of the Environment (DoE) for Northern Ireland to indicate the soundness of basalt and gritstone (greywacke) rock aggregates, the test being conducted on finely ground samples (Stewart & McCullough 1985), and it is under consideration by the British Standards Institution for inclusion in BS 812. A detailed and expanded test method was prescribed and used by Hills & Pettifer (1985) to test a range of basic igneous rock aggregates. Comparisons were made with the percentages of layer–lattice minerals and a swelling index, defined as the product of the percentages of expansive clay minerals and the expansion of basal spacing on glycolation using X-ray diffraction. They found that the correlation between the methylene blue value (MBV) and clay content was good for smectite but poor for mixed-layer chlorite–smectite and that there was a fairly good correlation between MBV and swelling index.

The test is sensitive to particle size, and Hills & Pettifer (1985) describe a method for obtaining a representative powdered sample passing the 75 μm test sieve. A representative test portion of 1 g taken from this sample is thoroughly dispersed in about 30 ml of water and then titrated by successively adding 0.5 ml increments of methylene blue solution that has been prepared by dissolving 0.1 g of methylene blue in 100 ml of distilled or de-ionized water. After each addition, the suspension is agitated for one minute and a drop removed with a glass rod and spotted on to a filter paper. This procedure is repeated at prescribed intervals until a definite pale blue halo is seen to have formed when the filter paper is held up to daylight. This is the end-point, and it indicates that the particles are no longer capable of further adsorption.

The methylene blue value (MBV), expressed as the percentage mass of methylene blue adsorbed by the test portion, is then:

$$\mathrm{MBV} = \frac{0.1\ V}{M}$$

where V (ml) is the total volume of methylene blue solution added to the suspension to reach the end-point and M (g) is the mass of the test portion.

With suitable facilities for sample preparation, including a power grinder, it is possible to test a large number of samples rapidly, as the titration can be completed within 15 minutes. Consequently, the MBV test provides a useful means of assessing the soundness of basic igneous rocks on a quality control basis, particularly when the samples are drawn from a single source.

The limiting values for potentially unsound rock may vary with the source and should also be related to aggregate usage. For basalt, a tentative maximum MBV of 1.0 has been set by the DoE, Northern Ireland, according to previous experience with aggregates from similar sources (Stewart & McCullough 1985). For gritstone, the value is 0.7, but as it appears that, apart from some specially selected shales, this rock does not contain the swelling clays to which the MBV test is sensitive, alternative test procedures for soundness may be more appropriate. For practical reasons it is suggested that up to 5% unsound materials may be tolerated in an aggregate for road construction.

The MBV of a powdered sample of rock should not be taken as the sole criterion for assessing the soundness of a source of basic igneous rock aggregate because it does not take account of the effect of petrological texture (the relationship between the grains of minerals forming a rock) on performance (Hills & Pettifer 1985). A secondary mineral rating system for indicating the durability of basalt road aggregate by accounting for the type, quantity, and distribution of clay and other secondary minerals by the examination of thin sections has been proposed by Cole & Sandy (1980), but the procedure is specialized and, although useful for an initial assessment of a source, it is not suitable for routine quality control.

To take account of some other test properties having a short test duration that might indicate the durability of basalt aggregates, Woodside & Woodward (1989) refer to a proposal by Woodward to combine the MBV with the 10% fines value and the porosity of the aggregate so as to yield:

$$\text{Basalt durability index} = 100 \times \frac{\text{Porosity}}{10\%\ \text{fines value}} \times \text{MBV}\ .$$

In this relationship, porosity is the apparent porosity P_a, which is defined in BS 1902 *Methods of testing refractory materials* as the ratio of open pores to the bulk volume of the material, the bulk volume being the volume of solid material plus the volume of sealed and open pores. Apparent porosity is given by:

$$P_a = 100 \times \frac{WR_s}{W+100}$$

where W is the percentage water absorption and R_s the relative density on a saturated and surface–dried (SSD) basis determined by the methods of BS 812 (Hosking 1974).

Relating the Index to assumed magnesium sulphate soundness loss limits, the authors suggest the following tentative limits for the Index:

Basalt durability index	Durability assessment	Assumed magnesium sulphate soundness loss limit
Less than 2.2	sound	less than 18%
2.2 to 4.4	marginal	18 to 30%
More than 4.4	unsound	more than 30%

The concept of combining a number of relevant test properties to estimate durability conforms with the established view that poor durability and unsoundness are normally a function of more than one variable. Before selecting the test properties for inclusion in the basalt durability index the authors carried out various laboratory tests on a wide range of weathered and unweathered types of basalt, and related the results to the magnesium sulphate soundness loss and methylene blue values (MBV) by linear regression equations and correlation coefficients. The coefficients relating magnesium sulphate soundness loss and MBV to porosity and water absorption were 0.90 and 0.92.

Not unexpectedly, however, the correlation between magnesium sulphate soundness loss and the dry strength values was not good. The correlation coefficients for ACV, AAV, and AIV were all less than 0.60, while that for 10% fines value was 0.74. It might, therefore, have been more appropriate to have incorporated in the Index the 24 hour soaked instead of the dry, or standard, 10% fines value. The 24 hour soaked test, used in association with the standard test, has a satisfactory record in South Africa, where comparable problems arise with basic igneous rock aggregate (Weinert 1980). It would not entail any increase in the overall testing period, as two days are already required for determining the porosity term in the Index.

With regard to the tentative limits, it is of interest to note that although most roadstone specifications would fail an aggregate with a magnesium sulphate soundness loss value in excess of 18%, a value of 30% (equivalent Index 4.4) is suggested by the authors as the limit for unsoundness of the Northern Ireland basalts, with 18% to 30% being rated as marginal.

The precision of the determination of the Index is not included with the proposal, but the repeatability and reproducibility of a combination of test results would be expected to be wider than those for an individual test. However, if these were found

Sec. 5.3] Tests and standards for aggregates in bituminous materials   239

to be acceptable for the proposed limits, this type of index could provide a useful and rapid indicator of the quality of a rock for road aggregates, and, depending on the frequency of testing, could justify the additional work entailed.

5.3.8 Adhesion of bitumen to aggregates

To perform satisfactorily as the binder in a mix, the bitumen needs to adhere well to the mineral aggregate. There must be good adhesion both at the mixing stage and subsequently during the service life of the asphalt.

Good initial coating and adhesion can be assured by using the hot-mix process where dry aggregate is mixed with molten bitumen at a temperature at which it has a low viscosity. The bitumen then wets the stone readily and good coating can be obtained. It is important that the surface of the aggregate should not be dusty as the presence of dust impairs the adhesion.

After laying, an open graded bitumen macadam mix may be susceptible to the effects of water. The high voids content of the mix gives interconnected voids that allow the ingress of water or water vapour. Then, although the permeability of bitumen to water is extremely low, the layer of bitumen coating the stone is very thin and so, given sufficient time, water molecules can diffuse through it. If the stone is acidic (see section 5.2.2), the water molecules can accumulate at the bitumen/stone interface to such an extent that the bitumen becomes detached from the stone (Hughes *et al.* 1960). This state of detachment of the bitumen film is not necessarily critical for the mix as the bond strength between the bitumen and the stone is equal to that in the dry state when the tensile stress is normal to the stone surface. A tangential stress, however, may disbond a detached film if the stiffness of the film itself is low enough. Thus, when traffic stresses are sufficiently severe, stripping of the binder from the aggregate can take place, resulting in a loss of cohesion in the mix. If the binder coating is not damaged by traffic stresses, it may re-adhere to the stone on drying out.

Dense graded mixes may be vulnerable to the effects of water if the fine aggregate or filler is unsuitable. In this case, although visible stripping may not occur, the ingress of water can result in a loss of strength of the mix by affecting the bitumen/stone bond at the large surface area presented by the fine aggregate. This problem can be prevented by using basic rocks for the fine aggregate, and limestone or Portland cement, both of which adhere well to the bitumen, for the filler. It is also beneficial to make the mix substantially impermeable by compacting it to a voids content of less than 5 or 6%.

Hydrated lime is also used in mixes as part of the filler, to promote adhesion. A further benefit of hydrated lime is that it reduces the age hardening of bitumen by oxidation — this effect depending on the source of the bitumen (Petersen *et al.* 1987).

Surface active compounds, such as fatty amines, are used as adhesion agents for bitumen (particularly cutbacks) to promote wetting of the aggregate and to prevent stripping.

5.3.8.1 Static water immersion tests

The total water immersion test (TWIT) is a static water immersion test developed by Shell (Van Asbeck 1964) that is intended to assess the extent to which a bituminous coating on an aggregate will resist displacement by water. The sample of aggregate, passing a ¾ in test sieve and retained on a ½ in test sieve (say 20 to 14 mm fraction) is prepared by washing it well and finally rinsing it with distilled water. It is then dried at 120°C for 5 hours, after which it is cooled for 24 hours in a closed container.

A test sample of 300 g of the aggregate at 40°C is mixed with 15 g of cutback bitumen at about 100°C, or 300 g of the aggregate at 60°C is mixed with 15 g of penetration grade bitumen at 120°C.

When the aggregate is completely coated it is immediately transferred to a glass jar. After being allowed to stand for half an hour, the sample is covered with freshly boiled distilled water and the jar placed in a water bath maintained at 40°C, where it remains for 3 hours. The jar is then removed and the area of the aggregate coated with binder is estimated visually to the nearest 5% or, when it applies, is reported as 'less than 30%'. The repeatability is 10%.

A stripping test employing static immersion is described in Appendix 4A of the PSA Airfields Branch Specification. After being coated with bitumen, 150 particles of 10 to 6.3 mm size aggregate are placed in trays that have been treated beforehand with a mixture of glycerol and dextrin to prevent adhesion of the binder to the tray. Each particle is completely separated from adjacent ones. After standing for one hour, the coated specimen is covered with distilled water at 18 to 20°C and maintained at that temperature. Following immersion for 48 hours, the water is decanted and the specimen allowed to dry at air temperature. The dried specimen is then examined particle by particle. The sample fails the test if there is evidence of the binder stripping from more than six of the total number of particles tested. This limit applies when the aggregate is for use in any of the specified mixes except friction course, for which the limit is three particles.

5.3.8.2 *Immersion wheel-tracking test*

As mentioned earlier, while the bitumen film coating the aggregate in a road surfacing can become detached because of the presence of water, stripping may occur only if the stresses due to traffic are sufficiently great and persistent. Some tests simulate traffic with a reciprocating loaded wheel that runs on the surface of the test specimen of compacted bituminous mix. An example of such a test is the immersion wheel-tracking test developed by TRRL (RRL 1962) in which the material tested is an open graded macadam with a specified grading and a binder content of 4.5%. Just enough material is mixed to fill a perforated metal mould when it is compacted by a hand roller in the prescribed manner to give a density of 1.75 g/ml. The material is intrinsically weak. Three specimens, each 12 in long, 4 in wide, and 1.25 in thick, are tested together while immersed in a water bath maintained at 40°C. The loading wheels are 8 in diameter and have 2 in wide, hard rubber tyres. The load on a wheel axle is 40 ± 0.5 lb, and there are 25 double passes of the wheel per minute.

The depth of penetration of the wheel into the test specimen is recorded against time. Initially there is a slow, steady rate of penetration. This is followed by a break in the curve on the recorder chart where the wheel starts to penetrate at a much greater rate. This critical point is adopted as the failure criterion, and the failure time is used as a measure of the resistance to stripping. An example of a failure time indicating a poor performance is 3 hours, while for a good performance the failure time might be 23 hours.

5.3.8.3 *Immersion mechanical tests*

Immersion mechanical tests are carried out on dense bituminous mixes, and may be either the compression test (ASTM D 1074 and ASTM D 1075) or the Marshall test.

In the ASTM test, cylindrical test specimens, 4 in by 4 in, are compacted statically at a pressure of 3000 lb/in^2, and cured in an oven at 60°C for 24 hours. The relative density of each specimen is determined before it is tested in compression at a constant speed of 0.2 in/min in an axial direction, at a temperature of 25°C.

Six specimens are sorted into two groups of three, each group having similar mean relative densities. One group of specimens is tested in the normal manner ('dry') and the other is tested after soaking in water for a specified time at a specified temperature — either 4 days at 49°C plus 2 hours at 25°C or 1 day at 60°C plus 2 hours at 25°C. The retained strength is reported as:

$$\text{Index of retained strength \%} = \frac{\text{average 'wet' strength}}{\text{average 'dry' strength}} \times 100 \ .$$

A commonly applied specification is that the index should not be less than 75%. The retained strength is low when the adhesion to the binder of the fine aggregate and/or filler is adversely affected by the presence of water.

The result depends on the amount of water absorbed by the specimens, and this, in turn, depends on the permeability and hence the voids content. It is therefore important to match the 'wet' and 'dry' batches of samples for average voids content. There are some aggregates that give a retained strength greater than 100%.

REFERENCES

American Association of State Highway and Transportation Officials (1986), Washington, DC. Standard Specifications for Transportation Materials and Method of Sampling and Testing. Test T103–86. *Soundness of aggregates by freezing and thawing*. Copyright 1986.

American Society for Testing and Materials. ASTM C88, 1983, *Standard test method for soundness of aggregates by use of sodium sulfate or magnesium sulfate*.

American Society for Testing and Materials. ASTM C127, 1988, *Standard test method for specific gravity and absorption of coarse aggregate*.

American Society for Testing and Materials. ASTM C128, 1988, *Standard test method for specific gravity and absorption of fine aggregate*.

American Society for Testing and Materials. ASTM C131, *Standard test method for resistance to abrasion of small size coarse aggregate by use of the Los Angeles Machine*.

American Society for Testing and Materials. ASTM C136–76, 1984, *Standard test method for sieve or screen analysis of fine and coarse aggregates*.

American Society for Testing and Materials. ASTM D692, 1988, *Standard specification for coarse aggregate for bituminous paving mixtures*.

American Society for Testing and Materials. ASTM D1073, 1988, *Standard specification for fine aggregate for bituminous paving mixtures*.

American Society for Testing and Materials. ASTM D1074, 1983, *Standard test method for the compressive strength of bituminous mixtures*.
American Society for Testing and Materials. ASTM D1075, 1988, *Standard test method for effect of water on cohesion of compacted bituminous mixtures*.
American Society for Testing and Materials. ASTM D1559, *Standard test method for resistance to plastic flow of bituminous mixtures using Marshall apparatus*.
American Society for Testing and Materials. ASTM D2041, 1990, *Standard test method for theoretical maximum specific gravity of bituminous paving mixtures*.
Asphalt Institute. *Mix design methods for asphalt concrete*. Manual MS-2, Maryland, USA.
Asphalt Institute. *Thickness design — full depth asphalt pavement structures for highways and streets*. Manual Series No 1, Maryland, USA.
Bonnaure, F., Gest, G., Gravois, A., & Uge P. (1977). A new method of predicting the stiffness of asphalt paving mixtures. *Proc. Assoc. Asphalt Paving Technologists*, **46**, 64–104.
British Standards Institution. BS 63: 1987. *Road Aggregates*.
   Part 1. *Specification for single-sized aggregate for general purposes*.
   Part 2. *Specification for single-sized aggregate for surface dressing*.
British Standards Institution. BS 410: 1986. *Specification for test sieves*.
British Standards Institution. BS434: 1984. *Bitumen road emulsions (anionic and cationic)*.
   Part 1. *Specification for bitumen road emulsions*.
   Part 2. *Code of practice for use of bitumen road emulsions*.
British Standards Institution. BS 594: 1985. *Hot rolled asphalt for roads and other paved areas*.
   Part 1. *Specification for constituent materials and asphalt mixtures*.
   Part 2. *Specification for the transport, laying and compaction of rolled asphalt*.
British Standards Institution. BS 598. *Sampling and examination of bituminous mixtures for roads and other paved areas*.
   Part 2: 1974. *Methods for analytical testing*.
   Part 3: 1985. *Methods for design and physical testing*.
   Part 101: 1987. *Methods for preparatory treatment of samples for analysis*.
   Part 102: 1989. *Analytical test methods*.
British Standards Institution. BS 812: 1960. *Methods for sampling and testing of mineral aggregates, sands and fillers*.
British Standards Institution. BS 812. *Testing aggregates*
   Part 1: 1975. *Methods for determination of particle size and shape*.
   Part 2: 1975. *Methods for determination of physical properties*.
   Part 3: 1975. *Methods for determination of mechanical properties*.
   Part 101: 1984. *Guide to sampling and testing aggregates*.
   Part 102: 1989. *Methods for sampling*.
   Part 103. *Methods for determination of particle size distribution*.
   Section 103.1: 1985. *Sieve tests*.
   Section 103.2: 1989. *Sedimentation test*.
   Part 105: 1985. *Methods for determination of particle shape*.
   Section 105.1: 1989. *Flakiness index*.
   Part 114: 1989. *Method for determination of the polished-stone value*.

Part 121: 1989. *Method for determination of soundness.*
British Standards Institution. BS 1047: 1983. *Specification for air-cooled blast-furnace slag aggregate for use in construction.*
British Standards Institution. BS 1447: 1988. *Specification for mastic asphalt (limestone fine aggregate) for roads, footways and pavings in buildings.*
British Standards Institution. BS 3690. *Bitumens for building and civil engineering.*
Part 1: 1989. *Specification for bitumens for roads and other paved areas.*
Part 3: 1990. *Specification for mixtures of bitumen with pitch, tar and Trinidad lake asphalt.*
British Standards Institution. BS 4987: 1988. *Coated macadam for roads and other paved areas.*
Part 1. *Specification for constituent materials and for mixtures.*
Part 2. *Specification for transport, laying and compaction.*
British Standards Institution. BS 5273: 1975 (1985). *Specification. Dense tar surfacing for roads and other paved areas.*
British Standards Institution. BS 5309: 1976. *Methods for sampling chemical products.*
Part 1. *Introduction and general principles.*
Part 4. *Sampling of solids.*
British Standards Institution. BS 6100. *Glossary of building and civil engineering terms.* Part 5. *Masonry.* Section 5.2: 1984. *Stone.*
British Standards Institution. BS 6577: 1985. *Specification for mastic asphalt for building (natural rock asphalt aggregate).*
British Standards Institution. BS 6925: 1988. *Specification for mastic asphalt for building and civil engineering (limestone aggregate).*
Cole, W. F. & Sandy, M. J. (1980). A proposed secondary mineral rating for basalt road aggregate durability. *Australian Road Research*, **10** (3), 27–37. Australian Road Research Board, Vermont South, Victoria 3133.
Collis, L. & Fox, R. A. (eds.) (1985). *Aggregates: sand, gravel and crushed rock aggregates for construction purposes.* Eng. Geol. Special Publication No. 1; The Geological Society of London.
Daines, M. E. (1986). *Pervious macadam: trials on trunk road A38 Burton by-pass, 1984.* Department of Transport, TRRL Report RR 57; Transport and Road Research Laboratory, Crowthorne, Berkshire, UK.
Davis, A. G. (1979). Discussion of paper 8144 'Skid-resistant road surfacings and tyre noise' by G. F. Salt. *Proc. Inst. Civ. Eng.*, **66** (Nov.), 699–713.
De Bats, F. Th. (1973). A computer simulation of Van der Poel's nomograph. *J. App. Chem. and Biotech.* **3**, p. 139.
Department of Transport (1976). *Specification for road and bridgeworks. Specification requirements for aggregate properties and texture depth for bituminous surfacings to new roads.* Technical Memorandum H 16/76. Department of Transport, London.
Department of Transport (1986). *Specification for Highway Works*, Parts 1–7. HMSO, London.
Department of Transport (1986). *Notes for Guidance on the Specification for Highway Works*, Parts 1–6, and Amendment No. 1, March 1988. HMSO, London.

Department of Transport, Highways and Traffic Directorate (1987). *Structural design of new road pavements*. Departmental Standard HD 14/87; Department of Transport, London.

Department of Transport, Highways and Traffic Directorate (1987) *Skidding-resistance of in-service trunk roads*. Departmental Standard HD 15/87 and Advice Note HA 36/87. Department of Transport, London.

Dhir, R. K., Ramsay, D. M., & Balfour, N. (1971). A study of the aggregate impact and crushing value tests. *J. Inst. Highway Eng.*, **XVIII**, 11, 17–27.

Dickinson, E. J. (1984) (reprinted 1985). *Bituminous roads in Australia*. Australian Road Research Board,, Vermont South, Victoria 3133.

DIN 4226, Blatt 3, (1971). *Zuschlag für Beton. Prüfung von Zuschlag mit dichtem oder porigem Gefüge*. Deutschen Normenausschusses, Berlin, 30.

Farrington, J. J. & Roberts K. (1986). Experiences with pervious macadams in Staffordshire. *J. Inst. Highways and Transportation*, **33**, 10, 8–16.

Fielding, B. J. & Maccarrone, S. (1982). Accelerated soundness tests for altered basalts. *Proc. 11th ARRB Conf.*, **11** (3), 129–144. Australian Road Research Board, Vermont South, Victoria 3133.

Fookes, P. G. & Higginbottom, I. E. (1980). Some problems of construction aggregates in desert areas, with particular reference to the Arabian peninsula. 1, Occurrence and special characteristics; 2, Investigation, production and quality control. *Proc. Inst. Civ. Eng.*, **68**, 39–90.

Fuller, W. B. & Thompson, J. E. (1907). The laws of proportioning concrete. *Trans. Am Soc. Civ. Eng.*, **59**, 67–172.

Geological Society of America (1963). *Rock-color chart*.

Hanson, F. M. (1934). Bituminous surface treatments of rural highways. *Proc. NZ Soc. Civ. Eng.*, **21**.

Hawkes, J. R. & Hosking, J. R. (1972). *British arenaceous rocks for skid-resistant road surfacings*. Department of the Environment, TRRL Report LR 488. Transport and Road Research Laboratory, Crowthorne, Berkshire, UK.

Heslop, M. F. W., Elborn, M. J., & Pooley, G. R. (1982). Recent developments in surface dressings. *J. Inst. Highway Eng.*, **29**, 7, 6–19.

Higgs, N. B. (1976). Slaking basalts. *Bull Ass'n Eng. Geol.*, **XIII**, 2.

Hills, J. F. (1973). The creep of asphalt mixes. *J. Inst. Petroleum*, **59**, 570, 247–262.

Hills, J. F. (1975). Creep tests for asphalt mix design. *Highways and Road Construction*, **43** (1782–3), 22–29.

Hills, J. F., Brien, D., & Van de Loo, P. J. (1974). *The correlation of rutting and creep tests on asphalt mixes*. Inst. of Petroleum Paper IP 74–001.

Hills, J. F. & Pettifer, G. S. (1985). The clay mineral content of various rock types compared with the methylene blue value. *J. Chem. Tech. and Biotech.* **35A**, 4, 168–180.

Hingley, C. E., Peattie, K. R., & Powell, W. D. (1976) *French experience with grave- bitume, a dense bituminous roadbase*. Department of the Environment, TRRL Report SR 242. Transport and Road Research Laboratory, Crowthorne, Berkshire, UK.

Hosking, J. R. (1974) *Synthetic aggregates of high resistance to polishing*. Part 3 — *Porous aggregates*. Department of the Environment TRRL Report LR 655. Transport and Road Research Laboratory, Crowthorne, Berkshire, UK.

References

Hosking, J. R. & Tubey, L. W. (1969). *Research on low-grade and unsound aggregates.* Ministry of Transport, RRL Report LR 293. Road Research Laboratory, Crowthorne, Berkshire, UK.

Hudec, P. P. (1983). *Aggregate tests — their relationship and significance.* Durability of Building Materials, 1 (1982/1983), pp. 275–300. Elsevier, Amsterdam, London, New York.

Hughes, R., I., Lamb, D. R., & Pordes, O. (1960). Adhesion in bitumen macadam. *J. App. Chem. (London)*, **10**, 433.

International Organization for Standardization (1982). ISO 6274–1982(E). *Concrete — Sieve analysis of aggregates.*

Jacobs, F. A. (1985). *An experiment to investigate rolled-asphalt wearing courses with different coarse aggregates on A30 Staines by-pass.* Department of Transport, TRRL Report RR 12. Transport and Road Research Laboratory, Crowthorne, Berkshire, UK.

Kennedy, C. K. (1978). *The development of slip-planes in rolled asphalt surfacings.* Department of the Environment, Department of Transport, TRRL Report LR 813. Transport and Road Research Laboratory, Crowthorne, Berkshire, UK.

Knight, B. H. & Knight, R. G. (1948). *Road aggregates; their uses and testing.* Edward Arnold & Co., London.

Larson, T. D., Cady, P. D., Franzen, M., & Reed, J. (1964). *A critical review of literature treating methods of identifying aggregates subject to destructive volume change when frozen in concrete and a proposed program of research.* Highway Research Board Special Report No. 80.

Lee, A. R. & Rigden, P. J. (1945). The use of mechanical tests in the design of bituminous road-surfacing mixtures. Part 1: dense tar surfacings. *J. Soc. Chem. Ind. (London)*, **64**, 133–161.

Leech, D. (1982). *A dense coated macadam of improved performance.* Department of the Environment, Department of Transport, TRRL Report LR 1060. Transport and Road Research Laboratory, Crowthorne, Berkshire, UK.

Leech, D. & Selves N. W. (1980). *Deformation resistance of dense coated macadams: effect of compaction and binder content.* Department of the Environment, Department of Transport, TRRL Report SR 626. Transport and Road Research Laboratory, Crowthorne, Berkshore, UK.

National Institute for Transport and Road Research (1986). *Surfacing seals for rural and urban roads.* Draft TRH3. CSIR, Pretoria.

Nunn, M. E. & Leech, D. (1986). *Substitution of bituminous roadbase for granular sub-base.* Department of Transport, TRRL Report RR 58. Transport and Road Research Laboratory, Crowthorne, Berkshire, UK.

Nunn, M. E., Rant, C. J., & Schoepe, B. (1987). *Improved roadbase macadams: road trials and design considerations.* Department of Transport, TRRL Report RR 132. Transport and Road Research Laboratory, Crowthorne, Berkshire, UK.

Nutt. P. E. (1987). *The implementation of heavy-duty macadam. Seminar on quality control in the blacktop industry and dense macadams of improved performance. Sutton Coldfield, BACMI.* British Aggregate Construction Materials Industries, London.

Perrin, R. M. S. (1976) *Manual of applied geology for engineers*; Section 5.2 — nature and origin of soils. Inst. Civ. Eng., London.

Petersen, J. C., Plancher, H., & Harnsberger, P. M. (1987). Lime treatment of asphalt to reduce age hardening and improve flow. *Proc. Ass'n. Asphalt Paving Technologists*, **56**, 632–653.

PIARC (1979). Technical Committee on Testing Road Materials. *XVIth World Road Congress*, Permanent International Association of Road Congresses, Vienna.

Pooley, G. R. & Clark, K. V. (1982). Full depth asphalt solves road reconstruction problems in Essex. *J. Inst. Asphalt Tech.*, *August*, **32**, 18–23.

Powell, W. D. (1987a). Bituminous roadbases and basecourses. *J. Inst. Highways and Transportation*, **34**, 3, 4–11.

Powell, W. D. (1987b). Collaborative research on improved bituminous macadams. *Seminar on quality control in the blacktop industry and dense macadams of improved performance. Sutton Coldfield, BACMI*. British Aggregate Construction Materials Industries, London.

Property Services Agency, Airfields Branch (1979). *Standard Specification Clauses for Airfield Pavement Works*. Part 4, *Bituminous Surfacing*. Croydon, UK.

Ramsay, D. M., Dhir, R. K., & Spence, J. M. (1973). Non-geological factors influencing the reproducibility of results in the aggregate impact test. *Quarry Managers' Journal*, London, **57**, 179–181.

Read, H. H. (1947). *Rutley's elements of minerology*. Thomas Murby & Co., London.

Rex, H. M. & Peck R. A. (1956). A laboratory test to evaluate the shape and surface texture of fine aggregate particles. *Public Roads*, December, US Government Printing Office, Washington DC.

Road Research Laboratory (1959). *Roadstone test data presented in tabular form*. Road Note 24. Department of Scientific and Industrial Research; HMSO, London.

Road Research Laboratory (1962). *Bituminous materials in road construction*. Department of Scientific and Industrial Research, HMSO, London.

Road Research Laboratory (1969). *Instructions for using the portable skid-resistance tester*. Road Note 27 (second edition). HMSO, London.

Salt, G. F. (1977). *Research on skid-resistance at the Transport and Research Laboratory (1927–1977)*. Department of the Environment, Department of Transport, TRRL Report SR 340. Transport and Road Research Laboratory, Crowthorne, Berkshire, UK.

Salt, G. F. & Szatkowski, W. S. (1973). *A guide to levels of skidding resistance for roads*. Department of the Environment, TRRL Report LR 510. Transport and Road Research Laboratory, Crowthorne, Berkshire, UK.

Shell International Petroleum Company Limited (1978). *Shell Pavement Design Manual*. London.

Shergold, F. A., & Hosking, J. R. (1959). A new method of evaluating the strength of roadstone. *Roads and Road Construction*, **37** (438), 164–167.

Shergold, F. A. & Hosking, J. R. (1963). Investigation of test procedures for argillaceous and gritty rocks in relation to breakdown under traffic. *Roads and Road Construction*, **41** (492), 376–378.

South African Bureau of Standards, SABS 1083–1976 (amended 1979). *Specification for aggregate from natural sources*. Pretoria.

Spence, J. M. (1979). Studies of strength and its determinative tests in roadstone aggregates. Thesis for the degree of Doctor of Philosophy in the University of Dundee. Unpublished.

Stewart, E. T. & McCullough, L. M. (1985). The use of the methylene blue test to indicate the soundness of road aggregates. *J. Chem. Tech. and Biotech.*, **35A**, 4, 161–167.

Szatkowski, W. S. & Hosking, J. R. (1972). *The effect of traffic and aggregate on the resistance to skidding of bituminous surfacings.* Department of the Environment, TRRL Report LR 504. Transport and Road Research Laboratory, Crowthorne, Berkshire, UK.

Szatkowski, W. S. & Jacobs, F. A. (1977). *Dense wearing courses in Britain with high resistance to deformation.* Colloquium 77: Plastic deformability of bituminous mixes. Technischen Hochschule Zurich, Mitteilung Nr. 37, 63.

Texas Highway Department (1978). *Open graded friction courses for highways.* National Cooperative Highway Research Program; Synthesis of Highway Practice 49. Transportation Research Board, Washington, DC.

Tran Ngoc Lan (1980). L'essai au bleu de méthylène. Un progrès dans la mesure et le contrôl de la propreté des granules. *Bull de liaison des Laboratoires des Ponts et Chaussées. Paris*, **107**, 130–135.

TRRL, (1979). *Final report of the working party on the slippage of rolled-asphalt wearing course.* Department of the Environment, Department of Transport, TRRL Report SR493, Transport and Road Research Laboratory, Crowthorne, Berkshire, UK.

United States Department of Transportation, Federal Aviation Administration, 1974 (revised 1985). AC No. 150/5370-10; *Standards for specifying construction of airports.* US Government Printing Office, Washington, DC.

US Army Corps of Engineers Waterways Experiment Station, Vicksburg, 1969. *Methods of testing stone for expansive breakdown on soaking in ethylene glycol.* Concrete Research Division, CDR-C 148–69.

Van Asbeck, W. F. (1964) *Bitumen in hydraulic engineering.* Vol. 2. 277 pp. Elsevier, Amsterdam/London/New York.

Van der Poel, C. (1954). A general system describing the visco-elastic properties of bitumens and its relation to routine test data. *J. App. Chem.*, **4**, Pt. 5, pp. 221–236.

Verbeck, G. & Landgren, R. (1960). Influence of physical characteristics of aggregates on frost resistance of concrete. *ASTM Proc.* **60**, 1063–1079.

Weinert, H. H. (1964) Basic igneous rocks in road foundations. CSIR Research Report No. 218. 47 pp. (*Nat. Inst. Rd. Res. Bull.* No. 5). Council for Scientific and Industrial Research, Pretoria.

Weinert, H. H. (1980). *The natural road construction materials of Southern Africa.* Council for Scientific and Industrial Research, Pretoria. Published by H. & R. Academica (Pty.) Ltd., Pretoria and Cape Town.

Williams, A. R. (1975). *Road Surfacing Materials.* British Patent Specification No. 1393885. HM Patent Office, London.

Williams, A. R. (1976). The importance of micro texture. *Symposium: Delugrip — the road surface designed for tyres.* Geological Society of London.

Williams, A. R. & Lees, G. (1969). Topographical and petrographical variations of road aggregates and the wet skidding resistance of tyres. *QJEG*, **2**, (3).
Woodside, A. R. & Peden, R. A. (1983). Durability characteristics of roadstone. *Quarry Management and Products*, August, 493–498.
Woodside, A. R. & Woodward, W. D. H. (1989). Assessing basalt durability — rapid alternative techniques. *Conference on durability and performance of bituminous highway materials*, Hatfield Polytechnic, UK.
Young, A. E. (1982). Road surface skid-resistance. *J. Inst. Asphalt Tech.*, **32**, 11–17.

6

Unbound aggregates

P. T. Sherwood (Section 6.1)
D. C. Pike (Section 6.2)

This chapter deals with two separate aspects of the unbound aggregates that are used in lower road pavement layers, i.e. sub-bases and roadbases. The first topic is frost-heave. A description is given of the various types of frost damage against the background of climatic conditions in the UK, and specifications to avoid frost-heave are considered. In particular the Transport and Road Research Laboratory (TRRL) frost-heave test, and the criteria for that test adopted by the Department of Transport (DTp), are reviewed.

The second topic is the stability of unbound sub-base under construction traffic. An outline of the problem of instability is given, and local experiences with various types of material are summarized. A programme of work at TRRL, including the development of a 300 mm shear-box test and a standard trafficking trial, is described.

Each of the two main sections is followed by its own list of literature references.

6.1 FROST SUSCEPTIBILITY

6.1.1 Types of frost damage

6.1.1.1 Frost-shattering

This occurs when water penetrates into the surface of a road and expands as it freezes in the voids of the road pavement layers. It is potentially the most common form of frost damage because conditions for it to arise occur every winter. However, because all public highways have a sealed surface there is no need for a freeze–thaw test to measure the susceptibility of materials to this kind of damage. Provided that the road is kept well-sealed and well-drained such damage should not occur, and if it does, it is a result of faulty construction or lack of maintenance, it is most common where a bituminous surfacing is nearing the end of its useful life, and the remedy is, of course, to ensure that the surface is well-sealed.

6.1.1.2 Frost-heave

More serious than frost-shattering is the problem of frost-heave. This can occur during prolonged periods of freezing when the temperature of the road pavement at depth falls below 0°C. When this occurs one of the conditions for frost-heave is met, and, if other conditions allow, water will be drawn from the sub-grade into the freezing zone where it will form ice-lenses. As these form, further water may be drawn into the freezing zone and more ice produced which can result in considerable expansion. It is the freezing of this transported water, rather than the 9% volume increase of water on freezing, which is the prime cause of frost-heave. The formation of ice-lenses in the lower layers of the road pavement may cause considerable heaving at the surface. After thawing, the foundations of the road are weakened by the excess moisture and, unless traffic is kept off throughout the period of thaw, the whole pavement structure may fail.

From the above discussion it is apparent that for frost-heave to occur three conditions must be satisfied:

(a) the materials in the frozen zone must be susceptible to frost-heave;
(b) water must be able to move freely from the water table into the freezing zone;
(c) frost must penetrate into the unbound layers of the road pavement and subgrade.

If any one of these conditions is absent, heave will not occur, and any procedure for avoiding heave damage should keep this in mind.

6.1.2 Measurement of frost-heave

6.1.2.1 The frost-heave test

A quantitative measure of the frost-susceptibility of granular base and sub-base materials can be obtained from the laboratory test developed at the Road Research Laboratory (now TRRL) and first described in RRL Report LR90 (Croney & Jacobs 1967). The test was developed at RRL originally as a research tool for examining the frost-susceptibility of soils, but it has been used in Great Britain since 1969 as a compliance requirement for unbound granular bases and sub-bases in the Specification for Road and Bridge Works (Department of Transport 1969 and later up-dates).

If a test is to be used as a comprehensive test the primary requirement is that there should be a written document setting out in full detail how test specimens should be obtained and prepared (British Standards Institution 1979). LR90 did not set out to be such a document, and when it was specified for this purpose the precision of the test was, not surprisingly, found to be extremely poor. As a result of this, extensive research was carried out at TRRL to improve the precision of the test. This research has led to the publication of the written document (Roe & Webster 1984) required for compliance purposes. This test specification, with minor modifications, has since been adopted as a British Standard (British Standards Institution 1989). The Standard gives details (see Table 6.1) of the precision of the test carried out by 19 laboratories on a flint-gravel sub-base. This shows that the precision of the test, although much improved, is still poor in relation to the criteria used for deciding whether or not a material is frost-susceptible.

Frost susceptibility

Table 6.1 — Precision estimates for frost-heave[†]

Material	Average (mm)	r_1 (mm)	R_1 (mm)	R_2 (mm)	$\sqrt{V_r}$ (mm)	$\sqrt{V_L}$ (mm)	$\sqrt{V_M}$ (mm)
Flint-gravel	13.5	7.0	8.5	9.0	2.5	1.8	1.0

[†] For definitions of precision terms see Chapter 2.

The precision data in the standard and in SR 829 have been criticized on the grounds that they give an optimistic impression because on both sets of tests the moisture contents and densities at which the specimens were prepared were nominated in advance. However, recent work at TRRL (Webster & West 1987, 1988) has shown that changes in moisture content and density of the frost-heave test specimens have little effect on the heave values recorded.

In the test, as it is now specified, nine cylindrical specimens 102 mm in diameter and 152 mm in height, are placed with their base in contact with water in a self-refrigerated unit (SRU). The SRU is designed so that the air temperature above the specimens can be maintained at $-17\pm1°C$ and the water in contact with the bases of the specimen can be maintained at $+4\pm0.5°C$. After compaction and extrusion the specimens are left to equilibrate in the SRU at room temperature for five days after which they are subjected to the temperature gradient given above. The heave of each specimen is recorded daily over a period of four days.

6.1.2.2 Interpretation of results

Normally, specimens are tested in sets of three. For a test to be valid, the temperature has to be within the specified limits throughout the duration of the test. The mean heave should not be suspiciously low (<2 mm) and, if the mean heave is less than 18 mm, the range of individual results has to be less than 5 mm. If a wide spread of results is obtained the test is repeated, using six specimens.

According to TRRL (Roe & Webster 1984) a material is classified as non-susceptible if the mean heave after four days of a valid test is less than 9.0 mm and as susceptible if the mean heave is more than 15.0 mm. Because of the poor precision of the test, if the mean heave is between 9.1 mm and 14.9 mm further tests are required at two other laboratories. The material is then classified as non-frost susceptible if the grand average heave is less than 12 mm. These criteria are based on the original observations by Croney & Jacobs set out in LR90 and modified to take into account of the fact that their work was largely based on observations made in soil subgrades rather than granular sub-bases and of changes made to the test procedure as described in LR90.

The classification procedure given in SR 829 is time-consuming and has been criticized because of the large amount of extra testing that is required for those materials which in the first test run have a heave in the range 9.1 mm to 14.9 mm. The need to re-test is a consequence of the poor reproducibility, and, to overcome this problem, proposals have been made (Pike *et al.* 1990) that, as part of each test, specimens of known heave should be included.

The proposal was a result of trials that showed that significant improvements in precision could be obtained if control specimens of known heave were included in each run. With this approach, the test results on the control specimens outside a range that allows for the reproducibility of the test would be a good indication that the results of the whole test run were of doubtful validity. The results of the work shows that this goal could be achieved by the use of reference aggregates of known frost heave, carefully selected to eliminate sampling and sample-reduction errors.

This approach is not new and there is already a precedent for it in another part of BS 812. The test for the determination of the polished–stone value (PSV) (described in BS 812: Part 114) also had poor precision when it was originally proposed as a BS test. In this case the early problems of poor precision were eventually overcome by requiring that samples of a stone of known PSV were included in each test run, in the procedure as now specified. If the PSV of the control stone is found to be outside a specified range, the results of the whole test run are rejected. If the results lie within the range, the results from the control stone are used to adjust those obtained with the material under test.

The proposals have been accepted in principle, and work is under way to establish the best method of introducing the use of reference samples into the test procedure. When this work has been completed an amendment to BS 812: Part 124 will be issued.

6.1.3 Availability of water

For frost-heave to become a problem, water must be able to travel freely from the water-table into the freezing zone where the ice lenses form. In heavy clay subgrades, water flow is restricted by low permeability; in coarse sands the suctions induced are too low to cause such flow. Hence, it is the intermediate materials such as gravels with fines, chalk, silts, and silty sands that are frost susceptible.

Dramatic reductions of frost-heave caused by a low water table or an impermeable subgrade have been demonstrated by tests at TRRL (Jacobs & West 1966, Burns 1977), in which full-scale freezing of road structures was simulated in 1.5 m deep test pits. On average, lowering the water table depth from 0.6 m to 1.4 m below the surface of the sub-base reduced the surface heave from about 70 mm to less than 10 mm. Laboratory tests on composite specimens, simulating frost-susceptible sub-bases placed on impermeable subgrades, extended this finding to a wider range of sub-base/subgrade combinations (Jones & Berry 1979).

Given the right kind of subgrade and good drainage, frost-heave should not therefore be a problem. However, the type of subgrade on which a road is to be built is not usually open to choice, and water may enter the pavement layers from the top or the sides to form a perched water table. Moreover, even if drainage is good to start with, it may deteriorate during the life of the road. Hence the design procedure currently in use in Great Britain makes the pessimistic assumption that the water table is always high and ignores the potential savings to be made when the chances of water entering the freezing zone are negligible.

6.1.4 Frost penetration

Under British climatic conditions, frost penetration into a road pavement in excess of 450 mm is almost unknown. Specifications therefore require that only those materials likely to be found in the top 450 mm of the surface of the road need to be non-susceptible to frost-heave as defined by the TRRL frost-heave test. Most roads have thicknesses of construction of at least 450 mm, so the frost-susceptibility of the subgrade is not usually a problem. It has been shown that stabilization with cement or bitumen is an effective method of dealing with frost-susceptible aggregates. Hence frost-susceptible aggregates used in an unbound form for base and sub-base construction are the only materials likely to be at risk from frost-heave. Unbound aggregates are frequently used within the top 450 mm of the road pavement, and British designs assume that, during the life of the road, frost penetration into these layers will occur no matter where in the country the road is situated (Croney & Jacobs 1967).

Frost penetration into a road depends on temperature levels, the duration of the cold spell, and the thermal properties of the layers. To measure geographical and yearly variations of penetration is impracticable, but air temperature records can be used, utilizing the concept of frost index (FI) which is a measure of the severity and duration of a cold spell. FI is measured in degree days below 0°C; for example a day with an average temperature of $-4°C$ which remains below zero all day would have an FI of 4 days. Successive days are aggregated for the period considered. Frost penetration (x) can be estimated (for average British conditions) from the relationship $x \simeq 40\sqrt{I}$ mm where I is the FI in degree days.

Using this concept, the general assumption of 450 mm of frost penetration for design has been re-examined recently (Sherwood & Roe 1986). The mean annual freezing index was obtained from nearly 50 meteorological stations throughout Britain for the period 1959–1979. The maximum freezing index for a single winter was 302 in 1962/63 (the coldest winter for 200 years). The grand mean was 50 degree days which corresponds to a frost penetration of about 280 mm. For three sites studied a freezing index exceeding 50 was not recorded in the study period, and for a further seven this figure was exceeded only in 1962/63. From this study it is clear that the present assumption of equal risk treatment regardless of location is untenable. For some sites there is little risk of penetration into the unbound layers of the road. For them, the 450 mm restriction cannot be justified and a thickness of about 200 mm would be more appropriate. There are areas where penetrations of 450 mm are common and no change to the present requirements can be contemplated, but for many other places there is a case for adjusting the thickness required to suit the likely penetration at a particular site on the basis of local meteorological records. Where such adjustments are made, non-susceptible material would still be required within the design depth, but savings might be achieved by using poorer materials below the level where frost penetration was unlikely to occur.

References

British Standards Institution (1979). *Precision of test methods — Part 1 Guide to the determination of repeatability and reproducibility for a standard test method.* BS 5497: Part 1: 1979.

British Standards Institution (1989). *Testing aggregates: Method for the determination of frost-heave.* BS 812: Part 124: 1989.

Burns, J. (1977). *The effect of the water-table on the frost-susceptibility of road making materials.* TRRL Supplementary Report SR 305.

Croney, D. & Jacobs, J. C. (1967). *The frost susceptibility of soils and road making materials.* RRL Report LR90.

Department of Transport (1969). *Specification for Road and Bridge Works (Fourth edition).*

Jacobs, J. C. & West, G. (1966). *Investigations into the effect of freezing on a typical road structure.* RRL Report No. 54.

Jones, R. H. & Berry, A. N. (1979). The influence of sub-grade properties on frost-heave. *Highways and Public Works 4* No. 1882, 17–22.

Pike, D. C., Sherwood, P. T., & Sym, R. (1990) The BS Frost-Heave Test: Development of the Standard and suggestions for further improvement, *Quarry Management* February 1990.

Roe, P. G. & Webster, D. C. (1984). *Specification for the TRRL frost-heave test,* TRRL Supplementary Report SR 829.

Sherwood, P. T. & P. G. Roe (1986). *Winter air temperatures in relation to frost damage in roads.* TRRL Research Report 45.

Webster, D. C. & West, G. (1987). *The effect of density on the heave of specimens in the frost heave test.* Transport and Road Research Laboratory, Materials Memorandum 168, 1987 (unpublished).

Webster, D. C. & West G. (1988) *The effect of density on the heave of specimens in the frost-heave test.* Supplement 1, Transport and Road Research Laboratory, Materials Memorandum 174, 1988 (unpublished).

6.2 STABILITY OF SUB-BASES

6.2.1 Patterns of use, and specifications

Large quantities of unbound aggregates are used in the lower layers of road pavements, partly to provide platforms strong enough to support lorries and other plant delivering and placing the materials that form the upper pavement layers. For many years sands and gravels were permitted aggregates in the lower layers, but occasional costly failures under construction traffic caused specifiers increasingly to demand crushed rock aggregates for this purpose. Suitable crushed rocks are scarce in the southeast of England, so considerable tonnages are now imported there over distances of 50 to 100 km, or more, and at high cost. The presumption against the use of sand and gravel increases the prices of approved aggregates in other areas too. If a means could be found of distinguishing satisfactory local materials, and perhaps of upgrading less stable aggregates, substantial savings could be made in haulage costs, perhaps amounting to several million pounds a year.

In a recent paper (Curtis & Loveday 1987) it was estimated that, of the 18.4 million tonnes of aggregates used in England and Wales in 1985/86 for granular sub-bases, 9.9 million tonnes consisted of crushed limestone, 5.2 million tonnes of 'hardstone' (i.e. igneous rocks and gritstones, etc.), and only 3.3 million tonnes of sand and gravel. An unknown tonnage of by-products and waste materials (such as slags from metal processing, colliery waste, crushed concrete, etc.) is also used.

From these figures it is clear that, currently, sand and gravel provide under 20% of the total market for sub-base aggregates. Twenty years ago the figure was nearer

50%. The leading specification, which has changed superficially very little over the past twenty years, is that of the DTp; this has recently been revised (Department of Transport 1986). For many years it has given requirements for two kinds of granular sub-bases, i.e. 'Type 1' and 'Type 2'. Type 1 materials are expected to be all-weather sub-base aggregates; no sands and gravels are allowed in Type 1. Type 2 materials have a looser specification and are not expected to give good performance under construction traffic in wet weather. Major changes in the market for aggregates have been caused by a growing tendency to exclude Type 2 sub-base materials from major road works.

The Notes for Guidance to the DTp's specification draw attention to the exclusion of sands and gravels from sub-bases for heavily-trafficked pavements, but suggest that, if local experience indicates that particular sands and gravels can be used successfully, there is a route for allowing those materials in DTp schemes. In a few restricted areas sub-bases made with sand and gravels have given consistently good performance; for example in southern Gloucestershire and northern Wiltshire millions of tonnes of sand and gravel have been used for this purpose after minimal processing from deposits in the upper Thames region which consist largely of fragments of Jurassic limestones.

To control the quality of this material Gloucestershire County Council imposed tight limits on particle size distribution ('grading'). The natural deposits are unusually consistent in grading; some favoured deposits have average gradings that lie at the centre of the envelope called up by the Gloucestershire C.C. specification. But the grading of these materials is not their most significant property: their inherent stability comes from the fact that they consist of a relatively soft mineral, i.e. calcite, which enables them to develop high shear strengths by interlocking when they are compacted. This property they hold in common with aggregates made by crushing Carboniferous limestone, which are favoured Type 1 sub-base materials. In contrast, siliceous sands and gravels, especially those that consist mainly of flint, have particles that are hard and glassy, and do not interlock well. It is the combination of surface hardness and smoothness that leads to low internal friction. This simple fact is not universally understood, and many engineers mistakenly believe that all sands and gravels, especially if their particles are well rounded, are intrinsically less suitable as road aggregates than crushed rocks.

Although there is a long history of use of sand and gravel in minor works, there have been few recent cases of their use on motorways and trunk roads. Enquiries have shown that local engineers feel the need of a firmer lead from the DTp. On its side, the DTp requires a strong case to be put by the local engineers, and thus, without a sustained programme, opportunities have been missed for an objective appraisal of all potentially suitable materials by a central agency. To counter this, a programme of research was carried out at TRRL with the cooperation of the Sand and Gravel Association (SAGA).

6.2.2 Performance requirements

Road sub-bases are usually laid directly on the prepared subgrade that results from earthworks carried out to achieve the required alignment. In some cases, e.g. where the soil is weak, a capping layer may be inserted below the sub-base. The quality of capping layers may not be as good as that of sub-bases, but their roles are broadly similar: (a) they must provide a working platform on which other paving materials can be transported, laid, and compacted — this implies a requirement in relation to shear strength; (b) they must insulate against freezing where the sub-grade is likely to be damaged by frost — hence the sub-base material must itself be frost resistant (see section 6.1); (c) they must make a contribution to the ability of the pavement to carry traffic during its working life — apart from adequate resistance to mechanical forces, sub-base materials must play a part in the drainage of those pavements that admit significant amounts of water.

This part of Chapter 6 is concerned with the first of these three functions, and describes an investigation into the resistance to rutting of granular materials, and the development of better methods of specifying them. Although this fact is not always recognized, some deformation at the surface of a sub-base subjected to construction traffic is almost inevitable, unless the aggregate is bound, e.g. with cement or bitumen. Bound sub-bases are used in only a minor proportion of roads in the UK and are beyond the scope of this book.

Deformations beyond a certain magnitude are unacceptable because, even if they are corrected at the sub-base surface, they can lead to damage to the subgrade, which cannot easily be repaired. There have been many cases in the UK where severe delays and extra costs have been caused by the use of sub-base materials that have been too weak to carry construction traffic, especially in wet weather and on weaker subgrades. The mode of failure is usually the accumulation of deep rutting under wheeled traffic. Fig. 6.1 shows the gradual accumulation of rut depth with traffic for a group of stronger materials, the virtual collapse of a weak material after only a few passes, and an intermediate result. The precise values of rut depth depend on a whole range of factors, including the nature of the sub-grade, the thickness and condition of the sub-base, and the type and condition of the vehicle, the extent and pattern of trafficking, etc. Additionally, there are several other considerations in the choice of an acceptable limiting value for rut depth.

As may be expected, the literature covering all these matters is extensive. It has been reviewed recently (Pike 1985) and, to summarize the available information on the tolerable extent of deformation, various authorities recommended a limit on rut depth somewhere 30 mm and 50 mm, measured from original level (i.e. not from peak to trough). Recent work at TRRL (Powell *et al.* 1984) lends support to a limit on deformation of this order.

Powell *et al.* proposed the following model of deformation:

$$\text{Log } N_{40} = \frac{h(\text{CBR})^{0.63}}{190} - 0.24 ,\qquad(6.1)$$

where N_{40} = number of standard axles to produce a deformation from original level of 40 mm in the surface of a sub-base; h = thickness of sub-base (mm); CBR = California bearing ratio (%) — a measure of the shear strength of the subgrade material.

Equation (6.1) was found by Powell *et al.* to fit several sets of data, and they used it to give recommended thicknesses of sub-base required to carry construction traffic — see Fig. 6.2.

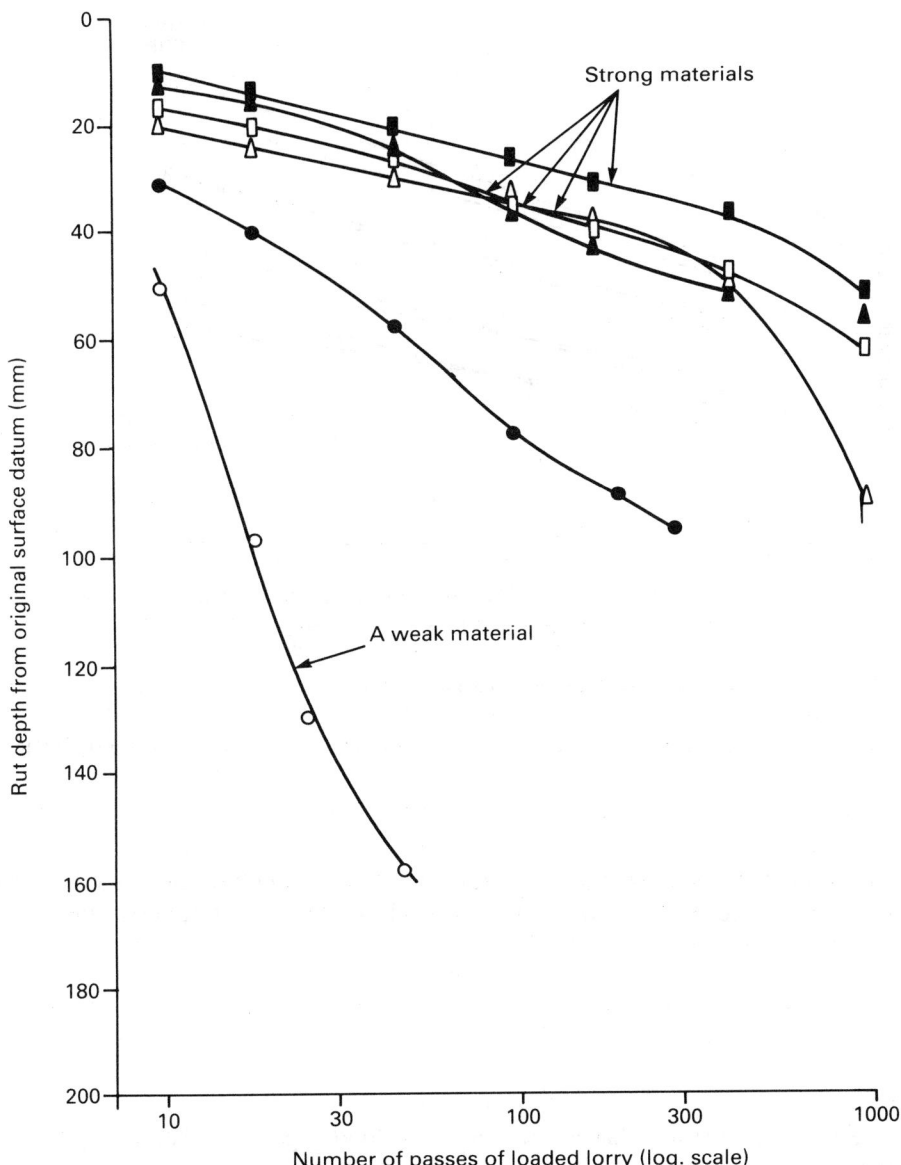

Fig. 6.1 — Average deformations measured in wheel-paths during trafficking trials of granular sub-bases materials.

6.2.3 Trafficking trials

In assessing the effects of traffic on road structures, highway engineers simplify the complex pattern of loads applied; for this purpose the concept of a standard axle has been introduced, and has become one of the pillars of road pavement design in the UK (TRRL 1970).

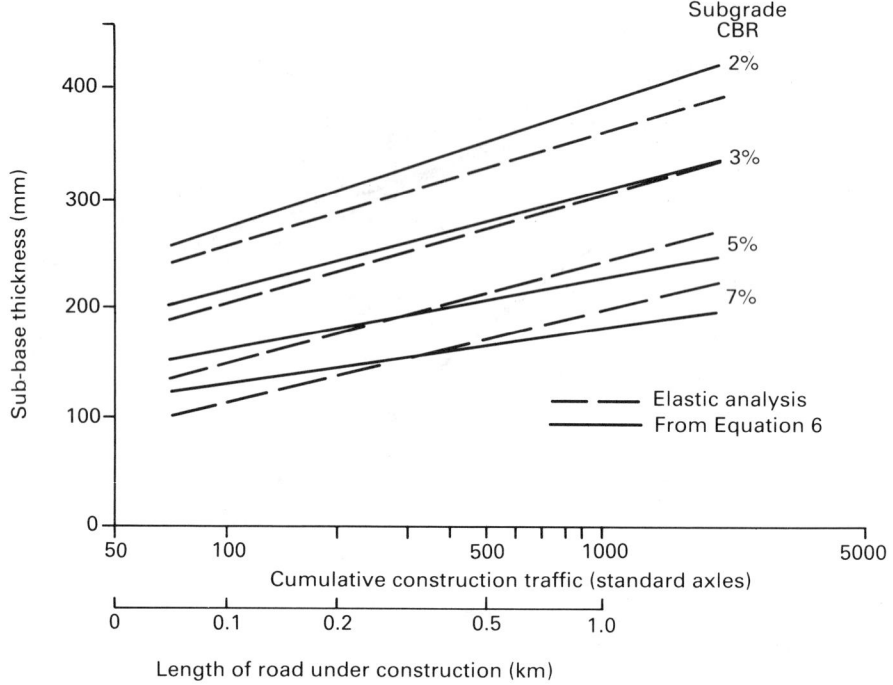

Fig. 6.2 — Thickness of sub-base required to carry construction traffic (after Powell *et al.* 1984).

One standard axle is a single pass of an axle weighing 8160 kg; other axles have a damaging power related to that of the standard axle by a fourth power law:

$$F = \frac{L^4}{L_s} \quad (6.2)$$

where F = an equivalence factor; L = any axle load (kg); L_s = standard axle load, 8160 kg.

Lorries carrying road construction materials can typically exert 3 to 4 standard axles per pass. Referring back to Fig. 6.2 and the work of Powell *et al.*, it will be seen that the cumulative construction traffic can be estimated at 1 standard axle per metre of road length under construction (i.e. between isolated access points). Lengths of several hundreds of metres are common, for example on motorway and trunk road projects.

This notion of predictable performance under standardized conditions suggested a basis for comparing the performance of various sub-base materials in simulative trafficking trials. Trials were carried out at TRRL, using a pit about 17.5 m × 5 m in plan and about 1 m deep filled with a clay of medium plasticity. The moisture content of the clay was adjusted so that its CBR was about 3%. The clay was covered with polythene sheeting, on top of which the experimental sub-base materials were laid and compacted.

For the first main series of trials, the sub-bases were laid about 230 mm thick. They were saturated with water after compaction, but allowed to drain overnight before trafficking. They were trafficked by a lorry that imparted 0.7 standard axles per pass; the wheel-tracks were canalized as closely as possible. Trafficking was continued until at least 50 mm of deformation from original level has been accumulated at the surface of the sub-base.

In later trials, capping layers (i.e. low-grade sub-base) were introduced between the subgrade and the experimental sub-base materials; the thickness of the experimental sub-bases was reduced to 150 mm; and heavier lorries were used for trafficking. In addition to trials at TRRL, larger experiments were conducted at specially constructed sites in Kent, Humberside, Surrey, and Scotland. Values for properties of the aggregates measured in the laboratory and in the field, including rut depths under given levels of traffic, were obtained for 41 sub-base materials, including both sands and gravels and crushed rocks, covering a wide range of petrology.

These trials gave an immediately credible test of performance; for example some sands and gravels performed well enough to be accepted for use on local contracts. But the trials were expensive to conduct, and thus it was obviously sensible to seek a laboratory-scale test that could reliably predict performance in the field.

6.2.4 The shortcomings of the CBR test
The CBR test was originally developed in the US to assess road-making gravels. As may be seen from Fig. 6.3, it is a punching shear test: a plunger is driven, via a load-measuring device, at a standard displacement rate into a sample of soil or aggregate compacted into a cylinder. Loads are recorded at standard levels of penetration. Alternatively, and with some modifications, the test can be applied to materials *in situ*. For many years this test has been a basic part of highway pavement design in the UK, and it might be thought to have immediate relevance to the suitability of sub-base materials. However, in the UK, it is recognized (TRRL 1970) that the CBR test is not applicable to sub-base materials. This is fairly obvious when it is seen that the diameter of the test plunger is only about 50 mm, hence a single large stone under the plunger could prevent an even development of shear under load. Nevertheless, the CBR test is still used in highway engineering in the UK because it offers a simple method of classifying soils in terms of their shear strength, even through the CBR ascribed to a particular soil is often chosen by considering other properties, e.g. plasticity and moisture content. It is therefore not surprising that some UK engineers persist in ascribing CBR values to sub-base materials, and, in some instances, try to modify the CBR test to make it applicable to coarse, granular materials. These attempts have not been conspicuously successful.

6.2.5 The shear-box test
The failure of unbound sub-bases by rutting under construction traffic is governed primarily by the shear strength of the aggregates. There are several established

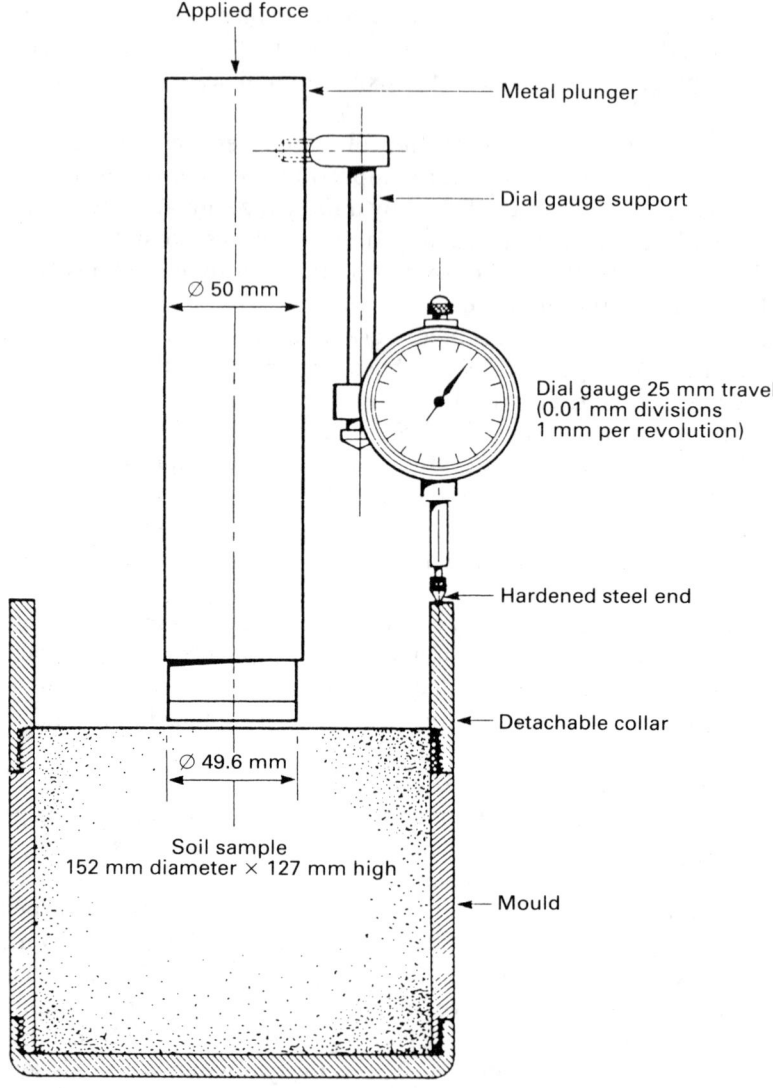

Fig. 6.3 — Arrangements for laboratory CBR test.

methods of testing the shear strength of aggregates having a maximum size of 40 mm or more, and the most convenient of these is the large shear-box.

Theories for shear strength and failure criteria are found in many texts; their common starting point is the equation:

$$T = \sigma_N . \tan \phi + c \tag{6.3}$$

where T = shear strength (kN/m^2); σ_N = normal stress (kN/m^2); ϕ = angle of internal friction (degrees); c = cohesion (kN/m^2).

For a densely-packed, well-graded aggregate, the shear stress rises with increase in displacement until a peak value is reached. It is usual to carry out the test at several levels of normal stress, so that ϕ and c can be measured.

This equation is usually attributed to Coulomb, and is derived from very early research on the frictional properties of solids. It is still widely used in soil mechanics, but it does not describe the behaviour of granular materials exactly. For example, over a extended range of normal stress, the relation between shear strength and normal stress is not linear. Thus it is not possible to ascribe a value of ϕ to a sub-base aggregate that is valid across the range of normal stress of interest in this investigation. A simple method of expressing the results of shear-box tests carried out in any standard condition is to quote the ratio of the peak shear stress to the normal stress. This peak shear stress ratio (PSSR) is equivalent to $\tan \phi$, if $c = 0$.

In 1968 a programme of tests was begun at TRRL using a standard 300 mm shear-box machine. This showed promise, and a substantial test programme was then undertaken, the first stage of which was the design and construction of a new shear-box machine, with additional features to make testing easier and more accurate. The general arrangement of the apparatus can be seen in Figs 6.4 and 6.5. With this machine a more detailed study was made of some of the factors that had been found to influence the results of shear-box tests on aggregate, i.e. normal stress, strain rate, grading, density, and moisture content. At this stage, the plasticity of fines was not considered. The following findings were published (Pike 1973): (a) the relations between peak shear stress and normal stress of non-plastic sub-base aggregates are markedly non-linear over an extended range of normal stress; (b) considerable differences in shear strength are found between aggregates of different type at low levels of normal stress, but these differences decrease with increase in normal stress; (c) changes in strain rate have no significant effect on shear strength; (d) the degree of compaction, moisture content, grading, and particle geometry all influence shear strength. The scale of influence was assessed via the effects of changes in density on shear strength, as follows:

(i) an increase in density of 1% caused by a change in grading alone leads to an increase in shear strength of 1 to 2%;
(ii) an increase in density caused by an increase in compactive effort or by optimizing the moisture content leads to an increase in strength of about 5%;
(iii) an increase in particle angularity and/or roughness leads to a decrease in density but to an increase in strength — for example, a 1% decrease in density from this cause was associated with a 3% increase in strength.

Limestones as a group showed higher levels of compactability and strength over a wide range of particle shapes than other groups.

A by-product of this work was an improved compactability test, which is described in section 6.2.8.

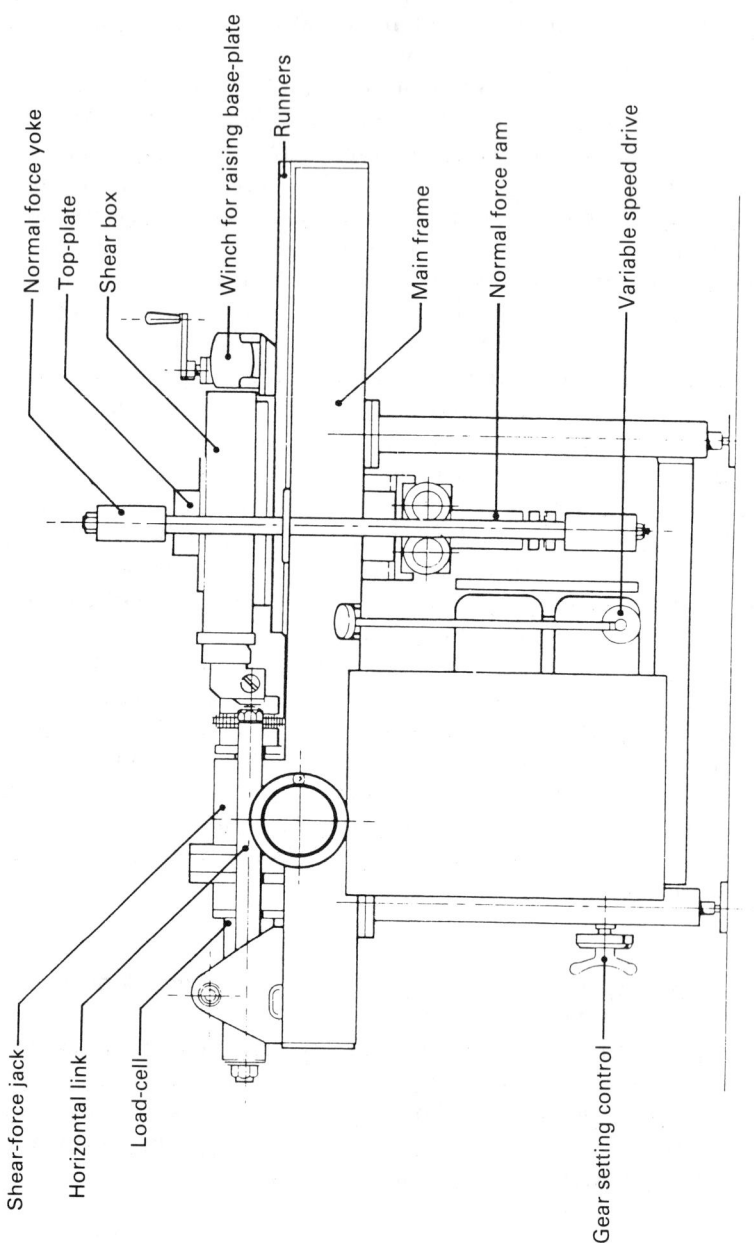

Fig. 6.4 — Diagram of shear-box machine.

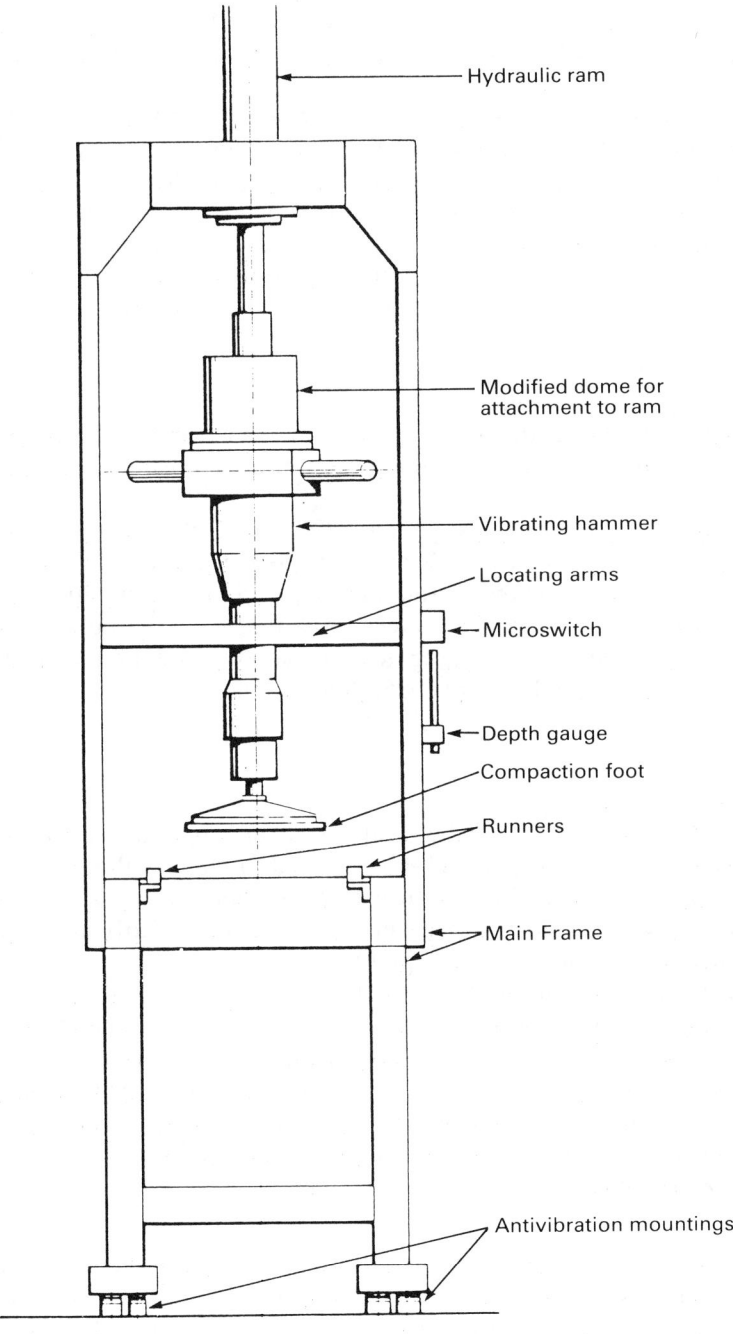

Fig. 6.5 — Diagram of heavy compaction rig used with shear-box.

6.2.6 A standard shear-box test

In 1980 a further programme of tests was put in hand by the Sand and Gravel Association and the Transport and Road Research Laboratory. Producers of sand and gravel were invited to nominate aggregates that they considered to be fit for use as high-grade sub-base materials.

After an initial period of weeding out materials that provided to be unstable when wet, or failed to meet the DTp's specification in some other way, e.g. by having an inadequate resistance to frost-heave, a list of aggregates for further investigation was prepared. Some of these aggregates were used in a set of shear-box tests carried out to determine the best combination of test conditions to suit the assessment of materials with strength properties affected by moisture. It was concluded from this study that the following test conditions were most applicable:

(a) normal stress: $10\,kN/m^2$;
(b) displacement rate: 1 mm/min;
(c) density: 97% of peak dry density determined by a compactability test (see 6.2.8);
(d) initial moisture content: 1% above the optimum moisture content.

The choice of a level of density set at 97% of peak dry density took into account that full compaction would not always be achieved in the field, and the choice of moisture content at 1% above optimum was based on a recognition that stability is often impaired by moisture. These test conditions gave the best separation of the weakest from the strongest aggregates and good correlation with the results of trafficking trials (Earland & Pike 1985, Pike 1985).

6.2.7 Correlations

Figs 6.6 and 6.7 bring together the results of shear-box tests and trafficking trials at two levels of standard axles, for sub-bases tested with and without capping layers. In each case there seems to be a critical level of shear strength, in terms of PSSR somewhere between 2 and 3, above which performance is fairly uniformly good, and below which performance is unacceptable.

The cut-off point may be fairly sharp in any given set of conditions, but, for practical engineering purposes, and when considering recommendations that might be made to cover a wide range of circumstances, it is prudent to take a rather conservative view. Thus, in Figs 6.6 and 6.7, curves A to D are drawn not through the centre of the plotted points to indicate the general trend, but at the upper limits of the deformations recorded for materials of given values of PSSR.

Three zones of strength can be postulated:

(a) *low strength* — PSSR less than 1.9: materials giving these values will not be suitable for use in sub-base layers but may be stable enough for capping layers;
(b) *medium strength* — PSSR 1.9 to 2.8: materials giving these values may be adequately stable in favourable conditions, but performance should be checked by a preliminary trafficking trial designed to simulate real conditions of service;
(c) *high strength* — PSSR above 2.8: materials giving these values should produce

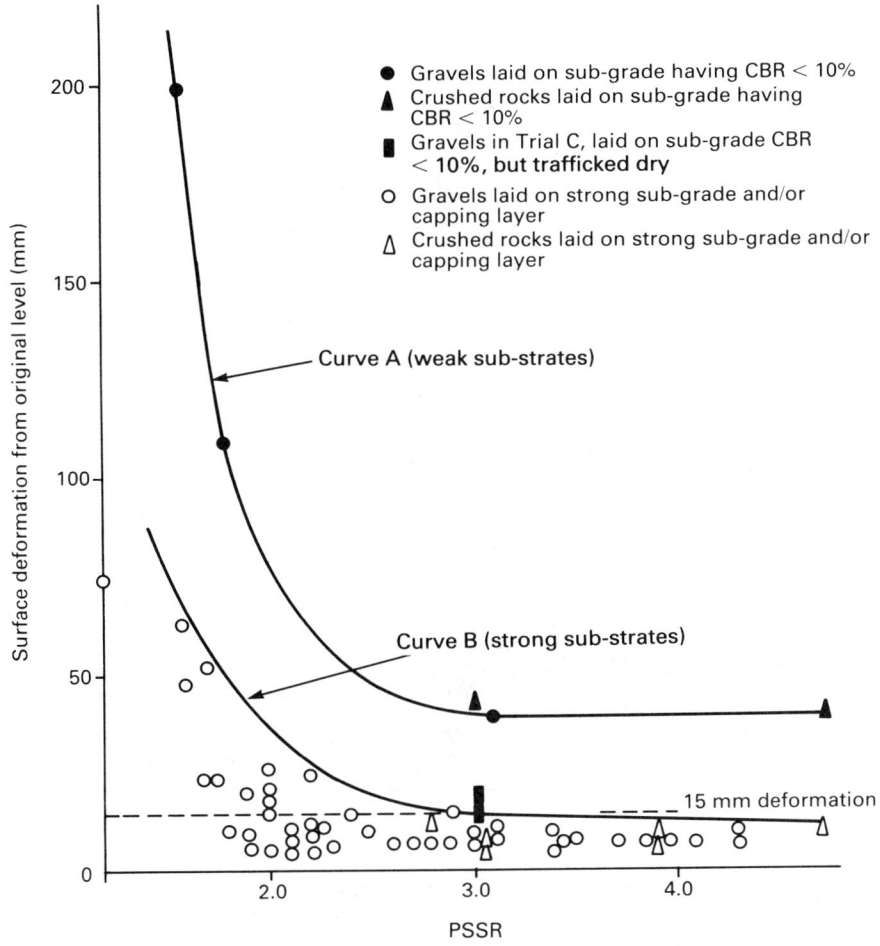

Fig. 6.6 — Relation between PSSR and surface deformations at 100 standard axles.

satisfactory stable sub-bases under normal construction traffic. Consideration should be given, however, to a trafficking trial if any of the following conditions may be encountered:

(i) exceptionally heavy lorries or other plant,
(ii) extensive working in continuously very wet weather or poor drainage conditions,
(iii) a requirement for very small deformations at the surface of the sub-base.

In addition to the shear-box test, several other possible indicators of shear strength were tried. It was confirmed that the CBR test was inapplicable, but other tests did give results that could be correlated with the results of trafficking trials.

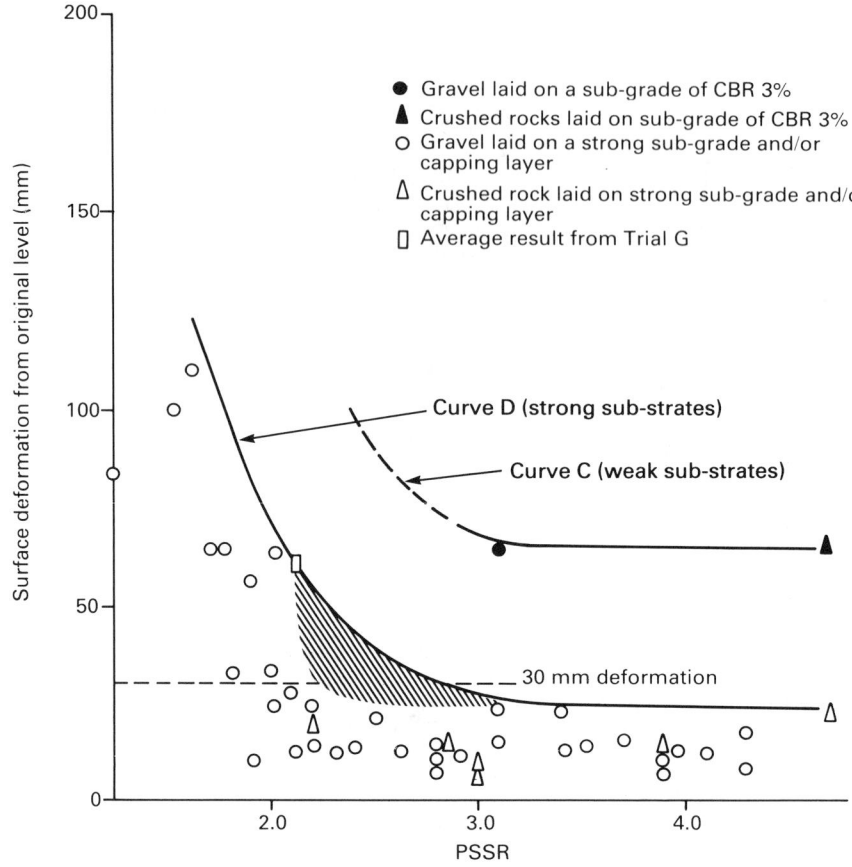

Fig. 6.7 — Relations between PSSR and surface deformations at 1000 standard axles.

These included the Clegg Impactometer, the Dynamic Cone Penetrometer, the Plate Bearing test, and, in limited circumstances, the Benkelman Beam. Further work is required to allow these *in situ* methods to be used reliably.

6.2.8 Compactability tests

Specifications for engineering soils and granular materials for roads often refer to two parameters, optimum moisture content and maximum dry density, on the premise that the compacted bulk density of such a material increases with increase in moisture content until a peak level of density is reached at an optimum moisture content (usually just before saturation is reached), after which further increase in moisture content leads to a decrease in density.

This is usually true for cohesive solids, but, while attempting to establish standard conditions of moisture content and density for the studies described in section 6.2.5, it was found not to apply to all sub-base materials. Indeed, it was found that almost the opposite effect was often achieved, i.e. high densities were found at low and high moisture contents, and low densities at intermediate moisture contents (Pike 1972). In many cases a pessimum moisture content was found. This effect may be explained

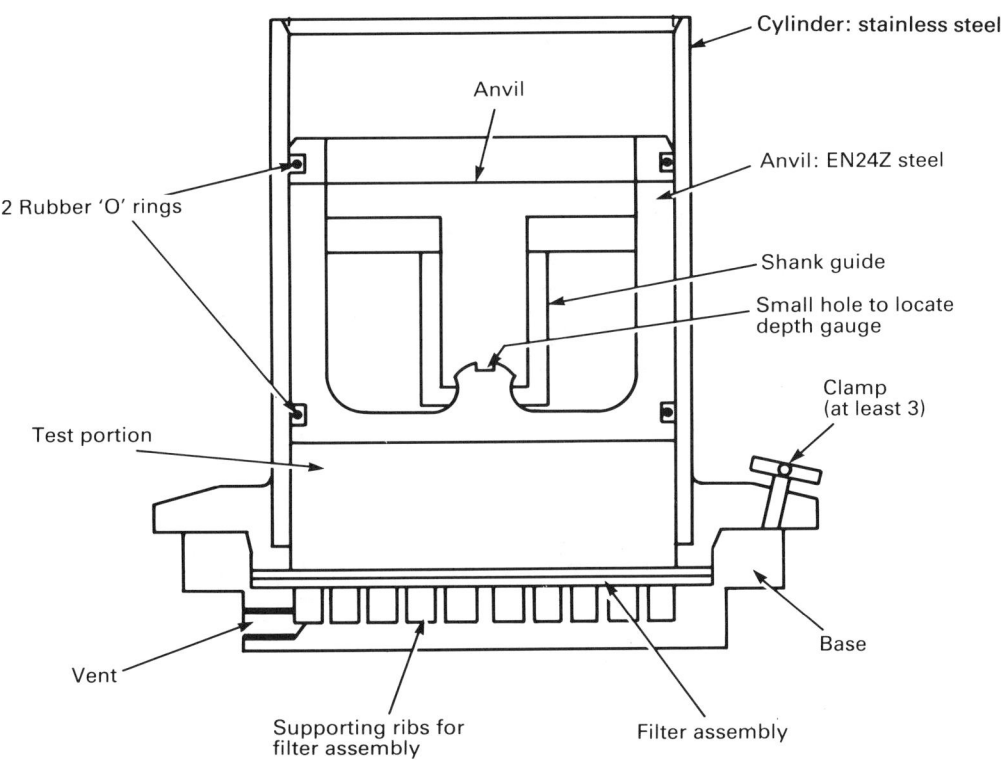

Fig. 6.8 — The compaction mould and anvil for use with the vibrated bulk density and optimum moisture content methods.

as follows. At moisture contents intermediate between the fully dry condition and saturation, aggregate particles are coated with a film of water bound by strong forces of surface tension. This film causes 'bulking' in building sands (see section 4.4), and it also comes into play during the compaction of granular materials. It effectively increases the volume of the aggregate particles, and hence reduces the bulk density. As more water is added, some free water is available to lubricate the particles, so, at or near saturation, the bulking effect is wholly overcome.

Improvements in apparatus and method produced a test adopted in British Standard BS 5835 (BSI 1980). The following improvements in the apparatus were provided:

(a) a specially designed mould having a filter base to allow the egress of air and water during the period of heavy vibration but to prevent the loss of fine solids, and a closely fitting anvil, sliding like a piston (see Fig. 6.8); the change in mould and

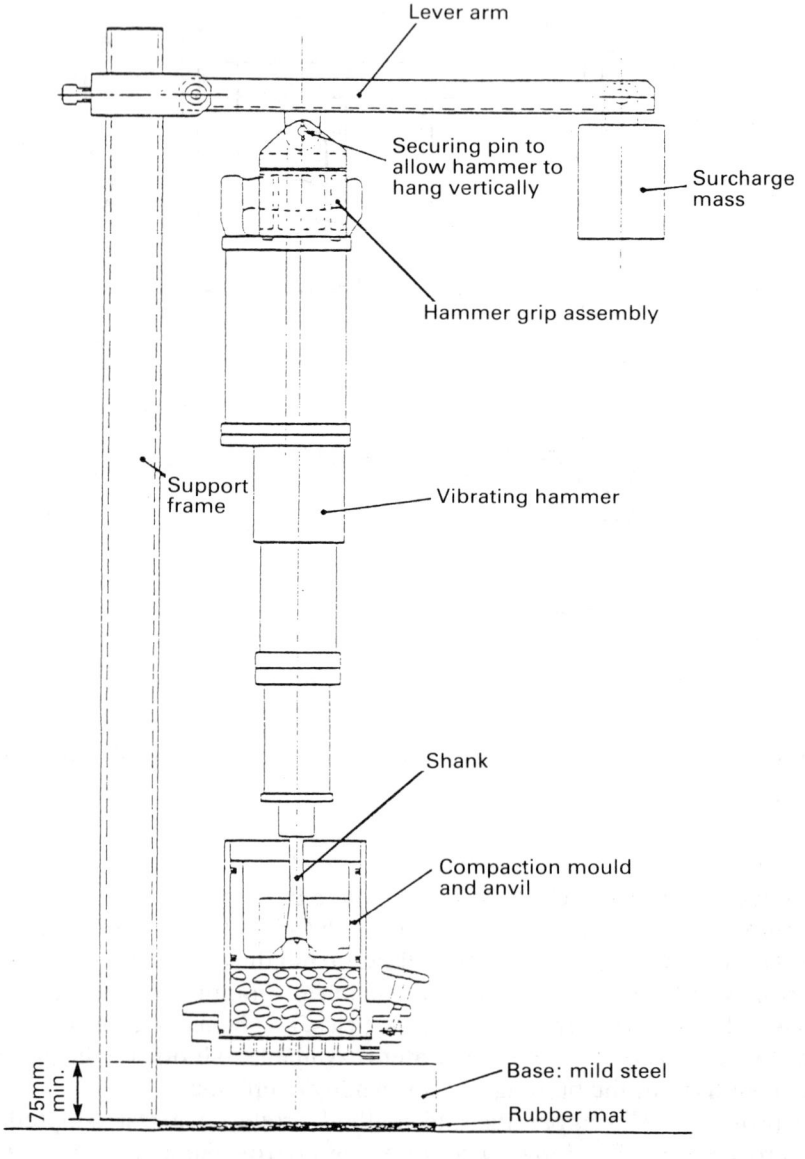

Fig. 6.9 — Compaction rig assembly.

associated equipment required the use of a vibrating hammer having a higher power rating;

(b) a simple, standard loading frame (see Fig. 6.9); previously there had been no effective control of the downward force on the hammer or the reaction to the mould.

This test is not yet in widespread use, but is preferred for determining the density and moisture content conditions for the improved frost-heave test — see section 6.1.

References

British Standards Instiututon. *Testing aggregates*. BS 812. 1975 *et seq.*

British Standards Instittion. *Precision of test methods*. Part 1. *Guide to the determination of repeatability and reproducibility* for a standard test method. BS 5497. 1979.

British Standards Institution. *Compactability tests for graded aggregates*. BS 5835: Part 1. 1980.

Burns, J. (1977) *The effect of the water table on the frost-susceptibility of roadmaking materials*. TRRL Supplementary Report SR305.

Croney, D. & Jacobs, J. C. (1967) *The frost susceptibility of soils and roadmaking materials*. RRL Report LR90. 1967.

Curtis, C. R. & Loveday, C. A. (1987) Granular materials in pavement foundations: the practicalities of production and supply. *Institution of Highways and Transportation National Workshop on Design and Construction of Pavement Foundations*. 1987.

Department of Transport. *Specification for highway works*. Her Majesty's Stationery Office. 1986 and earlier editions.

Earland, M. G. & Pike, D. C. (1985) Stability of gravel sub-bases. TRRL Research Report RR64. 1985.

Jacobs, J. C. & West, G. (1966) *Investigations into the effect of freezing on a typical road structure*. RRL Report LR54. 1966.

Jones, R. H. and Berry, A. N. (1979) The influence of sub-grade properties on frost-heave. *Highways and Public Works*. **47**, No. 1832. 17–22.

Pike, D. C. (1972) *The compactability of graded aggregates. 1. Standard laboratory tests*. Transport and Road Research Laboratory Report LR447.

Pike, D. C. (1985) The prediction of the performance of road sub-bases under construction traffic by shear-box tests. M.Phil thesis. Polytechnic of the South Bank.

Pike, D. C., S. M. Acott, & R. M. Leech, (1977) *Sub-base stability: a shear-box test compared with other prediction methods*. Transport and Road Research Laboratory Report LR185.

Pike, D. C., Sherwood, P. T., & Sym, R. (1988) *The TRRL frost-heave test: a further assessment of precision*.

Powell, W. D., J. F. Potter, Mayhew, H. C., & Nunn, M. E. (1984) *The structural design of bituminous roads*. Transport and Road Research Laboratory Report LR1132.

Roe, P. G. & Webster, D. C. (1984) *Specification for the TRRL frost-heave test*. TRRL Supplementary Report SR829. 1984.

Sherwood, P. T. & Pike, D. C. (1984) *Errors in the sampling and testing of sub-bases.* Transport and Road Research Laboratory Report SR831.

Sherwood, P. T. & Roe, P. G. (1986) *Winter air temperatures in relation to frost damage in roads.* TRRL Research Report RR45.

Transport and Road Research Laboratory. (1970) *A guide to the structural design of pavements for new roads.* Her Majesty's Stationery Office.

7
International and European standards

D. C. Pike

For many years it has been known that international standards for aggregates are being written and that they may eventually become applicable in the UK. Although there have been indications for about ten years that British Standards for aggregates will one day be replaced by European standards, there has always been uncertainty about whether or not the effects of such a change would bring serious problems. From time to time the question has been posed: 'Why should anyone in the UK be concerned about European Standards for aggregates — after all, the prospect of massive imports from the Continent is preposterous, so UK purchasers will have to keep buying from UK producers whatever happens to standards.' The honest answer to this is that it has been impossible to predict until very recently whether international or European standards will or will not pose problems, but experience shows that it is wise at least to take sensible precautions. In particular, there is a risk that unsuitable limits and/or test methods could be implemented unless all responsible parties take an active part in the drafting of standards.

The whole drift of legislation in the consumer field in the UK has been away from the ancient principle of 'buyer beware' and has reached a stage of general acceptance of the concept of 'products' fitness for purpose'. It may well go even further towards 'producers' strict liability', etc. As a result, specifiers have used and enforced standards increasingly, and aggregates producers find that more and more of their sales are tied to a contractual obligation to comply with a standard. This has led to a greater interest by the producers' trade associations (SAGA and BACMI) in British Standards, and a gradually growing awareness of the need to improve those Standards. in particular, considerable energies have been invested in improving the British Standards specifications for concrete aggregates and building sands, and on the complete overhaul of all the available test methods, with particular reference to precision. Some of the developments in British Standards are described in detail in other chapters of this book.

At the beginning of 1987 it became clear that the UK would have to take more notice of International and European Standards for aggregates. As a result of the determination by the member states of the EEC to complete their internal market by 1992, European legislation on several aspects of building materials is in preparation. An EEC Directive on construction products has been agreed; this covers only broad principles of 'essential requirements' and will be fleshed out by further EEC and national regulations. Among the secondary documents will be Eurocodes, one of which deals with structural concrete, and another covers masonry. New European Standards are required to support these Eurocodes. European Standards are written by CEN, the Committee for European Standardization, which is the European equivalent of BSI. Its members are the standards bodies of the eighteen member states of the EEC and EFTA. The rules of CEN mean that members of the EEC have to take European Standards seriously. There is, for example, a weighted majority voting procedure, so no single country can obstruct progress with a veto. There is also a standstill procedure that can prevent member states from drafting their own standards while CEN standards are being prepared.

A CEN Technical Committee, CEN/TC104, has been working for several years to produce a draft CEN standard, or Euronorm, for concrete (ENV206). The chairmanship and secretariat of CEN/TC104 are held by Germany. In 1987, CEN/TC104 decided to establish a Working Group (CEN/TC104/WG2) to write European Standards for concrete aggregates, and this was in place with a UK convenor and secretariat by May 1988. CEN/TC104 asked its WG2 to produce a specification for concrete aggregates by the end of 1989. This time schedule was, in normal BSI terms, impossibly tight. At its first meeting, CEN/TC104/WG2 agreed that it had to pay due regard to both technical and commercial factors in the aggregates industries in the member states. There would be no point in bringing out European Standards for aggregates that would cause technical problems, or increase costs unnecessarily to producers and users. Because of the very tight timescale imposed on WG2, it was accepted that it would not be feasible to write fully harmonized European Standards for concrete aggregates that could immediately replace national standards. Thus it was agreed to adopt a 'framework approach'. This would mean listing a draft CEN standard for concrete aggregates the requirements that must appear in the various national standards.

CEN/TC104/WG2 had not really started its work before the CEN Technical Board decided to set up a full Technical Committee for aggregates for all purposes (CEN/TC154). Its scope is 'Standardization in the field of natural and synthetic aggregates, by the preparation of a framework for specifying aggregates, performance requirements, sampling, and methods of tests, with the aim of developing harmonized standards afterwards'.

The UK has been given the chairmanship and secretariat of CEN/TC154, which has absorbed CEN/TC104/WG2. The first meeting of CEN/TC154 took place in London in November 1988. Thus, as this book was being written, a new era of standards-making for aggregates was beginning. The work of CEN/TC154 has been split among six sub-committees, covering:

SC1 — requirements for aggregates for masonry (chairmanship — Netherlands),
SC2 — requirements for aggregates for concrete (chairmanship — UK),

SC3 — requirements for aggregates for bituminous mixtures (chairmanship — FRG),
SC4 — requirements for unbound aggregates (chairmanship — France),
SC5 — lightweight aggregates (chairmanship — Denmark),
SC6 — test methods (chairmanship — UK).

It has been agreed within CEN/TC154 that work on a framework specification for concrete aggregates will proceed as arranged in CEN/TC104/W2 as a matter of highest priority, except that questions of tests will be referred to the new, specialist sub-committee SC6 of CEN/TC154. Framework specifications for aggregates for the other uses will also be drawn up. The detailed drafting of the frameworks is being done by four task groups (TG1 to TG4), all of which are convened by the UK.

CEN/TC154's sub-committeee SC1 on aggregates for masonry (e.g. building sands) will confer with a new CEN Technical Committee on masonry (CEN/TC125) which has been recently set up, also under a chairman and secretariat from the UK. Furthermore, SC3 and SC4 will discuss matters of common interest with a new CEN committee for road engineering, CEN/TC227, which has a chairman and secretariat from FRG.

Little progress towards full harmonization can be achieved until CEN/TC154/SC6 has produced test methods that are acceptable to the member states. Its programme requires over 40 such tests to be written in about three years. This will include the collection of precision data. This is a large programme of work; much of the burden falls on five task groups (TG5 to TG9) which are convened by FRG, France (which convenes two), Belgium, and Ireland.

Apart from CEN standards, consideration must be given to the publications of the International Standards Organisation (ISO). It is CEN's policy to accept existing ISO standards. If this is not possible, the drafters of CEN standards should try to proceed by small amendments to existing ISO documents. Only if the ISO documents are totally unacceptable should new CEN documents be written, and ISO should be informed.

There is no ISO committee for aggregates, but there is one for concrete (ISO/TC71) that has produced papers on concrete aggregates. Austria has the secretariat of ISO/TC71. Its sub-committee ISO/TC71/SC1 (test methods for concrete) produced several ISO test methods for concrete aggregates in the 1970s, but the UK and some other countries voted against all of these because they had serious technical defects. Then, in 1978, ISO/TC71 set up a working group (ISO/TC71/SC3/WG1) which has a convenor and secretariat from the USSR, to deal with aggregates. In the past ten years, ISO/TC71/SC3/WG1 has produced several drafts of a framework specification, and also several draft test methods, but none of these documents has yet proceeded to the stage of voting on a final draft ISO Standard.

The UK has contributed to the work of ISO/TC71/SC3/WG1, and supports the principle of using ISO documents as a basis for the drafts to be considered by CEN/TC154. Although the UK considers that the existing ISO documents are flawed, they provide at least a first step towards documents that could be accepted widely. The UK is seeking ways to draw the activities of ISO and CEN closer together.

Because the CEN work on aggregates has only just begun, it is too early to assess it critically. There is certainly a good deal of enthusiasm among the participants, and there is a good cross-section of UK interests. For successful results to be achieved it will be necessary for a wide range of generalized subjects to be dealt with, for example:

(a) differences between the rules of BSI, ISO, and CEN;
(b) differences in general approach between British Standards, other national standards, and ISO and CEN standards;
(c) the relation between specified requirements and test methods (the UK's view is that limits and test methods should appear in separate documents, and that advice should be given on the applicability, and limitations, of each test method);
(d) the concept of a hierarchy of test methods (the UK's view is that there should be only one standard method for determining any particular property to be specified, although subsidiary methods may be allowed to give a rapid, preliminary indication of compliance);
(e) the essential role of precision data;
(f) the elements of quality assurance and certification — UK aggregates producers would probably favour a simple system based on production control, self-certified by the producer, and compliance testing by the purchaser or his agent — there are some signs, however, of increasing pressure for third party certification of building products.

In addition to these policy issues, there are many technical topics to be considered in detail. The earlier chapters of this book deal with some of those topics, but it may still be worth noting here some of the points that seem likely to require specific attention.

7.1 SAMPLING

New Part 102 of BS 812 sets out a single method of obtaining a sample of average quality from a defined batch. The UK considers that the specification of the number of increments to be taken is part of a sampling standard, but other countries do not adopt this view: they see the question of increments as matter falling under 'quality control'.

The draft ISO Standard, ISO DIS4847, is based on the 1975 edition of BS 812, which is no longer acceptable in the UK. Special sampling arrangements may be needed in some circumstances, e.g. for petrological examination — see Chapter 2.

7.2 PARTICLE SIZE DISTRIBUTION

The sieve test in the new Part 103 of BS 812 differs little from the version in the 1975 edition of BS 812, except that it promotes washing-and-sieving, over dry sieving, as the normal method. This is to deal with the aggregation caused by fines.

Test sieves in general use suffer wear and tear, leading to alterations in aperture sizes, etc. Part 103 of BS 812 includes a calibration procedure to control wear, but this has not found universal favour, and the subject is being reconsidered.

The ISO sieve test method for aggregates (ISO 6274) was not accepted by the UK mainly because, although ISO 6274 attempted a harmonization of the sieve apertures, this was done in a way that did not achieve standardization. At the request of ISO/TC71/SC3/WG1, the UK has now provided a draft revision of ISO 6274. CEN/TC154/SC6/TG6 is also considering this matter.

As to the size distribution of fines, a revised sedimentation test has been published in Part 103 of BS 812. This uses the Andreasson pipette, as in the pre-existing version, but it extends the lower limit of sizes measured from 20 to 2 μm. ISO/TC71/SC3/WG1 has drafted a method that uses a pipette that is intended to give a rapid indiction of material finer than about 50 μm.

7.3 SHAPE AND SURFACE TEXTURE

Classification of shape and texture, using the terms 'irregular', 'rounded', and 'angular' extensively used in many textbooks on concrete mix-design, will be carried forward into Part 102 of BS 812. This is really a matter of initial description, rather than testing, which may not be familiar to other countries.

New Part 105 of BS 812 follows the previous text for the flakiness index, using shape gauges and slotted sieves. In some other countries, calipers are used to measure individual particles, and the criteria used to define flaky and elongated particles are sometimes different from those applied in the UK. There is surprisingly little evidence from the UK to link aggregate particle shape (or surface texture) with the performance of concrete, so it may be difficult to defend particular limits.

7.4 PARTICLE DENSITY, WATER ABSORPTION VALUE, AND BULK DENSITY

The term 'specific gravity' is in general use and widely understood, but, for some reason, the drafters of earlier editions of BS 812 used the term 'relative density'. They used this to mean density relative to water (which is what specific gravity means), but it has another meaning in materials engineering, i.e. a bulk density achieved in the field relative to a bulk density achieved in a standard test. This can lead to confusion, and, in future, the term 'particle density' will be used in BS 812. There are differences between the test methods for density used in the UK and those used in other countries, and these will have to be checked to ensure that the best methods are selected for European Standards.

Water absorption value is a rough indicator of porosity. The methods long used in BS 812 are crude, but there has been no concerted attempt to improve them, partly because water absorption is not frequently used to specify aggregates. Recent work at the Building Research Establishment indicates that porous aggregates can accelerate carbonation of concrete (Collins 1986). Also, porosity may in certain circumstances be linked with poor frost-resistance, soundness, and/or drying shrinkage. It is not surprising then that there are occasional suggestions in favour of using the water absorption test value as an indicator of these aggregate properties, not least because the direct tests can be lengthy and expensive. But there are counter-arguments to be weighed, partly because there are doubts about the precision of the water absorption test, and also because the question of limits to be applied to water absorption is complicated by petrology. Some aggregates give known problems at high water absorption values. For example, Roeder (1977) showed that white flint having a water absorption value of 10 to 12% is likely to cause spalling in concrete exposed to frost, whereas black or brown flint, with a water absorption value of 2 to 3%, is not likely to be associated with this problem. But Collins (1986) reported that there is no correlation between water absorption and frost resistance of concrete generally. Indeed he found that, in one set of experiments, the worst resistance to frost was shown by a concrete made with a coarse aggregate having a low water absorption value, i.e. under 1%. In yet other cases, aggregates with intermediate water absorption values give problems. The true position is that porosity alone is of limited use as an indicator of aggregate quality; the size of the pores, their distribution, and other factors, such as the degree of cementation of the grains within an aggregate particle, probably all have significant effects on performance.

7.5 RESISTANCE TO CRUSHING, ABRASION, AND POLISHING

There have been few problems with the resistance to crushing of natural aggregates used in the UK. This may be because there are abundant supplies of aggregates consisting of rock types of medium to high strength. There is a long-term interest, fostered by land-use planning considerations, in establishing whether lower strength materials, usually regarded as inferior, would be adequate for some purposes. Collins (1986) found that, for weak sandstones, aggregate strength as determined by the 10% fines value does not have much bearing on concrete strength. Obviously, when attempting to achieve higher levels of concrete strength, aggregate type can be a limiting factor, but this is only one of several relevant considerations, and it is usually more useful to proceed by trial mixes and concrete cube crushing tests than to limit testing to the aggregate.

Many other countries use attrition tests, such as the Los Angeles and Deval tests, to gain information about the strength and mechanical durability of aggregates, but they are not used much in the UK. The principle of these tests is that aggregate samples are put into a drum, with water and sometimes with metal balls, and then the drum is turned. After a specified number of revolutions, the sample is checked for weight loss. Some years ago, a wet attrition test was included in BS 812, but then it was dropped; British Rail still uses this test to assess the suitability of railway track ballast.

There is an abrasion test in BS 812 for coarse aggregates which is often called up, in conjunction with the polished–stone value test, for aggregates to be used in bituminous surfacings so as to promote skid-resistance and ensure its retention over a number of years.

In worn concrete pavements the abrasion resistance of the fine aggregates dominates the skid-resistance. Carbonate rocks (e.g. limestones and shell) polish more readily than other types, and there is a limit placed by the Department of Transport on the carbonate content of fine aggregate for pavement concrete. A simple acid-solubility test has been introduced as new Part 119 of BS 812 to give a standard basis of measuring carbonate content for this purpose.

7.6 FROST RESISTANCE AND SOUNDNESS

Frost damage to concrete is controlled mainly by a combination of environmental factors, especially by the degree of saturation and the number of freezing-and-thawing cycles to be withstood. Aggregate properties can be influential but Collins (1986) reports that there is no correlation between performance and the results of any of the tests available in BS 812. Direct freezing tests on aggregates or on concrete are used in some other countries, but not in the UK.

As has been noted, Roeder (1977) has drawn attention to a particular kind of spalling associated with weathered flint ('white flint' or 'cortex'), but no suggestion has yet been made that a test method to isolate this aggregate should be included in BS 812.

Unsoundness is a term used to describe those aggregates that may appear to be of adequate quality at the time of use but which subsequently degrade through environmental effects or as a result of chemical changes. Such aggregates are common overseas and are occasionally met in the UK. An investigation has recently been carried out into the applicability of several soundness tests, and the magnesium sulphate crystallization test has been drawn into BS 812. Both this and a similar test using sodium sulphate are used overseas, but there is conflicting evidence of the correlation between the results of that test and the performance of concrete, so it must be used with care. ISO/TC71/SC3/WG1 has recently concluded '.. this test is extremely severe, and can wrongly indicate that a good aggregate is not frost resistant. If an aggregate fails this test, it ought not be rejected immediately...'.

7.7 HARMFUL MATERIALS FINER THAN TWO MICROMETRES, ESPECIALLY CLAY MINERALS

For many years there has been a confusion in British Standards about clay minerals in aggregates. Terms like 'clay', 'clay, silt, and dust', etc., have been used rather loosely, and often without a proper link to a standard test method. Some improvements have been made to tests that measure particle size, but these cannot distinguish fines that are deleterious from those that are not. BSI has decided to investigate the possibility of a test for deleterious clay minerals; and a research project has been underway since 1988. At this stage of the project, the indications are that the methylene blue adsorption test should be applied for this purpose. It is already included in national standards in France and in other countries.

REFERENCES

Collins, R. J. (1986) Concrete using porous sandstone aggregate — results of recent research. *Quarry Management*. March, 33–38.

International Standards Organisation. (1984) *Concrete — sampling of normal weight aggregates*. ISO DIS4847.

International Standards Organisation (1982) *Concrete — sieve analysis of aggregates*. IS O6274.

International Standards Organisation. (1982) *Aggregates for concrete — determination of bulk density*. ISO 6782.

International Standards Organisation. (1982) *Coarse aggregates for concrete — determination of particle density and water absorption — hydrostatic balance method.* ISO 6783.

International Standards Organisation. (1987) *Fine and coarse aggregates for concrete — determination of the particle mass-per-volume and water absorption — pyknometer method.* ISO 7033.

Roeder, A. R. (1977) (1977) Some properties of flint particles and their behaviour in concrete. *Magazine of Concrete Research.* **29**, No. 99. pp 92–99.

Draft Part 120. *Methods for determination of drying shrinkage.* Published, with precision data.

Draft Part 121. *Soundness test (by magnesium sulphate crystallisation).* Published, with precision data.

Draft Part 122. *Methods of estimating the contents of deleterious materials in fine aggregates.* Circulated for public comment.

Draft Part 123. *Method for the assessment of alkali reactivity potential.* Circulated for public comment. Precision trial in progress.

Draft Part 124. *Method for determination of frost-heave.* Published, with precision data obtained. An amendment covering the use of reference aggregate is being considered.

Appendix 1

Status of new Parts of BS 812 at January 1990.

Part 100. *Definitions, symbols and calibration.* Draft circulated for public comment.

Part 101: 1984. *Guide to sampling and testing aggregates.* Published. Further work to provide advice on applicability of test methods is under consideration.

Part 102: 1984. *Methods for sampling.* Published.

Part 103: 1985. *Methods for determination of particle size distribution.* Section 1, specifying sieve tests, is published. Section 2, specifying a sedimentation method applicable to material finer than 75 µm, is also published. Precision data for both methods is included.

Draft Part 104. *Petrographical examination.* Circulated for public comment. A precision trial is in progress.

Part 105. *Methods for determination of particle shape:*

Part 105.1: 1985. *Flakiness index.* Published, with precision data.

Draft Part 105.2. *Elongation index.* Approved for publication. A precision trial is in progress.

Part 106: 1985. *Method for determination of shell content of coarse aggregate.* Published without precision data, but a precision trial is now in progress.

Draft Parts 107, 108, and 109. These cover tests for particle density, water absorption, bulk density, optimum moisture content, voids, bulking, and moisture content. Recommended for publication, with precision data.

Draft Parts 110, 111, 112, and 113. These cover tests for aggregate crushing value, 10% fines value, aggregate impact value, and aggregate abrasion value. Circulated for public comment in 1985. Precision trials are in progress.

Draft Part 114. *Methods of determination of the polished stone value.* Published with precision data.

Parts 115 and 116 — at present there are no plans to use these Part numbers.

Parts 117 and 118. *Methods for determinations of sulphate and chloride content.* Published, with precision data. A further section of Part 117, to cover modern methods including hand-held chloride meters, is being considered.

Part 119: 1985. *Method for determination of acid-soluble material in fine aggregate.* Published, with precision data.